Innovation in Forestry

Territorial and Value Chain Relationships

MIX
Paper from
responsible sources
FSC® C018575
FSC
www.fsc.org

COST – the acronym for European Cooperation in Science and Technology – is the oldest and widest European intergovernmental network for cooperation in research. Established by the Ministerial Conference in November 1971, COST is presently used by the scientific communities of 35 European countries to cooperate in common research projects supported by national funds.

The funds provided by COST – less than 1% of the total value of the projects – support the COST cooperation networks (COST Actions) through which, with €30 million per year, more than 30,000 European scientists are involved in research having a total value that exceeds €2 billion per year. This is the financial worth of the European added value which COST achieves.

A 'bottom-up approach' (the initiative of launching a COST Action comes from the European scientists themselves), 'à la carte participation' (only countries interested in the Action participate), 'equality of access' (participation is open also to the scientific communities of countries not belonging to the European Union) and 'flexible structure' (easy implementation and light management of the research initiatives) are the main characteristics of COST.

As precursor of advanced multidisciplinary research, COST has a very important role for the realization of the European Research Area (ERA) anticipating and complementing the activities of the Framework Programmes, constituting a 'bridge' towards the scientific communities of emerging countries, increasing the mobility of researchers across Europe and fostering the establishment of 'Networks of Excellence' in many key scientific domains such as: Biomedicine and Molecular Biosciences; Food and Agriculture; Forests, their Products and Services; Materials, Physical and Nanosciences; Chemistry and Molecular Sciences and Technologies; Earth System Science and Environmental Management; Information and Communication Technologies; Transport and Urban Development; Individuals, Societies, Cultures and Health. It covers basic and more applied research and also addresses issues of pre-normative nature or of societal importance.

Web: http://www.cost.esf.org

ESF provides the COST Office through an EC contract

COST is supported by the EU RTD Framework Programme

Innovation in Forestry

Territorial and Value Chain Relationships

Edited by

Gerhard Weiss

European Forest Institute Central-East European Regional Office (EFICEEC)
InFER – Institute of Forest, Environmental and Natural Resource Policy
University of Natural Resources and Life Sciences
Feistmantelstr. 4
1180 Vienna
Austria

Davide Pettenella

Università di Padova
Dipartimento Territorio e Sistemi Agro-Forestali
Via dell'Università 16
35020 Legnaro PD
Italy

Pekka Ollonqvist

Finnish Forest Research Institute
PO Box 18 (Jokiniemenkuja 1)
FI 01301 Vantaa
Finland

and

Bill Slee

The Macaulay Land Use Research Institute
Craigiebuckler
Aberdeen AB15 8QH
UK

www.cabi.org

CABI is a trading name of CAB International

CABI Head Office
Nosworthy Way
Wallingford
Oxfordshire OX10 8DE
UK

Tel: +44 (0)1491 832111
Fax: +44 (0)1491 833508
E-mail: cabi@cabi.org
Website: www.cabi.org

CABI North American Office
875 Massachusetts Avenue
7th Floor
Cambridge, MA 02139
USA

Tel: +1 617 395 4056
Fax: +1 617 354 6875
E-mail: cabi-nao@cabi.org

A catalogue record for this book is available from the British Library, London, UK.

Library of Congress Cataloging-in-Publication Data

Innovation in forestry : territorial and value chain approaches / edited by Gerhard Weiss, Pekka Ollonqvist, and Bill Slee.
p. cm.
Includes bibliographical references and index.
ISBN 978-1-84593-689-1 (alk. paper)
1. Forestry innovations. 2. Forests and forestry--Economic aspects. I. Weiss, Gerhard, 1966- II. Ollonqvist, Pekka. III. Slee, Bill. IV. Title.

SD387.I57I56 2011
333.75--dc22

2010035486

ISBN-13: 978 1 84593 689 1

Commissioning editor: Nigel Farrar
Production editor: Tracy Head

Typeset by SPi, Pondicherry, India.
Printed and bound in the UK by MPG Books Group.

Contents

Contributors vii

Preface xi

1 **The Study of Innovation in the Forest Sector: Relevance and Research Background** 1
Gerhard Weiss

2 **Theoretical Approaches for the Analysis of Innovation Processes and Policies in the Forest Sector** 10
Gerhard Weiss

3 **Networks of Small–Medium Enterprises Operating in Forestry: Some Theoretical Concepts and Empirical Evidence** 35
Davide Pettenella and Daria Maso

4 **Innovation-related Knowledge Flows: Comparative Analysis of Finnish and Estonian Wood Sectors** 48
Kadri Ukrainski and Miika Kajanus

5 **Institutional Innovation in European Private Forestry: the Emergence of Forest Owners' Organizations** 68
Américo M.S. Carvalho Mendes, Blaž Štefanek, Diana Feliciano, Diana Mizaraite, Dragan Nonić, Emil Kitchoukov, Erlend Nybakk, Gabriel Duduman, Gerhard Weiss, Liviu Nichiforel, Maria Stoyanova, Pekka Mäkinen, Rosário Alves, Vojislav Milijić and Zuzana Sarvašová

6 **The Contribution of Leader+ to the Implementation of Innovative Forest-related Projects** 87
Diana Feliciano, Bill Slee, Gerhard Weiss, Anne Matilainen and Thomas Rimmler

7 **How to Support Firm Competitiveness in Timber Industries? Clusters as Policy Means in Four European Countries** 101
Thomas Rimmler, Roger Coppock, Roland Oberwimmer, Andreja Pirc, Stjepan Posavec and Gerhard Weiss

8 **Innovation in Forest-related Territorial Goods and Services: an Introduction** 118
Bill Slee

9 **Innovative Market Opportunities Related to Carbon Sequestration in European Forests?** 131
Lorenzo Ciccarese, Peter Elsasser, Antonis Horattas, Davide Pettenella and Gregory Valatin

10 **The Role of Networks in Non-wood Forest Products and Services Market Development** 154
Daria Maso, Anne Matilainen and Davide Pettenella

11 **The Role of Cooperation in Enhancing Innovation in Nature-based Tourism Services** 169
Anne Matilainen, Gerhard Weiss, Zuzana Sarvašová, Diana Feliciano, Carmen Nastase and Monika Prede

12 **Innovations in Wood-based Enterprises, Value Chains and Networks: an Introduction** 189
Pekka Ollonqvist

13 **Role of Policies and National Programmes on Innovations in Timber-frame Construction** 204
Tomas Nord, Saana Tykkä, Denise McCluskey, Fahrudin Bajric, Laura Bouriaud, Mårten Hugosson, Anders Q. Nyrud, Pekka Ollonqvist, Anders Roos, Kadri Ukrainski and Kristian Bysheim

14 **Networks and Local Milieus as a Furniture Industry Innovation Platform** 233
Pekka Ollonqvist, Tomas Nord, Andreja Pirc, Kadri Ukrainski, Vuokko Takala-Schreib, Meelis Teder, Wladyslav Strykowski and Anne Viitala

15 **Innovation in the Wood Bio-energy Sector in Europe** 254
Erlend Nybakk, Anssi Niskanen, Fahrudin Bajric, Gabriel Duduman, Diana Feliciano, Krzysztof Jablonski, Anders Lunnan, Liana Sadauskiene, Bill Slee and Meelis Teder

16 **Policy and Market-related Factors for Innovation in Forest Operation Enterprises** 276
Laura Bouriaud, Edgar Kastenholz, Lukáš Fodrek, Zbigniew Karaszewski, Piotr Mederski, Thomas Rimmler, Arto Rummukainen, Liana Sadauskiene, Jaroslav Salka and Meelis Teder

17 **Innovation in EU Forestries: a Science–Policy Dialogue** 294
Filip Aggestam and Gerhard Weiss

18 **How to Support Innovation in the Forest Sector: Summary and Conclusions** 303
Gerhard Weiss, Pekka Ollonqvist and Bill Slee

Index 321

Contributors

Filip Aggestam, European Forest Institute Central-East European Regional Office (EFICEEC), InFER – Institute of Forest, Environmental and Natural Resource Policy, University of Natural Resources and Life Sciences, Feistmantelstr. 4, 1180 Vienna, Austria. E-mail: filip.aggestam@boku.ac.at

Rosário Alves, FORESTIS – Associação Florestal de Portugal, Porto, Portugal.

Fahrudin Bajric, The Economics, Energy and Prospective Department of the Technological Institute for Forestry, Cellulose, Construction Timber and Furniture, 10 avenue Saint Mandé, 75012 Paris, France.

Laura Bouriaud, Stefan cel Mare University of Suceava, str. Universitatii, nr. 13, 720229 Suceava, Romania. E-mail: bouriaud@usv.ro

Kristian Bysheim, Norsk Treteknisk Institutt, PO Box 113, Blindern, NO-0314 Oslo, Norway.

Américo M.S. Carvalho Mendes, Faculty of Economics and Management, Portuguese Catholic University, Porto, Portugal. E-mail: carvalho.mendes@sapo.pt

Lorenzo Ciccarese, Italian Institute for Environmental Protection and Research, Via Curtatone 3, 00185 Rome, Italy.

Roger Coppock, Head of Specialist Advisors, Forestry Commission GB, Silvan House, 231 Corstorphine Road, Edinburgh, EH12 7AT, UK.

Gabriel Duduman, Faculty of Forestry, 'The Stefan cel Mare' University, Suceava, Romania.

Peter Elsasser, Institut für Ökonomie, Bundesforschungsanstalt für Forst- und Holzwirtschaft, von Thuenen Institute, Leuschnerstrasse 91, D-21031 Hamburg, Germany. E-mail: peter.elsasser@vti.bund.de

Diana Feliciano, ACES – The University of Aberdeen & The Macaulay Institute, The School of Biological Sciences, Tillydrone Av., Aberdeen AB24 2TZ, UK. E-mail: d.feliciano@macaulay.ac.uk

Lukáš Fodrek, Researcher, Faculty of Forestry, Technical University in Zvolen, T.G. Masaryka 24, Zvolen, SK-96053, Slovakia.

Antonis Horattas, Department of Forests, 1414 Nicosia, Cyprus.

Mårten Hugosson, Department of Forest Products, Swedish University of Agricultural Sciences, Box 7008, SE-750 07 Uppsala, Sweden.

Krzysztof Jablonski, Department of Forest Technology, Poznan University of Life Sciences, Wojska Polskiego 71C, 60-625 Poznan, Poland.

Miika Kajanus, R&D Competence Centre for Entrepreneurship and Innovation, Savonia University of Applied Sciences, PO Box 72, FIN-74100 IISALMI, Finland.

Zbigniew Karaszewski, Researcher, The State Forests, Szczecinek, Czarnobor 3/2, Poland.

Edgar Kastenholz, Büro für Arbeitsschutz und Arbeitsgestaltung, Rütteberg 10, 79294 Sölden, Germany.

Emil Kitchoukov, University of Forestry, Sofia, Bulgaria.

Anders Lunnan, Department of Economics and Resource Management, Norwegian University of Life Sciences, N-1432 Ås, Norway.

Pekka Mäkinen, Helsinki University, Rural Entrepreneurship, Helsinki, Finland.

Daria Maso, Istituto Nazionale di Economia Agraria, Sede per il Veneto, Via dell'Università 14, 35020 Legnaro PD, Italy. E-mail: daria.maso@unipd.it

Anne Matilainen, Ruralia Institute, University of Helsinki, Kampusranta 9, 60320 Seinäjoki, Finland. E-mail: anne.matilainen@helsinki.fi

Denise McCluskey, Department of Forest Products, Swedish University of Agricultural Sciences, Box 7008, SE-750 07 Uppsala, Sweden.

Piotr Mederski, Senior Lecturer, Poznaÿ University of Life Sciences, ul. Wojska Polskiego 28, 60-637 Poznaÿ, Poland.

Vojislav Milijić, Serbian Federation of Private Forest Owners' Associations, Belgrade, Serbia.

Diana Mizaraite, Lithuanian Forest Research Institute, Kaunas, Lithuania.

Carmen Nastase, 'The Ştefan cel Mare' University, University Street nr. 13, 720 229 Suceava, Romania.

Liviu Nichiforel, Faculty of Forestry, 'The Stefan cel Mare' University, Suceava, Romania.

Anssi Niskanen, Faculty of Natural and Forest Sciences, University of Eastern Finland, Post-box 111, 80101 Joensuu, Finland.

Dragan Nonić, Faculty of Forestry, University of Belgrade, Belgrade, Serbia.

Tomas Nord, Industrial Marketing, Department of Management and Engineering, Linköping University, 581 83 Linköping, Sweden. E-mail: tomas.nord@liu.se

Erlend Nybakk, Norwegian Forest and Landscape Institute, Post-box 115, N-1431 Ås, Norway. E-mail: Erlend.Nybakk@skogoglandskap.no

Anders Q. Nyrud, Norsk Treteknisk Institutt, PO Box 113, Blindern, NO-0314 Oslo, Norway.

Roland Oberwimmer, Holzcluster Steiermark GmbH, Holzinnovationszentrum, 8740 Zeltweg, Austria.

Pekka Ollonqvist, Finnish Forest Research Institute, PO Box 18 (Jokiniemenkuja 1), FI 01301 Vantaa, Finland. E-mail: pekka.ollonqvist@metla.fi

Davide Pettenella, Dipartimento Territorio e Sistemi Agro-Forestali, Agripolis – Università di Padova, Via dell'Università 16, 35020 Legnaro PD, Italy. E-mail: davide.pettenella@unipd.it

Andreja Pirc, Asistant/internal doctorant, Faculty of Forestry, University of Zagreb Svetošimunska 25, 10000, Croatia.

Stjepan Posavec, Faculty of Forestry University of Zagreb, Svetošimunska 25, 10000, Croatia.

Monika Prede, Estonian Naturalists Society, 51003, Tartu, Estonia.

Thomas Rimmler, Finnish Forest Research Institute (Metla), Eastern Finland Research Unit/Joensuu, PL 6, Joensuu, Finland. E-mail: thomas.rimmler@metla.fi

Anders Roos, Department of Forest Products, Swedish University of Agricultural Sciences, Box 7008, SE-750 07 Uppsala, Sweden.

Arto Rummukainen, Senior Researcher, Finnish Forest Research Institute, Jokiniemenkuja 1, PB 18, FI 01301 Vantaa, Finland.

Liana Sadauskiene, Institute of Forestry, Lithuanian Research Centre for Agriculture and Forestry, Liepu street 1, Girionys, LT-53101, Kaunas distr., Lithuania.

Jaroslav Salka, Faculty of Forestry, Technical University in Zvolen, T.G. Masaryka 24, 960 53 Zvolen, Slovakia.

Zuzana Sarvašová, National Forest Centre, Zvolen, Slovakia.

Bill Slee, The Macaulay Land Use Research Institute, Craigiebuckler, Aberdeen, AB15 8QH, UK. E-mail: b.slee@macaulay.ac.uk

Blaž Štefanek, Forest Extension Service, Zagreb, Croatia.

Maria Stoyanova, Forest Research Institute, Bulgaria Academy of Sciences, Sofia, Bulgaria.

Wladyslav Strykowski, Wood Technology Institute Poznan Poland, Winiarska 1, Poznan, PL-60 654, Poland.

Vuokko Takala-Schreib, School of Culture and Design, Seinäjoki University of Applied Sciences, Kotikouluntie 1, FI-66300 Jurva, Finland.

Meelis Teder, Department of Forest Management, Institute of Forestry and Rural Engineering Estonian University of Life Sciences, Kreutzwaldi 5, 51014 Tartu, Estonia.

Saana Tykkä, EFICEEC, Department of Economics and Social Sciences, University of Natural Resources and Life Sciences, Vienna (BOKU), Feistmantelstr. 4, 1180 Vienna, Austria.

Kadri Ukrainski, Faculty of Economics and Business Administration, University of Tartu, 4-A211 Narva Rd, Tartu 51009, Estonia. E-mail: Kadri.Ukrainski@mtk.ut.ee

Gregory Valatin, Senior Economist, Forest Research, Northern Research Station, Roslin, Midlothian EH25 9SY, UK.

Anne Viitala, The Programme for the District of Kauhajoki, Furniture Industry, Nikkarikeskus Furniture Industry Development Center, FI-66300 Jurva, Finland.

Gerhard Weiss, European Forest Institute Central-East European Regional Office (EFICEEC), InFER – Institute of Forest, Environmental and Natural Resource Policy, University of Natural Resources and Life Sciences, Feistmantelstr. 4, 1180 Vienna, Austria. E-mail: gerhard.weiss@boku.ac.at

Preface

This book is the result of the COST Action E51 on 'Integrating Innovation and Development Policies for the Forest Sector', which studied innovation policies and processes in the forest sector in Europe. The central question of the Action was: to what extent do policies create an innovation-enabling environment for enterprises in forestry and the forest-based industries? Among others, the participants of the Action reviewed conceptual approaches for the study of innovation in the sector and conducted empirical case studies from the most important innovation fields.

This publication makes a significant contribution to the study of innovation in forestry and the forest-based industries. It presents the first comprehensive overview and assessment of the relevant research approaches for the study of innovation processes and policies in the forest sector and their application in the most relevant current innovation fields in Europe. It assesses the innovation-related particularities of two very different fields of production: territorial goods and services as well as wood value chains – two areas that differ significantly in their preconditions for innovation and that are conventionally tackled by separate research fields. It is hoped that this book will be a valuable resource for researchers, specialized students, interested stakeholders and policy makers.

As editors, we would like to thank, first of all, the contributing authors who put much effort into bringing their work from the COST Action into the required form of the book chapters. We would also like to thank all of the other participants of the COST Action. A special mention goes to Ewald Rametsteiner, who initiated the Action and, with his personal and professional engagement, led the Action until 2009. We would also like to take the opportunity to thank the COST officers who effectively, and in a friendly and supportive way, administered our work. Last but not least, we give our thanks to Saana Tykkä and Filip Aggestam, who, besides their scientific contributions, facilitated the COST Action and this book publication with their valuable skills. We sincerely acknowledge the financial contributions from the European Science Foundation for supporting the COST Action, and from the Austrian Ministry for Agriculture, Forestry, Environment and Water Management, which had been the main supporter of the Project Centre INNOFORCE and the Central-East European Regional Office of the European Forest Institute (EFICEEC) at the University of Natural Resources and Life Sciences, Vienna, Austria.

Davide Pettenella, Pekka Ollonqvist, Bill Slee and Gerhard Weiss
July 2010

1 The Study of Innovation in the Forest Sector: Relevance and Research Background

Gerhard Weiss

This book presents the results from innovation research in European forestry and forest-based industries. If we want to understand innovation processes, their conditions and drivers, patterns and outcomes, we have to see that innovation is conditional to spatial and sectoral, individual and policy factors. In practice, there are many ways how forestry is understood and what is produced. In this book, we span a very broad range of innovations: from business to institutional innovations, and from ecosystem services with public good characteristics to commodities on a global competitive market. We try to understand further the full range of factors that influence innovation processes: technological regimes, multiple actors, institutional frameworks and policy measures. The marketability of forest-based territorial goods and services is often limited to some degree: all the more innovation matters on firm and institutional levels. The many forest-based value chains differ strongly, whether regarding the traditional timber construction that does not look so traditional any more today, or the recent rise of the energy production on the basis of renewable sources that is still undergoing fast technological changes. Forestry and the forest-based industries thus look very colourful from an innovation perspective. The study of innovation in so many different fields and with so many influential factors quickly shows that there is by no means a uniform European forest sector: instead, there are European forestries.

1.1 The Relevance of Innovation Research in the Forest Sector

It has only been some 20 years since economists identified innovation as the key driving force behind economic growth. Since then, policy makers have increasingly turned their gaze to the means to induce and support innovation. The support of innovation has become a main concern of policy makers at various levels, national and European. Innovation policies as well as entrepreneurship-related policies are a key pillar of the EU 'Lisbon Strategy for Growth and Jobs', which is the economic development policy of the EU that was endorsed in March 2000. The central roles of innovation and competitiveness were reinforced since then and were confirmed very recently in the new economic strategy Europe 2020 that was launched by the European Commission in March 2010.

The name of the strategy, 'Europe 2020 – A European strategy for smart, sustainable and inclusive growth', which aims to lead Europe out of the current economic crisis and to prepare the EU economy for the next decade, identifies three major drivers for growth for Europe: (i) smart

 Innovation in Forestry: Territorial and Value Chain Relationships (eds G. Weiss *et al.*)

growth, developing an economy based on knowledge and innovation; (ii) sustainable growth, promoting a low-carbon, resource-efficient and competitive economy; and (iii) inclusive growth, fostering a high-employment economy delivering social and territorial cohesion. One of the headline EU-level targets is to invest at least 3% of the EU's GDP in research and development, which is a goal that was set already in the earlier economic policies. The strategy also includes other goals – related to employment, renewable energy, education and poverty, which were published already in earlier policies as well. These goals shall be reached by a number of concrete programmes to be implemented jointly by EU-level organization, the Member States and local and regional authorities. One of these is called 'Innovation Union' and aims to improve the framework conditions for innovation, to ensure that innovative ideas can be turned into products and services on the market.

The relevance of studying innovation in the forest sector is given in two directions: first, innovation and growth are as relevant as in any other sector, including forestry and the forest-based industries. There is growing international competition, and steady innovations are crucial to keep pace with global competitors, in terms of costs and quality. Secondly, the sector contributes much to the quality of life and to sustainable development. Forestry produces not only timber but also a range of ecosystem services and amenities. Forest-based products (including energy) use renewable sources and, therefore, contribute particularly to the sustainability aspect in the cited European goals and strategies. The forest sector furthermore provides income and employment opportunities in rural areas, which often face a decline in their economic significance.

These particular roles of the forest sector are increasingly recognized by policy makers and they are increasingly reflected in various policy programmes and initiatives. The new interest in renewable energy sources and renewable materials comes from various policy fields and the related opportunities are often not yet seen so clearly by actors within the sector. One example that illustrates this new interest is the Lead Market Initiative (LMI) for Europe: the forest sector is relevant for half of the named lead markets! The LMI for Europe was launched by the European Commission following the EU's 2006 broad-based innovation strategy. Six markets have been identified that are seen as highly innovative, that provide solutions to broader strategic, societal, environmental and economic challenges, that have a strong technological and industrial base in Europe and that depend more than other markets on the creation of favourable framework conditions through public policy measures. They are: eHealth, protective textiles, sustainable construction, recycling, bio-based products and renewable energies. The forest-based sector may contribute significantly to sustainable construction, bio-based products and renewable energies.

From this view, it can be said that forestry and the forest-based industries play an important role in rural economies and have a strong potential to contribute to a sustainable future of Europe, for instance in the following fields:

- recreation and tourism;
- nature protection, biodiversity conservation and landscape amenities;
- protective functions such as against natural hazards;
- bio-energy production and climate change mitigation; and
- bio-based products, including food, fibres, chemicals and wood construction.

The forest sector is often considered as a mature, 'low-tech' and declining industry. This understanding comes with the notion that our economy changes into a service economy, with the hype of information and communication technologies and with the popularity of high technologies as the source of future wealth. This provides research and high technology with attention in the media, in the public and in policy as if they were the only source for economic growth and innovation (Tunzelmann and Acha, 2005; Hirsch-Kreinsen and

Jacobson, 2008). This is not true: studies show that low- and medium-technology sectors still play a major role for employment and growth. Although these sectors invest less in research and development, they are still relevant for innovation. Innovations in mature sectors occur in different forms. Wood processing industries, for example, use sophisticated technologies in their production.

1.2 The Policy Background for Innovation and Development in the Sector

A number of policies are relevant for the support of innovation in forestry and the forest-based industries. At a pan-European level, Forest Europe/MCPFE (Ministerial Conference on the Protection of Forests in Europe) has adopted the Vienna Resolution on the 'Economic Viability of Sustainable Forest Management', calling for the strengthening of innovation and entrepreneurship in this sector. In the EU, a range of policy fields affects the forestry sector innovativeness. First of all, the public support for research and development plays an important role in the development of the sector. Innovation is an explicit objective in the creation of the European research area and in the EU Framework Programmes for Research. While the Fifth EU Framework Programme for Research hardly included the forest sector, this has since changed. The EC DG on research has a number of activities that are relevant for the forest sector, including the currently developed European Innovation and Research Plan, which will include the creation of European Innovation Partnerships between the EU and national levels, among others, with a proposal on Innovation Partnership in the Bio-Economy. The EC (DG RTD) will also prepare a Communication on Sustainable Bio-Economy, expected mid-2011, in turn providing a vision and action plan for a sustainable and innovative European bio-economy. It will aim at improving the framework conditions for innovation in the bio-economy sectors and at integrating the bio-economy approach into the forthcoming reform of the CAP. It will be interesting to see how far the forest sector will be able to take up those impulses and how far they will translate into new sectoral products.

A milestone in the development of the institutional framework for forest industry-related research was the creation of the Forest-based Sector Technology Platform (FTP) in 2004, by the European Confederation of Woodworking Industries (CEI-Bois), the Confederation of European Forest Owners (CEPF) and the Confederation of European Paper Industries (CEPI). Industry-wide technology platforms have been created by industry stakeholders to mobilize Europe's research, technological development and innovation efforts and to express the related goals *vis-à-vis* the European Commission. A central activity of the FTP was the development of a Strategic Research Agenda (SRA) that is aimed at supporting the wood-based value network in their further development. The National Research Agendas (NRAs) are the instrument to implement the SRA on the national level and to complement the relevant national policy documents. These NRAs, now available from 17 countries, transfer the SRA objectives for forest-based sector development in each participating country according to their specific circumstances and needs. They successfully influenced national policies but hardly reached out to include non-industry or other sector's actors in the development process (Niskanen *et al.*, 2010; Tykkä *et al.*, 2010).

Rural, regional and sustainable development policies are of specific importance for the sector, as they focus, among others, on the integrated sustainable development of rural regions in Europe and on the promotion of renewable materials and energy sources. These policies have developed strategies, have asked for national level programmes and provide concrete support instruments. The agricultural policy has been broadened to include the goals of an integrated cross-sectoral rural development and dedicates some of the financial resources for this goal. An interesting question here, of course, is how far the CAP effectively changes from an agricultural to a rural development policy and how far forestry may benefit. The rural development pillar (Pillar 2 of the policy) has so far only granted a minor share of the total budget and experience, and an evaluation shows that the benefits are still largely on the side of agriculture and not other rural sectors. A number of initiatives under the regional development policy address regional innovation strategies but, also here, it seems that very little of these funds have involved forest holdings and forest sector firms. The EU sustainable development goals have been, as can be seen in the above-mentioned economic development initiatives, included in various policies. In addition, the rural development guidelines consider the social aspects of rural economies, in terms of employment, biodiversity and quality of life in rural areas. The recent bio-energy boom, which is reflected in and supported by renewable energy policies on the European and national levels, has given forestry a new place in the sustainable development in practical terms.

Forestry and forest industries policies in Europe have reacted in various ways to the recent challenges experienced in the sectors. While the former has seen considerable environmental pressure and an increase in interest by society in recreation compared to the production of wood, many sectors in the forest industries in European countries have experienced commoditization of their products, globalization and industry concentration. In general, in all sectors, the number of jobs provided has declined and productivity has increased considerably. For quite a long time cost-cutting has been the main answer to economic viability, with the implementation of technological and organizational innovations. However, in recent years, awareness has risen across the industry that a main area of strategic competition is occurring *vis-à-vis* other substitute products, and in the field of product and service innovation. Societal changes do not necessarily only bring the sector under pressure but new societal demands also come along that provide new opportunities. Policy makers, however, do not always realize these new opportunities so quickly and often do not opt for proactively supporting the development of new solutions and offers by the sectors in order to take advantage of the new demands.

While there is a general understanding of the importance of innovation in both business and policy circles, neither innovation process is well understood, nor is it clear which kinds of policies are most effective in encouraging innovation and how to integrate innovation policies in the wider policy setting. Thus, despite the strong political interest in innovation and innovation-supporting policies, innovation-related research still leaves many questions unanswered. This is despite the fact that general innovation research has made considerable progress during the 1990s, with a shift from a linear understanding of innovation to a more complex systemic view (Edquist, 2001). Not yet well developed is the link between the innovation system research approach and its large implication on economic policy. Entrepreneurship research, despite its recent strong political support, has considerably less research tradition and an even weaker body of theory or consistent empirical research results (Audretsch, 2002).

With a linear understanding of innovation processes from research to the eventual marketing, the traditional science and technology approach to innovation policy focused strongly on the financing of public or private research and development activities. With the growing importance of systemic views of innovation processes, policy approaches are also changing. Lundvall and Borras (2005) describe how science policy changed into technology policy, which again developed into innovation policy, illustrated by the relevant OECD reports along the years. Since 1990, OECD has seen innovation as an interactive process. OECD (1997) links the possible innovation system-related policy responses to a range of possible systemic imperfections, such as informational failures, limited interaction between actors in innovation systems, institutional mismatches between (public) knowledge infrastructure and market needs or missing customer demand.

As seen in the example of the Europe 2020 strategy, policies are increasingly addressing framework conditions in addition to a mere financing-of-research approach. The Action Plans within the LMI are based on the following policy instruments: legislation, public procurement, standardization, labelling, certification and complementary actions such as training, and financial support, altogether representing a comprehensive and systemic approach to support innovation. Another prototypical example for a systemic innovation policy approach is the Leader instrument for rural development. The analysis of policy documents from relevant policy fields in the course of the COST Action E51 revealed that a systemic rhetoric is increasingly used but that measures often remain traditional (Weiss *et al.*, 2010).

1.3 Research on Innovation and Development in the Forest Sector

Many aspects of rural development have been covered by forestry-related research in a range of projects (COST E3, EU MULTIFOR. RD). In the field of social sciences, COST Action A12 has studied rural innovation. Sustainable forest management as a holistic concept for sustainable development has seen considerable effort put into operationalization, including forest sector research, e.g. EU projects on life cycle analysis or the IUFRO Task Force on SFM criteria and indicators. The EU Framework 6 Integrated Project EFORWOOD recently concluded with a prototype of an integrated sustainability impact assessment tool for the whole forest sector (ToSIA), which includes a policy component that shall be further developed into an interface by which the relations between sectoral production alternatives and policies can be assessed (Vogelpohl and Aggestam, forthcoming).

The EC commissioned a number of relevant overview studies, including the Study of the Effects of Globalization on the Economic Viability of EU Forestry (Nilsson *et al.*, 2007), the Study on the Development and Marketing of Non-Market Forest Products and Services (Mavsar *et al.*, 2008) and the Study on the Market Supply of Wood and Other Forest Products, in particular on obstacles to mobilization related to fragmented ownership structures (ongoing).

In recent years, several COST Actions and international research projects have addressed forest specific aspects of innovation, rural development or sustainable development/sustainable forest management. Innovation-related aspects have been analysed in the context of the European Forest Institute (EFI) Project Centre INNOFORCE (Rametsteiner *et al.*, 2005), the European Forest Institute Central–Eastern European Regional Office (EFICEEC) respectively.[1] The COST Action E30 on the Economic integration of urban consumers' demand and rural forestry production was especially dedicated to entrepreneurship aspects of small-scale forestry, the multifunctional use of forests as well as the timber and wood industries (Niskanen *et al.*, 2007). The area of innovations in forest services – without using the term innovation – were, and are being, addressed by COST Actions (E39 on forest-related health and well-being, E33 on forest recreation and nature-based tourism, or E45 on forest externalities) and EU research projects. The EU research project on Niche markets for recreational and environmental markets for forest enterprises was a landmark in the analysis of marketing possibilities of forest-related recreational and environmental services (Mantau *et al.*, 2001). The recently started EU research project NEWFOREX on New Ways to Value and Market Forest Externalities goes in the same direction and assesses several market-based methods for enhancing the provision of forest externalities, including e.g. payment schemes provision, certification or (re-) definition of property rights.

1.4 The Contribution of COST E51

The COST Action E51 'Integrating Innovation and Development Policies for the Forest Sector',[2] which provided the background for this book publication, particularly looked at the policy dimension of innovation and did this on two levels: the policy formulation and the impact of the policies 'on the ground'. The general objectives of the Action were to compare innovation, rural, regional and sustainable development policies and programmes that are relevant to forestry and forest sector enterprises, to assess their effects on innovation in forestry and forest sector firms in rural areas, and to develop approaches for a more integrated and coordinated implementation of innovation support in these policy fields. It aimed to identify and analyse key issues in strengthening cross-sectoral policy integration/coordination in those key development programmes relevant for forestry and forest sector firms in rural areas, in two major production fields:

- territory-based goods and services (e.g. the provision of recreational services, nature conservation or carbon sequestration); and
- vertical wood-related production chains (e.g. furniture, timber frame housing, bioenergy or timber harvesting operations).

Three guiding research questions were formulated:

- How far and in which way is the concept of innovation integrated into relevant policy fields?
- How far and in which way are the relevant policies coordinated?
- Do the policies create an innovation-enabling environment for forestry and forest-related industries?

In Phase I, the central policy documents from seven fields were analysed in 19 countries: forestry policy, forest-based industries policy, innovation policy, rural development policy, regional development policy, sustainable development policy and renewable energy policy. The results from the comparative analysis and a number of in-depth studies were published by Rametsteiner *et al.* (2010).

Phase II focused on innovation processes at the firm or local level. Groups of researchers conducted case study analyses from different European countries in some of the most important current innovation fields in forestry and the forest-based industries. A range of theoretical–conceptual approaches was used by the research groups. These approaches were compatible in that they all did not focus on innovation management within the companies but that they employed a broader, systemic view on innovation processes, appropriate to include institutional frameworks, policies and the aspect of actor-interrelations into the analysis. In Phase III, finally, conclusions and policy recommendations were derived. A science–policy dialogue, which was organized in course for the Final Conference of the COST Action E51 with the participation of key policy makers and stakeholders, was an important element.

The work of the COST Action E51 was carried out from 2006 to 2010 with the participation of approximately 40 research institutions from 20 countries. Various methods of data collection were applied, including desk research, document analysis, questionnaires and interviews with experts, carriers of innovation projects and involved stakeholders.

1.5 The Structure of the Book

The work carried out in phases II and III is presented in this book. The coordinated structure of the COST Action resulted in a fairly coherent structure of the book. The first part of the book presents the conceptual basis and a number of overarching themes. The empirical core of the book falls into two parts: one on the innovation field of territorial goods and services, and the other on wood value chains. Both parts start with a comprehensive conceptual examination of the innovation-related specificities of the two fields and include empirical analyses of some of the most important current innovation fields.

The major theoretical approaches that are available and relevant for this endeavour and that were used in the empirical work are presented in the first part of the book in Weiss's Chapter 2. After clarifying a number of central terms in the field of innovation research, a broad variety of mostly systemic innovation research approaches are described and the specific spatial (the regional dimension of innovation) and the sectoral aspects are discussed (the forest sector as a low-tech, traditional or mature sector and the importance to consider small and medium-sized enterprises). The first part of the book furthermore includes a number of conceptual articles and thematic analyses that are of overarching relevance. Pettenella and Maso (Chapter 3) take a closer look at the role of networks for innovations in forestry. They play a specific role for the small businesses that dominate the forest sector, even more pronounced in the increasingly important field of non-wood forest goods and services. While their chapter is more interested in informal business networks with a particular view towards cross-sectoral relations, Mendes *et al.* (Chapter 5) look at the enduring cooperation of forest owners within forest owners' associations. It frames the emergence of forest owners' organizations as an institutional innovation and compares the current situation in ten countries from northern, central-eastern and southern Europe.

The first section of the book further analyses two exemplary policy instruments with strong systemic features and that are of high relevance for innovation support, for both innovation fields: territorial goods and services, and wood-based value chains. The first analyses the Leader instrument of the EU Rural Development Programme. Feliciano *et al.* (Chapter 6) evaluate, on the basis of a questionnaire survey of Leader+ projects in four countries, how far Leader+ created enduring networks and how far it effectively supported bottom-up initiatives and risk-friendly and innovative projects (Leader+ was the name in the EU programming period 2000–2006). The second systemic innovation support instrument that is often applied in the wood sector is cluster initiatives. They can play a crucial role in the local networking among firms but also in the connection to international and/or scientific knowledge sources. Rimmler *et al.* (Chapter 7) compare cluster policies and initiatives in four European countries. Ukrainski and Kajanus (Chapter 4) look at a crucial part of innovation processes: the knowledge flows. Industrial firms gain ideas for innovation from various sources, internal and external. Their performance, and this is also true for smaller firms, increasingly depends on how successfully they are able to utilize external knowledge, such as from research. The analysis is based on data from the Community Innovation Survey (CIS).

The second part of the book focuses on innovation processes and policies in the field of forest-based territorial goods and services. The conceptual chapter by Slee (Chapter 8) includes a clear definition of territorial goods and services that points at two notions of the term that are both relevant for innovation: first, territoriality is connected with product differentiation where the locality of production is embodied in the character of the product. Secondly, multifunctional land use brings out multiple benefits, many of which are produced as externalities and have often public goods characteristics.

Three empirical studies shed light on important innovation fields that have quite different characteristics. Ciccarese *et al.* (Chapter 9), by asking which new market opportunities arise because of new demands for carbon sequestration in European forests, describe carbon markets as an example for an institutional innovation. They describe the opportunities related to the market-based mechanisms of the Kyoto Protocol, the European-regulated carbon markets and the voluntary carbon market, the latter being the most relevant at the moment. Maso *et al.* (Chapter 10) turn to another market for which business networks play an important role: non-wood forest products and services. Networks compensate for the typical problems that small enterprises face in introducing innovations. Another chapter particularly deals with forest-based tourism services. Matilainen *et al.* (Chapter 11), using cases from six western and eastern European countries, describe forms of cooperation between nature-based tourism enterprises and critical stakeholder groups.

The third part of the book is dedicated to wood value chains, spanning the timber harvesting and timber-based industries and bioenergy production. At the beginning, Ollonqvist's conceptual chapter (Chapter 12) describes how important actor interrelations and knowledge transmission through sectoral and regional innovation systems are for small businesses in this traditional sector.

Nord *et al.* (Chapter 13) then assess innovation processes in the timber construction industry and policy influences by using cases from eight countries. They particularly look at the position of the timber construction industry in the construction process and the sources of the innovations. Ollonqvist *et al.* (Chapter 14) study networks and local milieus as innovation platforms in the furniture industry. Using cases from five countries from western and eastern Europe, they analyse patterns of local furniture industry clusters and particularly focus on the role of knowledge and on patterns of knowledge transactions. The following chapter deals with a rather recent innovation, namely using wood resources for energy production on industrial scale. Nybakk *et al.* (Chapter 15) analyse the sources of innovation on the firm level and assess which policies supported or impeded the innovations. They use cases from eight countries from different European regions, including wood chips and pellets production, district heating and electricity generation. The last empirical chapter finally looks at an industry sector in the wood value chain that is often neglected, the timber harvesting operations. Bouriaud *et al.* (Chapter 16) study eight cases of innovations in micro-, small and medium-sized firms in central and eastern Europe, and identify market and policy-related conditions of innovation.

The final part of the book consists of two concluding chapters. Aggestam and Weiss (Chapter 17) present the results of a science–policy dialogue. The findings of the work in the COST Action E51 were presented and discussed at the final conference of the Action, 1 June 2010, in Brussels. This conference was designed to fertilize an exchange of experiences between the researchers and policy makers from the relevant fields. The moderated discussions were video recorded and then analysed. The final chapter (Weiss *et al.*, Chapter 18) eventually sums up all the experiences from the conceptual and empirical work and from the conference. It discusses a number of central issues that have arisen from the work: we need a differentiated understanding of entrepreneurship and innovation in this innovation field that is strongly related to the sustainable development of our society; we have to be aware of the specific demands for innovation in forestry, which is strongly connected to the territory and the multiple benefits that it provides to society; and we have to be aware of the specific requirements that the enterprises in the forest sector have, being small businesses that operate in a traditional sector.

Notes

[1] The EFI PC INNOFORCE existed 2000–2009 and worked on innovation processes and innovation policies in the European forest sector. It was then further institutionalized as EFICEEC that, among others, follows these research topics.

[2] For more information on the COST Action E51 – Integrating Innovation and Development Policies for the Forest Sector – see http://www.boku.ac.at/coste51.

References

Audretsch, D.B. (2002) Entrepreneurship: a survey of the literature. Prepared for the EC DG ENTERPRISE. London, Institute for Development Strategies, Indiana University and Centre for Economic Policy Research CEPR.

Edquist, C. (2001) The system of innovation approach and innovation policy: an account of the state of the art. Lead paper. *Nelson Winter Conference*, Aalborg.

Hirsch-Kreinsen, H. and Jacobson, D. (eds) (2008) *Innovation in Low-Tech Firms and Industries*. Edward Elgar, Cheltenham, UK.

Lundvall, B.-A. and Borras, S. (2005) Science, technology and innovation policy. In: Fagerberg, J., Mowery, D. and Nelson, R. (eds) *The Oxford Handbook of Innovation*. Oxford University Press, Oxford, pp. 599–631.

Mantau, U., Merlo, M., Sekot, W. and Welcker, B. (2001) *Recreational and Environmental Markets for Forest Enterprises: a New Approach Towards Marketability of Public Goods*. CAB International, Wallingford, UK.

Mavsar, R., Ramcilovic, S., Palahi, M., Weiss, G., Rametsteiner, E., Tykkä, S., Apeldoorn, R. v., Vreke, J., Wijk, M. v. and Janse, G. (2008) Study on the development and marketing of non-market forest products and services (FORVALUE). Study Report for DG AGRI, Study Contract No. 30-CE-0162979/00-21.

Nilsson, S., Rametsteiner, E., Böttcher, H., Havlik, P., Kraxner, F., Leduc, S., Obersteiner, M., Rydzak, F., Schneider, U., Schwab, D. and Willmore, L. (2007) Study of the effects of globalization on the economic viability of eu forestry. EC CONTRACT NUMBER 30-CE-0097579/00-89. Brussels.

Niskanen, A., Mederski, S., Nord, T., Bouriaud, L., Tykkä, S., Vennesland, B. and Vogelpohl, T. (2010) Policy diffusion in the forest-based sector in Europe. In: Rametsteiner, E., Weiss, G., Ollonqvist, P. and Slee, B. (eds) *Policy Integration and Coordination: the Case of Innovation and the Forest Sector in Europe*. OPOCE, Brussels.

Niskanen, A., Slee, B., Ollonqvist, P., Pettenella, D., Bouriaud, L. and Rametsteiner, E. (eds) (2007) *Entrepreneurship in the Forest Sector in Europe. Silva Carelica 52*, University of Joensuu, Faculty of Forestry.

OECD (1997) *National Innovation Systems*. OECD, Paris.

Rametsteiner, E., Weiss, G. and Kubeczko, K. (2005) *Innovation and Entrepreneurship in Forestry in Central Europe*. Brill, Leiden.

Rametsteiner, E., Weiss, G., Ollonqvist, P. and Slee, B. (eds) (2010) *Policy Integration and Coordination: the Case of Innovation and the Forest Sector in Europe*. OPOCE, Brussels.

Tunzelmann, N. v. and Acha, V. (2005) Innovation in ‘low-tech’ industries. In: Fagerberg, J., Mowery, D.C. and Nelson, R. (eds) *The Oxford Handbook of Innovation*. Oxford University Press, Oxford.

Tykkä, S., Bergseng, E., Weiss, G., Kleinschmit, A. and Rametsteiner, E. (2010) National Research Agendas of the Forest-based Sector Technology Platform. In: Rametsteiner, E., Weiss, G., Ollonqvist, P. and Slee, B. (eds) *Policy Integration and Coordination: the Case of Innovation and the Forest Sector in Europe*. OPOCE, Brussels.

Vogelpohl, T. and Aggestam, F. (forthcoming) Public policies as institutions for sustainability: potentials of the concept and findings from assessing sustainability in the forest-based sector. *European Journal of Forest Research*.

Weiss, G., Salka, J., Dobsinska, Z., Rametsteiner, E., Bauer, A., Aggestam, F. and Tykkä, S. (2010) Integrating innovation in forest and development policies: comparative analysis of national policies across Europe. In: Rametsteiner, E., Weiss, G., Ollonqvist, P. and Slee, B. (eds) *Policy Integration and Coordination: the Case of Innovation and the Forest Sector in Europe*. OPOCE, Brussels.

2 Theoretical Approaches for the Analysis of Innovation Processes and Policies in the Forest Sector

Gerhard Weiss

Abstract

This chapter gives an overview of theoretical and conceptual approaches that are relevant for the analysis of innovation processes and innovation policies in the forest sector. Research approaches are required that not only look at innovation processes within firms but that also include other relevant actors as well as institutional factors into the analysis. This chapter, therefore, briefly presents and discusses a range of available research approaches for these research questions: innovation systems, clusters, innovative milieus, regional governance, innovation governance and related approaches. The chapter particularly discusses the spatial and sectoral aspects that are of relevance: not only the production systems within forestry – and partly in the forest-based sector – have a territorial dimension, but also the related innovation processes and innovation systems. Specific sectoral aspects relate to the character of the forest-based industries as a traditional sector, and the fact that small enterprises play an important role in the sectoral production.

2.1 Introduction

Our ambition in this book is to analyse innovation processes in the forest sector and the policies that influence them. What we look at, therefore, are the development and implementation of new goods and services and new processes in the field of territorial goods or services of forestry, as well as in the forest-based industries. Furthermore, we explore the impact of institutional environments for these innovations. For the investigation of those questions, research approaches are required that are not confined to intra-firm processes but include other actors in the analysis, as well as the role of institutional and/or cultural factors. There are several theoretical–conceptual approaches available that look, for instance, at innovation systems, clusters, innovative milieus, innovation policies, and regional and innovation governance. Further specific issues that are particularly relevant when analysing the forest sector are the facts that: (i) the sector we are studying is classified as a low-tech and traditional sector, in which the majority of the firms are micro-, small or medium-sized enterprises (SMEs) that are often owned and run by families; and (ii) forestry has a strong territorial or place-specific dimension. The production of many goods and services of the sector and relevant innovation processes have thus a strong regional dimension.

In this book, different theoretical approaches are employed, however, within the common framework as described. This chapter on the theoretical approaches tries not to formulate or explicate one certain theoretical approach but rather tries to provide an overview of the various approaches that are available for our type of analysis and that – with changing foci – have been applied in the empirical chapters.

2.2 Definitions of Innovation

In this book project, we did not aim at using one single strict conceptual framework. We rather allowed for multiple conceptual approaches to be utilized. There was, however, still a common framework that guided our analyses and that acted as a common denominator: a broad understanding of innovation processes that include multiple actors and institutional frameworks in the analysis.

Throughout this book, the term 'innovation' is generally used from an economic and business perspective, meaning that in the centre of our interest are innovations on the market. It, therefore, is based on the work of Schumpeter, who defines innovation broadly as a discontinuously occurring implementation of new combinations of means of production (Schumpeter, 1934). This does not prevent us from taking a systemic view of innovation processes or looking at policy processes as well. The term 'innovation' is mostly reserved for business-related, market-related innovations, but for certain analyses, we also speak of institutional innovations that then refer to changes in the institutional or policy realm. We thus added to the commonly used types of innovation on the firm level as, for instance, defined by OECD (2005), the category of institutional innovations (Fig. 2.1). We decided to enlarge the understanding of innovation in this way in order to include further reaching innovations into our analyses that require institutional or policy changes. An analysis of the innovation frontiers for the forestry and forest-based industries in Europe reveals that, particularly in the field of territorial goods and services, this additional category is of high relevance also in practice (Rametsteiner *et al.*, 2010). In this book, Slee explores this aspect in detail in his conceptual chapter on innovations in territorial goods and services (Chapter 8). Mendes *et al.* (Chapter 5) describe forest owners' associations and Ciccarese *et al.* (Chapter 9) describe carbon markets as important examples for institutional innovations.

The term 'innovation' may be used to describe the process of innovation or the output of that process. Basically, by innovation

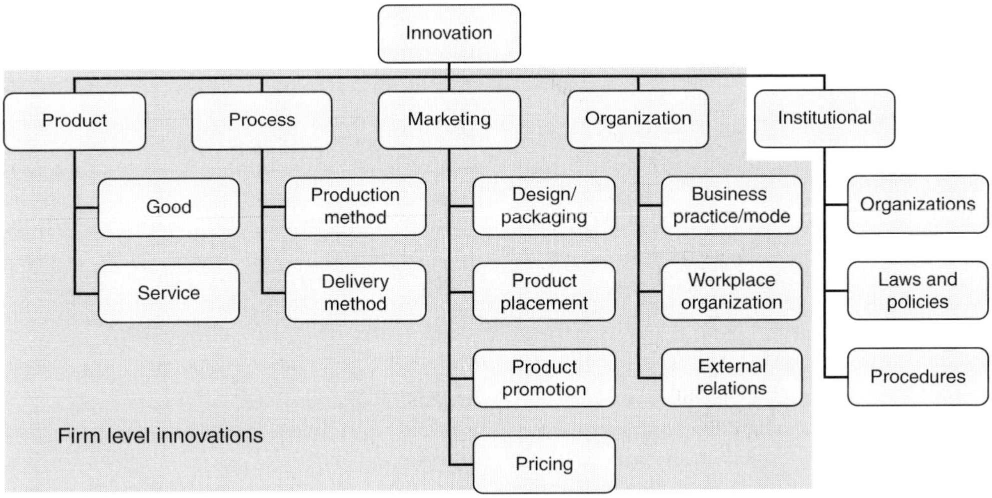

Fig. 2.1. Types of innovation (Weiss *et al.*, 2010a, modified from OECD, 2005).

we mean the introduction of novelties to the market (Rametsteiner *et al.*, 2005). Innovation is the first economic utilization of a new product or process. 'New' may mean new to the market, new to a sector or new to a firm. We thus speak of different degrees of novelty. Related to these different degrees is the concept of the diffusion of innovations (Rogers, 1995): not only is the implementation of a novelty by the first innovating firm relevant to society, but also the uptake/adoption of the innovation by others. Each adoption of a novelty by another firm is a non-trivial process within that firm that requires a smaller or larger adaptation of the novelty or the firm itself. A further distinction between different degrees of novelty is between radical and incremental innovations (Freeman, 1982). The first would be more profound change in a production process or product, the latter a small-scale adaptation that means an improvement of the still same product or process. Of course, radical innovations are particularly interesting, but most improvements in production processes or products are realized in smaller, incremental steps. In sum, the authors argue that many smaller, incremental innovations may in many cases be economically more important than the original radical innovation (Lundvall, 1992).

Innovations may be classified into new products (goods or services) and new processes (technological or organizational processes) (Edquist, 2005). It is noteworthy that innovations are not only new technologies but are also, for instance, new business models or new marketing methods that are important improvements in businesses. The OECD classification scheme makes that even more visible by classifying innovation into the following types (Fig. 2.1):

- product innovations – new or significantly improved goods or services (including technical specifications, components and materials, incorporated software, user friendliness or other functional characteristics);
- process innovations – new or significantly improved production or delivery methods (including significant changes in techniques, equipment and/or software);
- marketing innovations – new marketing methods (including significant changes in product design or packaging, product placement, product promotion or pricing);
- organizational innovations – new organizational methods in a firm's business practices, workplace organization or external relations.

It should be noted that, in the real world, the categories of innovations often overlap or are combined, i.e. that the implementation of new technologies often needs also changes in other fields, e.g. in workplace organization, etc. For the work of the COST Action E51, another type of innovation was added to these firm-level innovations (Weiss *et al.*, 2010a):

- institutional innovations – these are changes in the political–institutional framework of the sector. Many improvements or novelties cannot be implemented by market actors alone but depend on changes in the policy field or procedures (regulations or incentives) or joint action supported by public or semi-public organizations such as authorities or interest groups. They may be the creation of new markets (e.g. carbon trading) or joint actions (e.g. in the framework of forest owners associations, industry cluster initiatives, etc.).

2.3 Studying Innovation

One challenge in the study of innovations is the complexity of the processes that are involved in the development and implementation of novelties. It is widely recognized today that the conventional linear model of the so-called new product development process hardly pictures the real processes (Kline and Rosenberg, 1986). Instead of a linear process from the R&D department of a firm to its commercialization, innovation processes often involve repeated cycles and multiple actors, and are typically unpredictable in terms of costs and outcome (Kline and Rosenberg, 1986; Castellacci *et al.*, 2005; Pavitt, 2005). The innovation processes are complex within the firms, but go far beyond the firms, including a range of other actors, depending on institutional and cultural conditions.

The different levels that are relevant in innovation processes have been defined in different ways. In their overview article on innovation studies, Castellacci *et al.* (2005) distinguish organizational, systemic, sectoral and macro-economic levels. For the work of INNOFORCE, a different scheme was described: personal, enterprise, business-to-business and institutional levels (Kubeczko and Rametsteiner, 2002). Nybakk *et al.* (Chapter 15, this volume) employ another classification that defines an individual, organizational, inter-organizational and societal level that resembles the INNOFORCE approach. The overview presented below largely follows the two latter concepts and distinguishes the individual, organizational, inter-organizational and institutional–societal levels.

2.3.1 Individual level

On the individual level, innovation research deals with the personalities of owners and managers of businesses, but also other staff that are relevant for the innovation processes. Rogers (1995) classifies individual agents according to their innovation-related behaviour as innovators (the venturesome), early adopters (respected local opinion leaders), early majority (deliberate followers), late majority (the sceptical) and laggards (traditionalists). It is often posited that certain innovative individuals play a crucial role in the innovation processes, who obviously are the masterminds and driving agents behind an innovation. Schön or Jenssen and Jørgensen refer to them as 'innovation champions' (Schön, 1963; Jenssen and Jørgensen, 2004). The present book's chapter on new forest-based bio-energy systems (Nybakk *et al.*, Chapter 15) takes a firm-level perspective with a particular attention to the role of the entrepreneur. In the case of owners as managers of small businesses, the personal and organizational levels are closely linked.

Innovators are essentially entrepreneurial persons: innovativeness is linked to entrepreneurial attitudes and behaviour as a personal trait. An entrepreneurial attribute is mostly connected with risk-taking, but also responsibility: Morris describes entrepreneurship as the willingness to 'embrace new opportunities and take responsibility for effecting creative change' (Morris, 1998). Within Cultural Theory, the 'entrepreneur' is described as an individualist cultural style of behaviour. For an individualist, nature is benign and resilient – able to recover from any exploitation, and man is inherently self-seeking and atomistic (Thompson *et al.*, 1990).

2.3.2 Organizational level

On the organizational level, the study of innovation focuses on the issue of managing innovation. Innovative capabilities of a firm are its capability to generate customer value by developing and introducing new products and services to the market or by reducing the costs (Ollonqvist and Rimmler, 2010): the innovative capabilities depend on organizational structures and management practices. The firm's innovation system is embedded in the firm's resource base, management system and organizational structure. The rate of innovations depends mostly on continuous business routines, and spontaneous innovations are rather the exception than the rule. Innovative firms are entrepreneurial firms that engage in 'product-market' innovations, 'undertake somewhat risky ventures, and are first to come up with pro-active innovations' (Wiklund, 1998).

Important features for the innovativeness of firms relate to their structure, culture and strategies. Rogers defines a number of variables that explain organizational innovativeness (Rogers, 1995): he assumes centralization and formalization as having a negative impact, and complexity, interconnectedness, organizational slack and system openness as positive. Mintzberg (1979) relates innovativeness to five typical organizational archetypes: (i) machine bureaucracy and (ii) divisionalized form with low degrees of innovativeness; (iii) simple structure and (iv) professional bureaucracy with a medium degree of innovation potential; and (v) 'adhocracy', which works on the basis of mutual adjustment with a high innovation potential (Mintzberg, 1979). Other scholars have coined different typologies of innovative firms. One conclusion is that different types of organizational structures have different innovation potential but highly innovative firms may also have different organizational structures. While adhocracy is a highly dynamic and adaptive organizational structure found in Silicon Valley-type companies, another organizational form of innovative companies would be the 'J-form' of typical Japanese companies with a highly organized innovation management (Lam, 2005). Freeman empirically discovered the following innovation strategies that allow one to assess the innovation-orientation of a firm: offensive and defensive innovation strategy, imitation strategy, traditional and niche strategies, as well as dependent satellites (the latter of which do not innovate) (Freeman, 1982).

Scholars have taken different perspectives from which to suggest innovation strategies for firms. Cooper developed the stage-gate model (Cooper, 2006) for structuring the new product development process within a firm 20 years ago. As a more market-oriented model for innovation management, Hippel at about the same time presented the demand-pull driven lead-user model (Hippel, 1986). These different approaches highlight that different aspects are relevant for the management of innovation. A more comprehensive view is taken by Bean and Radford, who, in their innovation management model, define four activity clusters within firms, each essential to the comprehensive understanding of innovation and each requiring management (Bean and Radford, 2002): the first and most overreaching cluster is that of providing a nurturing and hospitable environment for innovation (nurture/build). The second (create/develop) is the creation of the capacity for innovation. The implement/commercialize cluster is the everyday blocking and tackling. Finally, the exploit/manage cluster focuses on carefully reaping the benefits from the effort.

2.3.3 Inter-organizational level

Engaging in inter-organizational collaboration seems to be of increasing importance for successful companies in recent decades (Castellacci *et al.*, 2005; Powell and Grodal, 2005; Smith, 2005): studies show that collaboration is positive for the firm's innovative capabilities and innovative firms tend to cooperate more strongly than others. It seems to be particularly beneficial to have diverse ties and to be centrally placed within a network. Various forms of organizational cooperation and collaboration have been described, among others distinguishing between horizontal or vertical cooperation. Major contributions to research and policy came from network theory and Porter's cluster concept. Most literature focuses on the role of cooperation among firms (business-to-business), but from an innovation research point of view the relations to any other organizations or actors may also be important, as for instance described by Freeman's stakeholder theory. Studies on inter-organizational cooperation and networking often focus on regional relations but the importance of global-wide cooperation is increasingly discussed.

The network concept is a key element in many innovation models on the business-to-business level, including the approaches of the innovation district, the creative milieu or the learning region. Networks are intermediate organizational forms between markets and firms, when these fail in efficiency and efficacy. They find themselves between the flexibility and autonomy of markets and the force and control of organizational authority (Powell and Grodal, 2005). Networks offer benefits such as a division of labour or information exchange and learning. Grabher characterizes networks by four basic features: reciprocity, interdependence, loose coupling and power (Grabher, 1993). Forms of networks range from formal contractual relations to informal ties such as a common membership in professional associations. Strong and weak ties (Granovetter, 1972) have different functions within networks and exert various effects on the innovativeness of firms. The analysis of firm networks plays an important role in this book, even if our studies do not apply highly structured network analyses. Pettenella and Maso (Chapter 3) explain further the study of networks in multifunctional forestry and the case study analyses by Maso *et al.* (Chapter 10) and Matilainen *et al.* (Chapter 11) on various territorial goods and services pay particular attention to actor networks.

Clusters, as described by Porter (1998), are based on competition rather than trust (as would be typical of networks). For Porter, they can be an important source for competitive advantages of a nation. The positive effects of agglomeration in specialized clusters are related to information flows and relatively low transaction costs. Firms benefit, among other factors, from specialized education and training as well as from a specialized labour market. By use of his 'diamond model', Porter characterizes clusters by way of four key elements: a geographical concentration; a specialization of several industries in closely related technologies or markets; a presence of companies together with other institutions such as training and research institutions, regulatory institutions, public bodies, intermediaries, financing institutions, etc.; and the interrelations between these actors.

Empirical studies of successful clusters and Porter's conclusions and recommendations triggered many policy initiatives to support or create industry clusters as a means for development. The cluster concept thus exists as a concept for empirical research as well as a concept for policy intervention. Clusters as a policy means are often regionally based service organizations aimed at supporting the collaboration within specific sectors and knowledge development, in other words, functions of innovation systems. It is, therefore, important to be clear if statements or studies on clusters refer to the empirical phenomenon or policy initiatives. Cluster policies can be regarded as systemic innovation policy approach as framework conditions are taken into account and non-firm actors are given important roles (Nauwelaers, 2003). In this book, Rimmler *et al.* (Chapter 7) and Ollonqvist *et al.* (Chapter 14) present analyses of clusters as policy means in wood industry contexts.

Stakeholder theory argues that for the long-term success of a firm, not only are investors, employees, suppliers and customers important, but there are also many other parties involved, including public administration, political groups, trade associations and unions, communities, any interest groups and the public at large. This analysis approach is rooted in a strategic management approach that was originally presented by R.E. Freeman in 1984 and further developed by many researchers (Freeman, 1984/2010; Freeman *et al.*, 2010). The stakeholder view tries to find out which stakeholder groups are crucial and tries to accommodate the different demands that are expressed towards the company. Mitchell *et al.* (1997) give three attributes that define critical stakeholder groups: power – in how far a party has means to impose its will; legitimacy – in how far structures or behaviours are socially accepted; and urgency – the time sensitivity of stakeholders' claims. Matilainen *et al.* ask in Chapter 11 (this volume) how nature-based tourism firms manage critical stakeholder groups.

Inter-organizational cooperation is one of the major crucial issues in innovation, including in the forest sector, and its better support has been recommended by earlier studies (Rametsteiner *et al.*, 2005; Mavsar *et al.*, 2008). Cooperation may be seen both as an innovation in itself (a new organizational model) and as a supporting factor for innovations (e.g. information exchange in sector meetings or maintaining cross-sectoral relations). The chapter in this book by Mendes *et al.* (Chapter 5) is dedicated to forest owners' organizations as an institutionalized form of cooperation in forestry.

Stronger cooperation is considered important in different ways: horizontal cooperation of forest holdings and forest sector firms in order to gain stronger market power or to divide production functions, vertical cooperation along production chains in order to coordinate better the production processes, or cross-sectoral relations in order to develop new business opportunities.

2.3.4 Institutional–societal level

On the institutional–societal level, factors come into the picture that go beyond the firms and their interaction: the significance of institutional arrangements such as property rights, the roles and functions of public and semi-public actors such as governments or interest groups, and characteristics of societies. Such analyses may be a means for analysing innovation processes and performances at different levels, particularly the macro-level, and they offer conceptual frameworks for the governance of innovation on the policy and firm level. Both government policy making as well as formulating innovation strategies of firms may rely on these analyses.

An array of disciplines, including economics, political sciences and sociology, have contributed to innovation research on the institutional–societal level. In the narrower sense, institutional level innovation research can be understood as systemic approaches to the analysis of innovation processes. Systemic approaches are mostly based in new institutional economics and evolutionary economics. These analyses look at the role of institutions in furthering or preventing economic growth and innovation. They look at transaction costs, political economy, property rights, hierarchy and organization, public choice, growth, industrial dynamics, interdependencies, competition and resource constraints within the economy.

Systemic approaches include elements such as: actors, institutions, and their interactions, as well as the specific role of knowledge and learning. Social systems are seen as consisting of interrelated and interacting components – actors and institutions – working towards common objectives (innovation systems). Actors hereby are any relevant actors from the business field, research and education, and from the political system. Institutions should in this context be understood as any 'humanly devised constraints that structure political, economic and social interactions' (North, 1991). They include formal and informal rules of our society and they constrain or open up chances for innovation. Formal institutions are written rules such as laws or standards as well as organizations. Informal institutions are unwritten rules, such as routines or established practices, and become visible when, for instance, travelling through different cultures. In this book, an institutional level perspective prevails in most chapters as they analyse the innovation processes from a systemic perspective. Slee and Ollonqvist (Chapters 8 and 12, respectively) further explore societal level and systemic aspects in their conceptual introductions to the two central parts of this volume.

Knowledge and learning are key to any innovation and become increasingly important in our society ('knowledge society'). Both explicit and tacit types of knowledge are important in the production processes and in innovations (Polanyi, 1958). Explicit knowledge such as scientific knowledge can be formalized, while tacit knowledge is in the pre-logical phase and cannot or has not been formalized and is, therefore, communicated through learning by doing or word of mouth. The economy in industrialized countries is increasingly seen as a 'learning economy', with learning as the most important factor in the production process (Lundvall and Johnson, 1994). Almost all learning processes are social and interactive rather than being solely individual. Learning has to be understood more broadly than what happens in education and training: firms engage in cooperative interactive learning relationships with a wide range of actors. Learning takes place not only within specific knowledge-producing sectors but to a large extent as a by-product of ordinary economic activities. These learning processes are extensively described, among others, under the concept of the learning region (Morgan, 1997). In this book, the role of learning in innovation processes is touched upon in various ways. Ukrainski and Kajanus (Chapter 4) particularly analyse how research results feed into companies' innovation processes. The challenge of combining internal and external knowledge sources for innovations is further explored by Ollonqvist (Chapter 12) and reflected in the chapters on the various wood value chains.

Technological change is seen as an evolutionary process, which is – in analogy to biology – based on diversity creation and selection mechanisms. The processes include path dependencies and possible lock-in situations (Arthur, 1989). As social growth processes are historically specific and institutions are self-reinforcing, a standard that is first-to-market can become entrenched even if later, better advanced standards would emerge. Because of high switching costs, lock-in situations may arise where a non-optimal situation cannot be changed.

2.4 Systemic Approaches to the Study of Innovation

2.4.1 Systemic models: clusters, industrial district, innovative milieus and innovation systems

How innovations actually occur and how they can be supported has been analysed and interpreted in very different ways. Schumpeter was one of the first to study the phenomenon of innovation. He connected it strongly to the role of entrepreneurs as innovators, a view on the innovation process as termed 'Schumpeter Mark I'. In his later work, he acknowledged a much greater complexity of the process ('Schumpeter Mark II'). He was the first to describe the phenomenon that innovations tend to 'cluster' in certain industries and in certain periods. The cluster concept was subsequently further developed by, among others, Porter for industry or business clusters. Porter compared the competitive advantages of different nations and formulated that the core factor is the role of the nation in stimulating competitive improvement and innovation. In his diamond model, he explains the innovative clusters by four sets of conditions that need to be fulfilled: factor conditions (availability of skilled labour, capital, knowledge or infrastructure), demand conditions (primarily home demand), related and supporting industries (e.g. suppliers), and firm strategy, structure and rivalry (how firms are managed in a nation or region) (Porter, 1990).

From this model of clusters, the model of industrial districts (Marshall, 1920; Harrison, 1992) differs in how it explains innovations and what factors it values as more important. The similarities between the two models are the observed collocation of the firms and the important role of labour and suppliers in the region. For both, the competition between the firms is also crucial. However, the industrial district model strongly emphasizes the collaboration between the firms. SMEs, in particular, may gain advantage from collaboration, as was shown on the often-cited example of northern Italy, e.g. in Bianchi and Giordani (1993).

While scholars of the industrial district see the advantage of geographical proximity primarily in economies of scale through the collaboration (division of labour and joint forces in procurement, production and marketing), the innovative networks, innovative milieu and creative milieu approaches emphasize again another factor – the collective learning processes (Camagni, 1991; Cooke and Morgan, 1994; Maillat, 1996). This research field focuses on the quality of transactions, alliances and partnerships between enterprises in a region. They focus less on ties between businesses and more on the collective environment for innovation and pay specific attention to institutions other than the firms (Moulaert and Sekia, 2003). The authors emphasize the role of socio-cultural conditions for collective learning processes among a set of diverse actors within a production system and a certain culture.

In sum, the authors assume that innovative regions have to be supported by three network systems: business, social and political networks (Weber, 2002; Fornahl and Brenner, 2003). The regional actors' adaptability and their ability to learn is the central question of the study of learning regions (Morgan, 1997; Scheef, 1999; Butzin, 2000). With that, studies of innovations at the regional level strongly turned to emphasize the analysis of knowledge and learning processes (Asheim and Cooke, 1999; Breschi and Lissoni, 2001).

Another systemic research approach is termed 'systems of innovation' and was developed by authors such as Freeman (1987), Lundvall (1992), Nelson (1993) and Edquist (1997). In this systems view, actors, institutions and their relationships are defined as the elements of analysis. It concentrates on the relationship between the actors and the institutions in the innovation process rather than concentrating on certain elements. The authors look at the innovation process as an institutional process. Innovating firms do not act on their own, but are embedded in a system of other actors and institutions that may be supportive or hindering. In its first phase, innovation systems were described and compared on the national level. Later on, the authors also recognized local–regional as well as sectoral specificities. Innovations are often particularly supported by sets of actors on the regional level, called regional innovation systems (RIS; Asheim and Isaksen, 2002), or within certain sectors, called sectoral innovation systems (Breschi and Malerba, 1997). The two latter are specifically relevant for the analysis of innovations in the forestry and forest industry sectors, whereby the regional approach has a specific significance when looking at territorial-based forest ecosystem services such as recreation or environmental services, the sectoral approach when analysing innovations in the wood value chain (Rametsteiner *et al.*, 2005). It should be noted that both – regional and sectoral – approaches should be used for the analysis of any forest- and wood-related innovation because both dimensions are highly relevant in any case. From earlier studies, the sectoral innovation systems rather focus on timber-related innovations and other forestry-related services are rather supported by RIS (Kubeczko *et al.*, 2006). These specializations, however, are interpreted as deficiencies.

2.4.2 A broad understanding of innovation systems

The study of innovation systems has developed quickly in the recent decades. The concept, however, is used in different understandings and with different foci. Besides the above-described foci on national, regional or sectoral innovation systems, they also have different research questions in the centre of their analyses and differ in their understandings of the innovation system. What the systemic approaches have in common is that they: put innovations, institutions and learning into the focus of their analyses; include product and process innovations; look at the innovation processes from an evolutionary perspective; strive to be holistic and interdisciplinary; and emphasize the interdependence of factors and the non-linearity and non-optimality of innovations (Edquist, 1997). The approach provides a conceptual framework rather than formal theories and draws on several theoretical strands within institutional economics and related fields. The above-described innovation models are not distinctly different research fields but overlap to some extent; for instance, studies of innovative milieus and learning regions are seen as part of systemic approaches of the study of innovation. One of the available definitions of innovation systems reads as follows (Freeman, 1995):

> ... the network of institutions in the public and private sectors whose activities and interactions initiate, import, modify and diffuse new technologies.

In this definition, institutions are to be understood in a broader sense, including organizations that act as agents, and also technologies would – in a broad understanding – comprise products, services or organizational changes. This definition includes a strong reference to the innovation process and the functions of the innovation system. Functions and activities, however, have been described in very different ways and a common understanding does not yet exist (Edquist, 2001; Kubeczko *et al.*, 2006). The functions or services that the institutions provide within innovation systems were described by Edquist and Johnson (1997) in the following way:

- to provide information in order to reduce uncertainty;
- to manage conflicts and cooperation; and
- to provide pecuniary and non-pecuniary incentives.

The authors, however, also notice that institutions often may act as innovation brakes rather than accelerators. This is in line with what Schumpeter (1934) described – that innovations or innovators often have to fight against established structures or ways of doing things. Established power structures in fact may often hinder technological change or the development of new products or services, as described by Buttoud *et al.* (forthcoming) in their article on barriers to innovation in the forest sector.

The concept of innovation systems is used in different ways and for different purposes, as Castellacci *et al.* (2005) observed. The authors distinguish two distinct traditions within the innovation systems research, which they call the historical–empirical approach and the interactive learning-based approach. The first applies a rather narrow definition of the innovation system that mainly includes the institutions and organizations involved in the innovation process. An emphasis is put on the historical evolution of national institutions. A main function of the concept in this approach is to serve as a benchmarking tool for analysing national innovation systems and policy instruments, as seen in the work of, e.g. Freeman and Nelson. According to these authors, the second tradition employs a broader understanding of the innovation system that includes more parts of the economic structure and institutional set-up, which affect learning processes. Authors such as Lundvall and scholars in the field of RIS (Asheim and Gertler, 2005), creative milieus (Cooke and Morgan, 1994) or learning regions (Florida, 1995) strongly focus their analyses on the role of knowledge and they understand learning as an interactive social process.

Innovation systems may be characterized in at least three dimensions: spatially, sectorally and in terms of activities (Edquist, 2005). This applies to all types of innovation systems that have been described – national, sectoral and RIS. Edquist states that the definition of the boundaries of the systems in these three dimensions is not easy and he acknowledges that there are narrower and broader approaches. His quest for searching 'boundaries', however, reflects a narrower understanding of innovation systems in which the boundaries would be predefined. Edquist, with much relevance, calls it a weakness of the innovation systems approach that it is still conceptually diffuse and used differently by different authors. A broader understanding would leave that consciously open for each case and for the empirical work, as expressed by Lundvall (1992) when saying 'that a definition of the system of innovation must be kept open and flexible'. Nelson and Rosenberg (1993) express it in the following way:

> There is no presumption that the system was, in some sense, consciously designed, or even that the set of institutions involved work together smoothly and coherently. Rather, the 'system' concept is that of a set of institutional actors that, together, play the major role in influencing innovative performance.

In our view, the openness as demanded by Lundvall should be understood as meaning the system is defined in a procedural way but not predefined in, for instance, geographical terms. The strength in this would be that it makes it possible to define empirically in each case the boundaries of any specific innovation system that is studied. Looking at the system in the way Nelson and Rosenberg do is more fruitful for understanding all the supporting and hampering processes.

Most authors are interested in established and enduring systems that explicitly aim to generate innovations, whether at the national, regional or sectoral levels. They sometimes assume the innovation systems as being consciously created by policy makers or not (Edquist, 1997). However, in forestry at least, many innovations come about without the purposeful, systemic and coordinated work of a well-defined innovation system (Kubeczko *et al.*, 2006). This has been observed for the forest sector in the work of INNOFORCE (Rametsteiner *et al.*, 2005): many important innovations are actually born 'between' established sectoral and/or RIS in a more *ad hoc* project structure within an arena of various actors and institutions (Kubeczko *et al.*, 2006).

From this is concluded that for the analysis of innovations in the fields of forestry and the forest-based industries, a broader understanding of the innovation system approach is appropriate. The analysis would then, first of all, not only look at purposefully created and enduring innovation systems but primarily at the innovation processes. This would allow one to answer then the question of how far established innovation systems influence the innovation. Secondly, the existing innovation systems should be empirically described, including their boundaries, functions and specific sectoral specifications. This would allow for a better understanding of the many ways innovation systems actually look in the real world.

2.5 Innovation Policy and Governance

Most of the chapters of the present book include analyses of the role of the political–institutional level and public policies in innovation processes. One relevant theoretical field is, of course, policy analysis, but such questions are also tackled by innovation research approaches that often include the issue of policy intervention (Lundvall and Borras, 2005). Within policy analysis, implementation research (Pülzl and Treib, 2007) would particularly be relevant as the questions often regard the impact of policies on the industry and on innovation processes.

Political science scholars increasingly use the term and concept of governance in their work and hereby refer to the growing complexity of the political world and actions (Benz, 2004). The concept of governance follows from the central insights of policy studies from recent decades – the reality does not follow the early conceptions of policy making: in a democracy, the relationship between state and non-state actors is not a unidirectional hierarchy, as power is distributed among many stakeholders in a policy making process; public administration follows not only formal but also informal goals of organizational growth and power; the policy-making process does not follow the logical model of the policy cycle with the stages of formulation, implementation and evaluation, etc. The conventional governmental view of political control is increasingly replaced by the broader governance concept that sees political actors interrelated in networks of negotiations. Under this paradigm, different modes of coordination among various public and private actors – ranging from mutual adaptations, competition on markets to hierarchical control – are conceptualized as complementary (Kooiman, 1993; Mayntz and Scharpf, 1995; Mayntz, 1998).

The term 'governance' was first introduced as a scientific concept in economy and particularly refers to coordination mechanisms and institutional arrangements of the market (Williamson, 1985). In this understanding, it is compatible with its use in political studies (Frey, 2002; Benz, 2004).

2.5.1 Innovation governance

Boekholt (2004) relates the concept of governance to innovation systems: an interactive and systemic approach pays much more attention to multiple sources of new ideas and knowledge creation, to the bridges and linkages of various actors in the system, and to widening the 'pipeline perspective' of innovation than was common in the linear innovation approaches. This insight has induced a gradual process of reshaping the system and rethinking the role of public and private actors in this system.

In practice, innovation governance is often related to the concepts of innovation systems and the triple helix model that was developed by Etzkowitz and Leydesdorff (2000), in which the innovation system is divided into three main subsystems: the industrial system, the education and research system, and the system of government and governance.

Innovation policy is no longer understood to be limited to traditional research and technology policy but is considered a policy issue with large interfaces with other policy areas. A number of studies have supported the search for better coherence in the area of innovation policy (e.g. Arnold and Boekholt, 2003; Edler *et al.*, 2003). Furthermore, innovation is not solely understood any more as a means for reaching economic growth, strengthening competitiveness and employment. Innovation is increasingly understood as a means to contribute to the solution of a range of societal problems and goals, most prominently in environmental protection.

The traditional science and technology policy approach prevailed in most OECD countries in the post-war period and the systemic innovation policy approach has gained increasing importance during the last two decades. The former focuses on fostering scientific and technological advance, and enhancing the flow of knowledge down along the innovation chain. Systemic innovation policy is seen as a complex process, taking place in an environment of interacting actors and institutions (innovation system); having multiple sources (apart from research activities); and running through multiple feedback loops between the different stages. Policy addresses the systemic environment in which innovation takes place in ways that can better inform decisions about research, commercialization, technology adoption and implementation (Weiss *et al.*, 2010a). Although a systemic view of innovation policy gains in importance in research and practice, the analysis of relevant policies revealed that the use of systemic rhetoric in policy documents is still often only symbolic and the concrete measures often remain traditional (Weiss *et al.*, 2010b).

2.5.2 Regional governance

The concept of governance in innovation processes has been elaborated in greater detail in local or regional level studies (regional development studies) under the term of 'regional governance' (Frey, 2002; Adamaschek and Pröhl, 2003; Benz and Fürst, 2003; Böcher, 2006, 2008; Scherer, 2006). In a similar way, it has been taken up as a central concept in the RIS approach (Braczyk *et al.*, 1998). Böcher shows how, for instance, in the Leader instrument, hierarchical top-down steering (government) and bottom-up initiatives (governance) work in combination (Böcher, 2006).

Scholars have used the concept of regional governance from different theoretical starting points, including systems theoretic (Kooiman, 1993, p. 258) and actors' approaches (Rhodes, 1996, 1997). Regional development studies have also taken up insights from public governance research that follows the actor-centred institutionalism (Mayntz and Scharpf, 1995). A recent example that presents a well developed conceptual basis was provided by Scherer (2006). He connects the concept of regional governance with innovation networks and learning regions. On the basis of three detailed case studies in Austria, Germany and Switzerland, he concludes that regional governance systems are increasingly emerging and typically take over development tasks (Scherer, 2006).

2.6 Spatial Aspects

The forestry and forest-industry sector is characterized by a number of particularities.

First of all, the forest that produces timber and other goods and services is a natural resource; the products are renewable. The production is connected with the land base, with the implication that the production is distributed across the land, the producing land base cannot be moved to somewhere else, and the product has to be transported from the site of production to the processing and the place of use. While this transportation is possible in most provisioning services of forest ecosystems (goods such as wood, mushrooms and, to some extent, drinking water), for the regulating, supporting and cultural ecosystem services, this is typically not the case. These services are required or have to be enjoyed at the very location, such as in the case of the protection against natural hazards, biodiversity conservation or landscape amenities. The production further depends on natural site and climate conditions, implying a certain yearly production potential and the types and quality of the products (e.g. the tree species or ecosystem services). Together with site-dependent harvesting technologies, they are important naturally defined factors for the profitability. A further important specificity is that forests produce at the same time timber and other goods and services with joint production and trade-off relations (Glück, 2000a,b). Put differently, the timber production affects in various ways other ecosystem services of forests such as biodiversity or its recreational value, which are socially demanded and often secured by regulations but can be marketed only to a limited degree (Mantau *et al.*, 2001; Mavsar *et al.*, 2008). They are often externalities and have public good character. The territorial dimension of forest goods and services is comprehensively discussed in Slee's conceptual chapter (Chapter 8, this volume).

Not only forestry production but also innovation processes have a spatial dimension that has been discussed in the broader fields of economic geography or geography of innovation, and under several headings, including the cluster, industrial district, innovation network, innovative milieu, learning region or regional innovation system. Innovation processes are not only linked to the spatially available resources, but also to local or regional actor networks and they depend on regional or national institutional frameworks. Spatial aspects of innovation processes are therefore discussed also by Ollonqvist's chapter on innovation in wood value chains (Chapter 12, this volume).

2.6.1 The geography of innovation: clusters, networks or RIS?

Important elements of a geography of innovation (Asheim and Gertler, 2005) are the proximity of various types of actors, the role of tacit knowledge and interactive learning – three elements that are closely related to each other. Tacit knowledge does not travel easily and is best shared through face-to-face interaction. The geographical consequence is that the actors who are involved in the innovation or learning process should be located close to each other – as is observed in networks, clusters or RIS.

Which of these different approaches are suitable for our analysis? Martin and Sunley (2003) deconstruct the cluster concept as a fashionable policy label without sound conceptual basis. In fact, many authors remark on the weak theoretical foundation of that concept as well as on the fact that, in practice, differing kinds of policy interventions are labelled clusters or are inspired by the aim to create local industry clusters, e.g. in the paper collections edited by Breschi and Malerba (2005) and Asheim *et al.* (2006). As Simmie (2006) critically states, Porter assumes that the clustering causes innovation, productivity gains and competitiveness. Simmie argues that it is the reverse: the functioning of local systems of innovation is required for a high level of innovations. RIS may resemble cluster structures but, according to Simmie, the explanation for the innovativeness is to be found in the innovation systems. Studies show that competitive sectors may be clustered but many clustered sectors are not competitive (Simmie, 2006). Asheim *et al.* (2006) conclude that the cluster concept has been overestimated and oversold as a policy concept. Clusters are often not as successful as a few frequently reported best practice cases suggest. Besides the promises and advantages, the drawbacks and limitations should also be acknowledged (Asheim *et al.*, 2006). Breschi and Malerba (2005) observe that public policy attempts to form new clusters through top-down intervention such as technopoles, science parks or firm incubators are often ineffective. Instead, government policies play an important role in cluster development by accommodating the formation of new firms, investments in education and the provision of support infrastructure (Breschi and Malerba, 2005). What successful cluster initiatives do seems to be precisely that: providing a supportive infrastructure for local firms or, in other words, fulfilling RIS functions (compare Rimmler *et al.*, Chapter 7, this volume).

The concept of RIS differs from the cluster approach by using different causality relations and also in the way it may incorporate a broader network of actors into the innovation model. As shown above, a narrow understanding of innovation systems focuses on public and private organizations that are more directly concerned with the generation and use of new knowledge in the production system. A broader understanding, however, would also include civil society groups, neighbours and the public. The broader understanding reflects a bottom-up, interactive innovation model such as behind the learning regions concept (Asheim and Gertler, 2005). What differs in the innovation system approach from network approaches is that it includes the institutional dimension.

2.6.2 A broader view on RIS

Asheim (1998) and Cooke (1998) defined three types of RIS that coincide and that illustrate the variety in the relationships between the production structure and the institutional set-up in regions (Asheim and Gertler, 2005):

- The territorially embedded regional innovation system (Asheim, 1998) or 'grassroots RIS' (Cooke, 1998) implies a broader definition of innovation system and can be explicated by what was described as industrial districts. In this type, innovation activities of firms are based on localized learning processes that are stimulated by geographical, social and cultural proximity.
- In regionally networked innovation systems (Asheim, 1998) or 'networked RIS' (Cooke, 1998), the organizations are embedded in a specific region and are characterized by localized learning, but policy interventions intentionally strengthen the region's institutional infrastructure by providing, for instance, regionally based R&D institutes or vocational training organization. This type is regarded as ideal because the local resources and activities are embedded in a supportive institutional environment. The localized learning and informal knowledge of the firms (that are often small businesses) are complemented by knowledge from external sources such as scientific research that is accessed through institutional actors including technology transfer agencies or services centres.
- In regionalized national innovation systems (Asheim, 1998) or 'dirigiste RIS' (Cooke, 1998), exogenous actors and relationships play a stronger role. These narrower innovation systems primarily incorporate the R&D functions of universities, research institutes and corporations and follow a stronger linear innovation model. Examples such as science parks or technopoles, according to the authors, have mostly failed to develop innovative networks within the parks or with local firms.

The approach of RIS – especially if understood in a broad sense – seems particularly relevant for the forest sector in connection to regional development. For the support of an integrated endogenous rural development and to generate new income possibilities, the inclusion of many stakeholders in a cross-sectoral perspective is required (Böcher, 2006). Sectoral innovation systems have deficiencies in these respects because of their uni-sectoral orientation and their focus on rationalization. While sectoral innovation systems concentrate on technological improvements, studies show that new goods and services of the forest sector are rather generated through local cross-sectoral *ad hoc* networks (Kubeczko *et al.*, 2006). Such local networks may be interpreted as grassroots RIS. This model thus is an appropriate approach to analyse innovation processes in small innovation projects in forestry or forest industry. RIS then have a complementary function to sectoral innovation systems and, in the analysis, both models should be used in combination.

Cluster initiatives and Leader or similar regional development structures can be seen as network RIS as defined above. Regional development policies today take into account insights from regional development, regional government and regional innovation studies. Policy instruments, such as cluster initiatives or regional and rural development agencies, increasingly employ endogenous development models. Their success then depends on how far they follow the regional governance/regional innovation principles in the implementation process. In fact, group interests may stand against the effective realization of bottom-up approaches and the participation of cross-sectoral local stakeholder groups such as envisaged by the Leader instrument.

2.7 Sectoral Aspects

In addition to the characteristics of the forest sector that are described under the spatial dimension, there are some more sectoral particularities that are important for the innovation processes and innovation policies. The forest-based sector is what is often termed 'low-tech' but what we prefer to call a traditional sector. Furthermore, most firms are SMEs, often run by family owners.

Sectors differ in their innovation patterns, such as the sources of innovations and the actors, institutions and policies that play major roles in the innovation processes (Breschi and Malerba, 1997; Malerba, 2005). In an industry life cycle view, innovation patterns of an industry may be divided into what is often called a Schumpeter Mark I or Mark II pattern. The term 'Schumpeter Mark I' refers to Schumpeter's early work, where he focused strongly on individual entrepreneurs. Industries, in their early stages, reflect conditions of high technological opportunities, low appropriability and low cumulativeness at the firm level. Rapidly changing knowledge, high uncertainty and low entry barriers make new firms the main innovators. By contrast, high appropriability and high cumulativeness lead to a Schumpeter Mark II pattern, when the industry matures and technological change rather follows well-defined trajectories. In his later work, Schumpeter turned to study innovations in large firms, which dominate the mature stage of an industry (Breschi and Malerba, 1997; Malerba, 2005).

An influential distinction between sectors was made by Pavitt with accord to the sources of innovation and the appropriation mechanisms (Pavitt, 1984):

- in supplier-dominated sectors, new technologies are embodied in new equipment and new technologies are diffused through learning-by-doing and similar mechanisms;
- in scale-intensive sectors, innovations particularly focus on rationalization, and they come from internal as well as external sources and may be protected by secrecy or patents;
- specialized suppliers focus on the quality of performance, reliability and customization; and
- science-based sectors are characterized by high rates of product as well as process innovations, coming from internal and external R&D.

In reality, these types are not found in pure forms. The forest sector would mostly be categorized as supplier-dominated, applying innovations that were developed in other sectors, although parts would fall under the relatively innovative scale-intensive category. How innovation differs between sectors has more recently been studied from an innovation system perspective by Breschi and Malerba (1997), who describe different types of sectoral innovation systems on the basis of their different technological regimes, actors and institutions.

2.7.1 Technological regimes

As Nelson and Winter (1982) have shown, the technological environment in terms of opportunity and appropriability conditions affects innovation processes and outcomes. Opportunities for innovations differ across sectors in their level, variety, pervasiveness and sources such as university research or endogenous learning (Breschi and Malerba, 1997). The appropriability differs in the level and means, such as patents, secrecy, continuous innovations, or the control of complementary assets. If innovations cannot be strongly protected from imitation, the level of knowledge externalities is higher (spillovers). If the knowledge in a sector is of cumulative character, building on existing knowledge, the level of appropriability is higher (Malerba, 2005). The nature of knowledge in a sector can be described in terms of specificity, tacitness, complexity and independence. Another related characteristic of a sector's knowledge base is the typical means of knowledge transmission – how a firm can gain access to knowledge (Breschi and Malerba, 1997): by informal means such as face-to-face talk, personal teaching and training, mobility of personnel (important for tacit knowledge), or formal means such as publications, licences or patents (relevant for standardized/codified knowledge).

Breschi and Malerba (1997) analysed several types of sectoral innovation systems: the forest industry belongs to what they term 'traditional sectors'. The technological regimes in traditional sectors are characterized by low degrees of innovation opportunities, appropriability and cumulativeness (on firm level). It does not take long until other firms imitate improved production processes or new products. The knowledge base is relatively simple, generic and often embodied in equipment and materials. The relatively simple and codified knowledge base makes spatial proximity relatively less important. However, the low level of the firms' technological competencies limits their ability to exploit distant sources of knowledge. As a result, producers possibly develop symbiotic relationships with local suppliers and public agencies. Opportunities to innovate are mainly related to the search for lower production costs; the ability to innovate consists in the effective incorporation of new generic and codifiable knowledge into existing products. Competitive advantage may rest upon the use of less conventional means, such as trademarks or aesthetic design (Breschi and Malerba, 1997).

2.7.2 Actors and their relationships

Heterogeneous actors are part of sectoral systems, such as firms as the key actors, but also governmental and non-governmental actors. The firms of a sector are characterized by specific beliefs, expectations, goals, competencies and organization, and are continuously engaged in processes of learning and knowledge accumulation (Malerba, 2005). Users and suppliers may have different close relationships with the producers. In dynamic and innovative settings, suppliers and users affect innovation processes and continuously redefine the boundaries of the sectoral systems. Other actors such as universities, financial organizations or public agencies influence innovation processes in different ways. Malerba further mentions that the appropriate unit of analysis may – instead or in addition to the firms – often be individuals or firms' subunits.

In the view of sectoral systems of innovation, all kinds of stakeholders are part of the analysis. Knowledge on the forest sector may therefore come from policy or governance studies that often well describe all types of actors, their rationales, values and relationships. In the triple helix model, the three main subsystems are the industry, education and research, and policy (Etzkowitz and Leydesdorff, 2000). This model reflects a top-down government-oriented approach to innovation systems. Arnold and Kuhlmann (2001) have developed a more detailed model with the most relevant groups of actors of a national innovation system from which a model of a forestry innovation system (Fig. 2.2) was derived by Rametsteiner *et al.* (2005). Although including already quite extensive groups of actors, this model is still restricted as to which stakeholders are included: other than sectoral interest groups, a local or broader public is not shown.

Rametsteiner *et al.* (2005) applied this model of a sectoral innovation system in their comparative analysis of forestry innovation systems in seven Central European countries and in a more in-depth analysis on the example of Austria. Their conclusions include that only a few actors seem relevant in innovation governance in the sector, that there is virtually no interaction of sectoral and national innovation system actors, and that there is a lack of interaction with other sectors where currently important relevant innovations are occurring, for instance, in bio-energy, climate change mitigation or tourism (Rametsteiner and Weiss, 2006).

2.7.3 Institutions

Institutions such as patent systems, property rights or innovation support infrastructures play important roles for the innovation activity and technological change in a sector. They also include social norms, routines, habits or established practices that shape agents' cognition, actions and interactions (Malerba, 2005). They are binding or less binding, formal or informal.

Because of the strong externalities and public goods character of the forest resources, the forest policies have long had important impacts on the sector. This specificity includes, for instance, the fact that most forestry faculties in Central Europe have a forest policy chair – something that hardly exists for other sectors. Institutions include the unspoken understanding of good forest management and what 'sustainable forest management' means. The latter has undergone profound changes in the past (Kennedy *et al.*, 2001). They also include the education and training structures in countries. In the example of the Austrian provinces, it can be observed how important the differences in

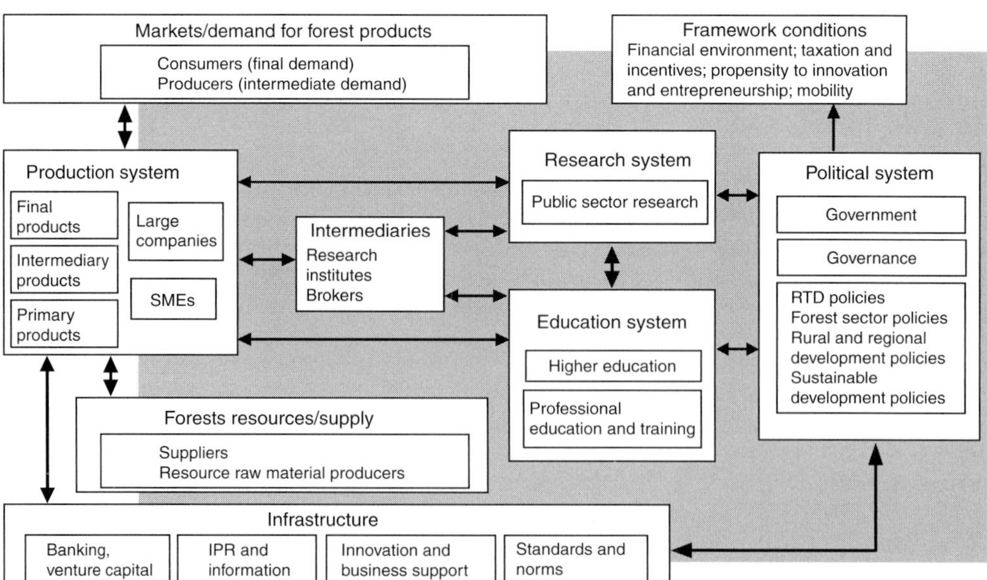

Fig. 2.2. Forestry sectoral innovation system. Source: Rametsteiner *et al.* (2005) (adapted from the NIS model in Arnold and Kuhlmann, 2001).

the sector-specific institutional structures may be: the forestry sections within the chambers of agriculture differ greatly in their capacities between the provinces with the implication that the chambers were able to support the establishment and maintenance of forest owners' organizations and cooperatives in those provinces where they were disposed of larger budgets and staff numbers. The well-staffed and well-budgeted provincial chambers, furthermore, were able to provide resources for the further development and diffusion of bio-energy systems such as wood biomass-based district heating plants (Weiss, 2004). In Austria, the majority of rurally located biomass heating plants have been established and are run by forest owners' cooperatives.

Institutional factors affect the sector's development in the timber market (see the example of clusters in Rimmler *et al.*, Chapter 7, this volume), as well as in recreational or environmental services of forests (for the example of tourism, see Matilainen *et al.*, Chapter 11, this volume), often through constituted power structures (Kubeczko *et al.*, 2006; Buttoud *et al.*, forthcoming).

The system change in the former communist countries in eastern Europe had profound impacts on their economies, including the innovation systems. The transformation of socialist to post-socialist innovation systems has been analysed by Radosevic (1997). One challenge hereby is how to restructure the domestic sectoral knowledge bases that otherwise are prone to erode (Radosevic, 1997). The institutional changes in eastern European countries are highly visible in the technological development of the forest sector and ongoing innovation activities (Rametsteiner *et al.*, 2005; Glück *et al.*, 2010).

2.7.4 Low-tech or traditional?

An often-used and very simple distinction of sectors is into low technology and high technology. With the growing attention to the concepts of the knowledge society (Drucker, 1994) and learning economy (Lundvall and Borras, 1997), policy makers have turned to an enforced support of research, development and innovation in high-tech sectors. In line with this particular focus is the aim to make the European Union the world's 'most competitive and dynamic knowledge-based economy' (Lisbon Strategy). Research-intensive, science-based industries are seen as the key drivers of future economic prosperity. Such industries receive high political attention, whereas low-technology industries are neglected. This view, however, is challenged by recent research (Hirsch-Kreinsen *et al.*, 2005).

The OECD uses the ratio of R&D expenditure to turnover as the indicator for defining high- and low-technology sectors: high-tech sectors are those investing more than 5% of their turnover in R&D and include, for instance, pharmaceuticals or computing technologies. Medium-high-tech industries invest more than 3%, medium-low-tech more than 0.9% and low-tech even below that. Low-tech industries include, for instance, the broad group of wood, pulp, paper, printing and publishing sectors, as well as food or textiles (OECD, 2002).

Following the above logic, the mention of innovations in low-tech industries seems a contradiction in terms. According to the OECD data, the high-tech sectors, however, account for only around 3% of value added in the OECD countries (Hirsch-Kreinsen *et al.*, 2003). This is an indication that: (i) the high-tech sector is overestimated in terms of their economic impact; (ii) innovation processes include more than just R&D; and (iii) the focus of innovation policy and innovation research on the high-tech sectors and on R&D is misleading because it does not appropriately understand innovation processes in the economy (Tunzelmann and Acha, 2005; Hirsch-Kreinsen *et al.*, 2008). Even more explicitly, Hirsch-Kreinsen *et al.* (2008) state that the use of the low-tech/high-tech definition perpetuates a number of errors in understanding the role of technologies in innovation processes and economic development. Low- and medium-tech sectors:

- achieve respectable growth in productivity;
- draw heavily on high-tech sectors;

- generate substantial innovations themselves; and
- are important elements in the innovativeness of regional and transnational industrial value chains.

A systemic understanding of innovation, in contrast to the described reduced view, puts much broader relationships of different actors and sectors into the picture: very different kinds of knowledge may contribute to innovations and the process of diffusion is at least as important as the development of radical innovations in the first place. As Arundel *et al.* (2008) have analysed from a European innovation survey, low-tech firms innovate differently.

The OECD (2005), in its so-called Oslo Manual on collecting and interpreting innovation data, gives an account on the systemic characters and complex processes connected with innovation (Smith, 2005). There are many aspects of innovation processes that should be noted when speaking of the so-called high- and low-tech industries. Innovation is mostly not a linear process from R&D to marketing, but a learning process that involves a complex interaction of various actors and feedback loops in the different stages of knowledge creation. Kline and Rosenberg (1986) explicate that innovation does NOT depend on invention processes as a source for new ideas (!) but that R&D is rather to be seen as problem-solving technologies as a part of ongoing innovation processes. Research and invention are hardly initiating factors for innovation but have a serving function in the realization of an innovation. The Community Innovation Survey (CIS) programme tries to incorporate increasingly more of these systemic aspects of the innovation process (Smith, 2005).

The forest sector (forestry and forest-based industries) is a very diverse sector but also certainly one with a very limited research input. Following the given considerations, at least two aspects should be noted:

- important innovation processes do not necessarily include R&D but rely rather on tacit knowledge and related forms of knowledge transfer; and
- important parts of the forest sector actually use quite sophisticated technologies in production, whether the bio-energy, paper, pulp or timber value chains.

On the basis of the CIS methodology and data, the critique of the 'high-tech bias' can be substantiated as studies show that innovation is prevalent across all sectors and R&D is not the most important innovation input (Smith, 2005). A detailed analysis of the wood sector in Finland and Estonia is given in Ukrainski and Kajanus' chapter (Chapter 4, this volume), which confirms the relatively lower importance of research organizations as input for innovation processes.

Arguably, the term 'low-tech industries', and the allocation of the forest sector into this class, is at the same time correct but also misleading. An alternative concept would be the notion as a traditional sector as described earlier, following Breschi and Malerba (1997).

2.7.5 Small is beautiful?

The overwhelming majority of firms in the forest sector are SMEs, which employ the most persons and contribute the most to the value-added of the sector. We need to place particular attention on this fact because: (i) small enterprises differ in their structure, organization and management significantly from large enterprises; and (ii) these differences are often neglected in policy and research. It is an interesting fact that it is not only a specificity of the forest sector that we need to call for more attention to the specific requirements of small enterprises, but it is true for technology and innovation policy in general: most research and policy are oriented on the model of large enterprises but most enterprises in general are small. An indication that underlines this fact is that, although small enterprises dominate our economy, there is a special term for them (SME) as well as special policies that try to balance out the distorted attention towards larger companies ('Think small first' in the title of the EU Small Business Act).

The EU defines SMEs by using the number of staff and the yearly turnover as indicators (EC Recommendation 2003/361/EC). A company qualifies as an SME if employing fewer than 250 persons and reaching a turnover up to €50 million. A micro-enterprise is one with fewer than ten persons employed and a turnover up to €2 million.

SMEs account for 99% of all enterprises in the EU, and 90% are micro-enterprises. SMEs provide two-thirds of the private sector jobs (approximately 65 million) and contribute to more than half of the total value-added created by businesses in the EU. With specific SME-related policies such as the Small Business Act, the EU acknowledges that they are primarily responsible for wealth and economic growth, and that they have a key role in innovation (COM/2008/394).

Schumacher's annotation on the diseconomies of scale under the keywords 'small is beautiful' are well-known (Schumacher, 1973); there have been many authors since that specifically studied the innovation behaviour of small businesses and as a result it can be said that these have advantages and disadvantages for innovation. Because of the smallness, small businesses may have more flexible organizational structures and a stronger personal leadership. With owner managers, the strategic decisions typically also include personal and family-related criteria versus purely economic ones. Small firms have a number of advantages when it comes to innovations, including the lower degree of bureaucracy and the better environment for the excitement and engagement of the personnel in developing novelties (Rothwell, 1989; Acs and Audretsch, 1990; Kowol, 1998; Arundel *et al.*, 2008). According to Rothwell and Dodgson (1994), large firms' innovatory advantages lie primarily in material factors such as their greater financial and technological resources, while small firms rather have behavioural advantages such as their greater entrepreneurial dynamism, internal flexibility and responsiveness to changing circumstances. Rothwell and Zegveld (1982) conclude that SMEs play highly significant roles at the early, fluid stages of development in new technological industries. Rothwell and Dodgson (1994), however, note that the role played in innovation by SMEs is strongly sectorally influenced – the small size of forest sector companies alone obviously do not make them exceptionally innovative. They also note that innovations are often pursued in collaboration between small and large firms. The particular importance of collaboration for small firms, which was mentioned earlier in this chapter, is emphasized by Ollonqvist in his chapter on innovations in the wood-based industries (Chapter 12, this volume), and is further elaborated by Pettenella and Maso in Chapter 3 (this volume).

Innovations in small low-tech firms are typically performed in cooperation with other firms and providers of knowledge intensive business services (see also Ollonqvist, Chapter 12, this volume). The specificity of small low-tech firms implies that they need specific innovation support infrastructures (Ollonqvist and Rimmler, 2010). That is also confirmed by the study of Ukrainski and Kajanus (Chapter 4, this volume) on the knowledge flows in innovation processes in the Finnish and Estonian wood sector. They conclude that ensuring knowledge creation alone is not enough to support small firms in their innovation activities; this also requires the improvement of the firms' capabilities in innovation management as well as the improvement of the capabilities of universities and other research organizations to meet the needs of small firms.

If small firms, because of their limited capacities, are not able to fully utilize knowledge-intensive consultancy services from the market, it is rational that public or semi-public organizations step in. They exist in many forms and should be further developed or made relevant for the forest sector: clusters, technology centres and regional development offices are just examples of infrastructures that have been established for the support of SMEs in rural areas. Voluntary or obligatory sectoral interest organizations also have functions in supporting the development and implementation of innovations. Rametsteiner *et al.* (2005), however, have found that in the forest sector, innovation policies hardly exist within the sectoral innovation system and that there are huge gaps in the provision of relevant information and intra- and cross-sectoral networking. Innovation incentives focus on technological novelties and rationalization rather than the development of new products and often lack support for the early phases and risky projects. Weiss (2004) and Kubeczko *et al.* (2006) confirm that new ideas are rather supported from the outside than from the inside, and the forest sectoral system becomes active only in the later phases in order to disseminate the innovations. The creation of sectoral–regional clusters is suggested as a promising policy tool and empirically examined in this book in the chapters by Rimmler *et al.* (Chapter 7) and Ollonqvist *et al.* (Chapter 14). The support structures that have been established within the Leader programmes contain good potential and are also empirically investigated (Feliciano *et al.*, Chapter 6).

References

Acs, Z. and Audretsch, D.B. (1990) *Innovation and Small Firms.* MIT Press, Cambridge, Massachusetts.

Adamaschek, B. and Pröhl, M. (eds) (2003) *Regionen erfolgreich steuern. Regional Governance - von der kommunalen zur regionalen Strategie.* Verlag Bertelsmann Stiftung, Gütersloh.

Arnold, E. and Boekholt, P. (2003) Research and innovation governance in eight countries. Technopolis, Brighton, UK.

Arnold, E. and Kuhlmann, S. (2001) RCN in the Norwegian Research and Innovation System. Background Report No. 12 in the Evaluation of the Research Council of Norway. Royal Norwegian Ministry for Education, Research and Church Affairs, Olso.

Arthur, W.B. (1989) Competing technologies, increasing returns, and lock-in by historical events. *Economic Journal* 97, 642–665.

Arundel, A., Bordoy, C. and Kanerva, M. (2008) Neglected innovators: how do innovative firms that do not perform R&D innovate? Results of an analysis of the Innobarometer 2007 survey. INNO-Metrics Thematic Paper 125.

Asheim, B. (1998) Territoriality and economics: on the substantial contribution of economic geography. In: Jonsson, O. and Olander, L.-O. (eds) *Economic Geography in Transition. The Swedish Geographical Yearbook 74.* Lund.

Asheim, B. and Cooke, P. (1999) Local learning and interactive innovation networks in a global economy. In: Malecki, E. and Oinas, P. (eds) *Making Connections: Technological Learning and Regional Economic Change.* Ashgate, Aldershot, UK, pp. 145–178.

Asheim, B. and Gertler, M.S. (2005) The geography of innovation: regional innovation systems. In: Fagerberg, J., Mowery, D.C. and Nelson, R. (eds) *The Oxford Handbook of Innovation.* Oxford University Press, Oxford, pp. 291–317.

Asheim, B. and Isaksen, A. (2002) Regional innovation systems: the integration of local 'sticky' and global 'ubiquitous' knowledge. *The Journal of Technology Transfer* 27, 77–86.

Asheim, B., Cooke, P. and Martin, R. (2006) The rise of the cluster concept in regional analysis and policy: a critical assessment. In: Asheim, B., Cooke, P. and Martin, R. (eds) *Clusters and Regional Development. Critical reflections and explorations.* Routledge, London, pp. 1–30.

Bean, R. and Radford, R. (2002) *The Business of Innovation: Managing the Corporate Imagination for Maximum Results.* AMACOM, New York.

Benz, A. (2004) Einleitung: Governance – Modebegriff oder nützliches sozialwissenschaftliches Konzept? In: Benz, A. (ed.) *Governance – Regieren in komplexen Regelsystemen. Eine Einführung.* Verlag für Sozialwissenschaften, Wiesbaden, pp. 11–28.

Benz, A. and Fürst, D. (2003) Region – Regional Governance – Regionalentwicklung. In: Adamaschek, B. and Pröhl, M. (eds) *Regionen erfolgreich steuern. Regional Governance – von der kommunalen zur regionalen Strategie.* Bertelsmann, Gütersloh.

Bianchi, P. and Giordani, M. (1993) Innovation policy at the local and national level: the case of Emilia-Romagna. *European Planning Studies* 1, 25–42.

Böcher, M. (2006) The concept of regional governance and the promotion of sustainable rural development processes in Germany. Insights from the implementation of Regionen Aktiv und Leader+. In: Rametsteiner, E. (ed.) *Proceedings of the 1st COST Action E51 Joint MC and WG Meeting*. University of Natural Resources and Applied Life Sciences, Vienna, pp. 49–66.

Böcher, M. (2008) Regional governance and rural development in Germany: the implementation of Leader+. *Sociologia Ruralis* 48, 372–388.

Boekholt, P. (2004) Ensuring policy coherence by improving the governance of innovation policy. *EU Trendchart Workshop: Ensuring policy coherence by improving the governance of innovation policy*. Brussels.

Braczyk, H.-J., Cooke, P. and Heidenreich, M. (eds) (1998) *Regional Innovation Systems: the Role of Governance in a Globalized World*. UCL Press, London.

Breschi, S. and Lissoni, F. (2001) Knowledge spillovers and local innovation systems: a critical survey. *Industrial and Corporate Change* 10, 975–1005.

Breschi, S. and Malerba, F. (1997) Sectoral innovation systems: technological regimes, Schumpeterian dynamics and spatial boundaries. In: Edquist, C. and Johnson, B. (eds) *Systems of Innovation. Technologies, Institutions and Organizations*. Pinter, London, pp. 130–156.

Breschi, S. and Malerba, F. (2005) Clusters, networks and innovation: research results and new directions. In: Breschi, S. and Malerba, F. (eds) *Clusters, Networks and Innovation*. Oxford University Press, Oxford, pp. 1–26.

Buttoud, G., Kouplevatskaya-Buttoud, I., Slee, B. and Weiss, G. (forthcoming) Barriers to institutional learning and innovation in the forest sector in Europe: markets, policies and stakeholders. *Forest Policy and Economics*.

Butzin, B. (2000) Netzwerke, Kreative Milieus und Lernende Region: Perspektiven für die regionale Entwicklungsplanung? *Zeitschrift für Wirtschaftsgeographie* 3/4, 149–166.

Camagni, R.P. (ed.) (1991) *Innovation Networks: Spatial Perspectives; GREMI – Groupe de Recherche Européen sur les Milieux Innovateurs*. Belhaven, London.

Castellacci, F., Grodal, S., Mendonca, S. and Wibe, M. (2005) Advances and challenges in innovation studies. *Journal of Economic Issues* XXXIX, 91–121.

Cooke, P. (1998) Introduction: origins of the concept. In: Braczyk, H.-J., Cooke, P. and Heidenreich, M. (eds) *Regional Innovation Systems*. UCL Press, London, pp. 2–27.

Cooke, P. and Morgan, K. (1994) The creative milieu: a regional perspective on innovation. In: Dodgson, M. and Rothwell, M. (eds) *The Handbook of Industrial Innovations*. Edward Elgar, Aldershot, UK, pp. 25–32.

Cooper, R.G. (2006) Formulas for success – the seven principles of the latest Stage-Gate® method add up to a streamlined, new-product idea-to-launch process. *Marketing Manager* 2006(March/April), 18–24.

Drucker, P.F. (1994) *Post-Capitalist Society*. Harper, New York.

Edler, J., Kuhlmann, S. and Smits, R. (2003) New governance for innovation. The need for horizontal and systemic policy co-ordination. Fraunhofer Institute Systems and Innovation Research, Karlsruhe, 28.

Edquist, C. (1997) Systems of innovation approaches – their emergence and characteristics. In: Edquist, C. (ed.) *Systems of Innovation. Technologies, Institutions and Organizations*. Pinter, London, pp. 1–35.

Edquist, C. (2001) The system of innovation approach and innovation policy: an account of the state of the art. Lead paper. Nelson Winter Conference, Aalborg.

Edquist, C. (2005) Systems of innovation: perspectives and challenges. In: Fagerberg, J., Mowery, D.C. and Nelson, R. (eds) *The Oxford Handbook of Innovation*. Oxford University Press, Oxford, pp. 181–208.

Edquist, C. and Johnson, B. (1997) Institutions and organizations in systems of innovation. In: Edquist, C. (ed.) *Systems of Innovation. Technologies, Institutions and Organizations*. Pinter, London, pp. 41–63.

Etzkowitz, H. and Leydesdorff, L. (2000) The dynamics of innovation: from National Systems and 'Mode 2' to a triple helix of university–industry–government relations. *Research Policy* 29, 109–123.

Florida, R. (1995) Towards the learning region. *Futures* 5, 527–536.

Fornahl, D. and Brenner, T. (2003) *Cooperation, Networks and Institutions in Regional Innovation Systems*. Edward Elgar, Cheltenham, UK.

Freeman, C. (1982) *The Economics of Industrial Innovation*. Pinter, London.

Freeman, C. (1987) *Technology Policy and Economic Performance: Lessons from Japan*. Pinter, London.

Freeman, C. (1995) The national system of innovation in historical perspective. *Cambridge Journal of Economics* 19, 5–24.

Freeman, R.E. (1984/2010) *Strategic Management*. Reprint. Cambridge University Press, Cambridge.

Freeman, R.E., Harrison, J.S., Wicks, A.C., Parmar, B. and de Colle, S. (2010) *Stakeholder Theory. The State of the Art.* Cambridge University Press, Cambridge.
Frey, R.L. (2002) Regional governance. Input Paper to the Workshop. Avenir Suisse, Basel, p. 22.
Glück, P. (2000a) Policy means for ensuring the full value of forests to society. *Land Use Policy* 17, 177–185.
Glück, P. (2000b) Theoretical perspectives for enhancing biological diversity in forest ecosystems in Europe. *Forest Policy and Economics* 1, 195–207.
Glück, P., Avdibegović, M., Čabaravdić, A., Nonić, D., Petrović, N., Posavec, S. and Stojanovska, M. (2010) The preconditions for the formation of private forest owners' interest associations in the Western Balkan Region. *Forest Policy and Economics* 12, 250–263.
Grabher, G. (1993) Rediscovering the social in the economies of interfirm relations. In: Grabher, G. (ed.) *The Embedded Firm.* Routledge, London, pp. 1–32.
Granovetter, M.S. (1972) The strength of weak ties. *American Journal of Sociology* 78, 1360–1380.
Harrison, B. (1992) Industrial districts: old wine in new bottles? *Regional Studies* 26, 469–483.
Hippel, E. v. (1986) Lead users. A source of novel product concepts. *Management Science* 32, 791–805.
Hirsch-Kreinsen, H., Jacobson, D., Laestadius, S. and Smith, K. (2003) Low-tech industries and the knowledge economy: state of the art and research challenges. Paper written within the context of the research project PILOT: Policy and Innovation in Low-Tech.
Hirsch-Kreinsen, H., Jacobson, D. and Robertson, P. (2005) 'Low-tech' industries: innovativeness and development perspectives. A summary of a European research project. PILOT Project Consortium, Dortmund, p. 33.
Hirsch-Kreinsen, H., Hahn, K. and Jacobson, D. (2008) The low-tech issue. In: Hirsch-Kreinsen, H. and Jacobson, D. (eds) *Innovation in Low-Tech Firms and Industries.* Edward Elgar, Cheltenham, UK, pp. 3–22.
Jenssen, J.I. and Jørgensen, G. (2004) How do corporate champions promote innovations? A literature review. *International Journal of Innovation Management* 8, 63–86.
Kennedy, J.J., Thomas, J.W. and Glück, P. (2001) Evolving forestry and rural development beliefs at midpoint and close of the 20th century. *Forest Policy and Economics* 3, 81–95.
Kline, S.J. and Rosenberg, N. (1986) An overview of innovation. In: Landau, R. and Rosenberg, N. (eds) *The Positive Sum Strategy: Harnessing Technology for Economic Growth.* National Academy Press, Washington, DC, pp. 275–305.
Kooiman, J. (1993) Findings, speculations and recommendations. In: Kooiman, J. (ed.) *Modern Governance. New Government–Society Interactions.* Sage, London, pp. 249–262.
Kowol, U. (1998) Innovationsnetzwerke. Technikentwicklung zwischen Nutzungsvisionen und Verwendungspraxis. PhD thesis, Deutscher Universitätsverlag, Wiesbaden.
Kubeczko, K. and Rametsteiner, E. (2002) Innovation and entrepreneurship – a new topic for forest related research? Discussion Paper P/2002-1. Institute of Forest Sector Policy and Economics, Vienna.
Kubeczko, K., Weiss, G. and Rametsteiner, E. (2006) The role of sectoral and regional innovation systems in supporting innovations in forestry. *Journal of Forest Policy and Economics* 8, 704–715.
Lam, A. (2005) Organizational innovation. In: Fagerberg, J., Mowery, D.C. and Nelson, R. (eds) *The Oxford Handbook of Innovation.* Oxford University Press, Oxford, pp. 115–147.
Lundvall, B.-Å. (ed.) (1992) *National Systems of Innovation. Towards a Theory of Innovation and Interactive Learning.* Pinter, London.
Lundvall, B.-Å. and Borras, S. (1997) *The Globalising Learning Economy.* European Communities, Luxembourg.
Lundvall, B.-Å. and Borras, S. (2005) Science, technology, and innovation policy. In: Fagerberg, J., Mowery, D.C. and Nelson, R. (eds) *The Oxford Handbook of Innovation.* Oxford University Press, Oxford, pp. 599–631.
Lundvall, B.-Å. and Johnson, B. (1994) The learning economy. *Journal of Industry Studies* 1, 23–42.
Maillat, D. (1996) From the industrial district to the innovative milieu: contribution to an analysis of territorialized productive organisations. Working Paper IRER Nr. 9606b. University Neuchatel, Neuchatel.
Malerba, F. (2005) Sectoral systems: how and why innovation differs across sectors. In: Fagerberg, J., Mowery, D.C. and Nelson, R. (eds) *The Oxford Handbook of Innovation.* Oxford University Press, Oxford, pp. 380–406.
Mantau, U., Merlo, M., Sekot, W. and Welcker, B. (2001) *Recreational and Environmental Markets for Forest Enterprises: A New Approach Towards Marketability of Public Goods.* CAB International, Wallingford, UK.

Marshall, A. (1920) *The Principles of Economics.* McMillan, London.
Martin, R. and Sunley, P. (2003) Deconstructing clusters: chaotic concept or policy panacea? *Journal of Economic Geography* 1, 5–35.
Mavsar, R., Ramcilovic, S., Palahi, M., Weiss, G., Rametsteiner, E., Tykkä, S., Apeldoorn, R. v., Vreke, J., Wijk, M. v. and Janse, G. (2008) Study on the development and marketing of non-market forest products and services (FORVALUE). Study Report for DG AGRI, Study Contract No. 30-CE-0162979/00-21.
Mayntz, R. (1998) New challenges to governance theory. Jean Monnet Chair Paper 50. Robert Schuman Centre of the European University Institute, Florenz.
Mayntz, R. and Scharpf, F.W. (eds) (1995) *Gesellschaftliche Selbstregelung und politische Steuerung.* Schriften des Max-Planck-Instituts for Gesellschaftsforschung Köln, Frankfurt a.M. Campus.
Mintzberg, H. (1979) *The Structuring of Organization.* Prentice Hall, Englewood Cliffs, New Jersey.
Mitchell, R.K., Agle, B.R. and Wood, D.J. (1997) Towards a theory of stakeholder identification and salience: defining the principle of who and what really counts. *The Academy of Management Review* 22, 853–866.
Morgan, K. (1997) The learning region: institutions, innovation and regional renewal. *Regional Studies* 31, 491–504.
Morris, M. (1998) *Entrepreneurial Intensity: Sustainable Advantages for Individuals, Organisations, and Societies.* Quorum Books, Westport, Connecticut.
Moulaert, F. and Sekia, F. (2003) Territorial innovation models: a critical survey. *Regional Studies* 37, 289–302.
Nauwelaers, C. (2003) Background paper on cluster policies. Prepared for the Trend Chart Policy Workshop Innovative Hot Spots in Europe: Policies to promote trans-border clusters of creative activity, 5–6 May, Luxembourg, p. 48.
Nelson, R. (ed.) (1993) *National Systems of Innovation.* Oxford University Press, Oxford.
Nelson, R. and Rosenberg, N. (1993) Technical innovation and national systems. In: Nelson, R. (ed.) *National Systems of Innovation: A Comparative Study.* Oxford University Press, Oxford, pp. 3–27.
Nelson, R. and Winter, S. (1982) *An Evolutionary Theory of Economic Change.* Belknapp Press, Cambridge, Massachusetts.
North, D.C. (1991) Institutions. *Journal of Economic Perspectives* 5, 97–112.
OECD (2002) The measurement of scientific and technological activities. Proposed Standard Practice for Surveys on Research and Experimental Development: Frascati Manual. OECD, Paris.
OECD (2005) *Oslo Manual: Guidelines for Collecting and Interpreting Innovation Data.* OECD, Paris.
Ollonqvist, P. and Rimmler, T. (2010) Policy integration supporting innovation policy design for SMEs. In: Rametsteiner, E., Weiss, G., Ollonqvist, P. and Slee, B. (eds) *Policy Integration and Coordination: the Case of Innovation and the Forest Sector in Europe.* OPOCE, Brussels.
Pavitt, K. (1984) Sectoral patterns of technical change: towards a taxonomy and a theory. *Research Policy* 13, 343–373.
Pavitt, K. (2005) Innovation processes. In: Fagerberg, J., Mowery, D.C. and Nelson, R. (eds) *The Oxford Handbook of Innovation.* Oxford University Press, Oxford, pp. 86–114.
Polanyi, M. (1958) *Personal Knowledge: Towards a Post Critical Philosophy.* Routledge, London.
Porter, M.E. (1990) *The Competitive Advantage of Nations.* McMillan, London.
Porter, M.E. (1998) Clusters and the new economics of competition. *Harvard Business Review* (November–December), 77–90.
Powell, W.W. and Grodal, S. (2005) Networks of innovators. In: Fagerberg, J., Mowery, D.C. and Nelson, R. (eds) *The Oxford Handbook of Innovation.* Oxford University Press, Oxford, pp. 56–85.
Pülzl, H. and Treib, O. (2007) Implementing public policy. In: Fischer, F., Miller, G.J. and Sidney, M.S. (eds) *Handbook of Public Policy Analysis.* CRC Press, Boca Raton, Florida, pp. 89–108.
Radosevic, S. (1997) Systems of innovation in transformation: from socialism to post-socialism. In: Edquist, C. (ed.) *Systems of Innovation: Technologies, Institutions and Organizations.* Pinter, London, pp. 371–394.
Rametsteiner, E. and Weiss, G. (2006) Innovation and innovation policy in forestry: linking innovation process with systems models. *Forest Policy and Economics* 8, 691–703.
Rametsteiner, E., Weiss, G. and Kubeczko, K. (2005) *Innovation and Entrepreneurship in Forestry in Central Europe.* Brill, Leiden.
Rametsteiner, E., Weiss, G., Ollonqvist, P. and Slee, B. (eds) (2010) *Policy Integration and Coordination: the Case of Innovation and the Forest Sector in Europe.* OPOCE, Brussels.

Rhodes, R.A.W. (1996) The new governance: governing without government. *Political Studies* XLIV, 652–667.

Rhodes, R.A.W. (1997) *Understanding Governance: Policy Networks, Governance, Reflexivity and Accountability.* Open University Press, Buckingham, UK.

Rogers, E.M. (1995) *Diffusion of Innovations.* The Free Press, New York.

Rothwell, M. (1989) Firm size, innovation and industrial change. *Small Business Economics* 1, 51–64.

Rothwell, M. and Dodgson, M. (1994) Innovation and size of firm. In: Dodgson, M. and Rothwell, M. (eds) *Handbook of Industrial Innovation.* Edward Elgar, Cheltenham, UK.

Rothwell, R. and Zegveld, W. (1982) *Innovation and the Small and Medium Sized Firm: their Role in Employment and in Economic Change.* Kluwer and Nijhoff, Dordrecht.

Scheef, J. (1999) *Lernende Regionen, Regionale Netzwerke als Antwort auf globale Herausforderungen.* Linde Verlag, Vienna.

Scherer, R. (2006) *Regionale Innovationskoalitionen. Bedeutung und Erfolgsfaktoren von regionalen Governance-Systemen.* Haupt, Bern.

Schön, D.A. (1963) Champions of radical new inventions. *Harward Business Review* 41(March–April), 77–86.

Schumacher, E.F. (1973) *Small is Beautiful: a Study of Economics as if People Mattered.* Vintage, London (re-edition 1993).

Schumpeter, J.A. (1934) *The Theory of Economic Development.* Harvard University Press, Cambridge, Massachusetts.

Simmie, J. (2006) Do clusters or innovation systems drive competitiveness? In: Asheim, B., Cooke, P. and Martin, R. (eds) *Clusters and Regional Development. Critical Reflections and Explorations.* Routledge, London, pp. 164–187.

Smith, K. (2005) Measuring innovation. In: Fagerberg, J., Mowery, D.C. and Nelson, R. (eds) *The Oxford Handbook of Innovation.* Oxford University Press, Oxford, pp. 148–179.

Thompson, M., Ellis, R. and Wildavsky, A. (1990) *Cultural Theory.* Westview Press/Praeger, Boulder, Colorado/Westport, Connecticut.

Tunzelmann, N. v. and Acha, V. (2005) Innovation in 'low-tech' industries. In: Fagerberg, J., Mowery, D.C. and Nelson, R. (eds) *The Oxford Handbook of Innovation.* Oxford University Press, Oxford, pp. 407–432.

Weber, M.K. (2002) Innovation networks and the transformation of large socio-economic systems: the case of combined heat and power technology. In: Pyka, A. and Küppers, G. (eds) *Innovation Networks: Theory and Practice.* Edward Elgar, Cheltenham, UK, pp. 133–168.

Weiss, G. (2004) Die Rolle von Innovationssystemen in der Entwicklung und Verbreitung von Biomassefernwärmeanlagen in Österreich. *Centralblatt für das gesamte Forstwesen* 121, 225–242.

Weiss, G., Salka, J., Dobsinska, Z., Rametsteiner, E., Bauer, A., Aggestam, F. and Tykkä, S. (2010a) Integrating innovation in forest and development policies: comparative analysis of national policies across Europe. In: Rametsteiner, E., Weiss, G., Ollonqvist, P. and Slee, B. (eds) *Policy Integration and Coordination: the Case of Innovation and the Forest Sector in Europe.* OPOCE, Brussels.

Weiss, G., Ollonqvist, P., Rametsteiner, E. and Slee, B. (2010b) Summary results and conclusions. In: Rametsteiner, E., Weiss, G., Ollonqvist, P. and Slee, B. (eds) *Policy Integration and Coordination: the Case of Innovation and the Forest Sector in Europe.* OPOCE, Brussels.

Wiklund, J. (1998) Small firm growth and performance: entrepreneurship and beyond. PhD thesis, Jönköping International Business School. JIBS Dissertation Series 003.

Williamson, O.E. (1985) *The Economic Institutions of Capitalism. Firms, Markets and Relational Contracts.* The Free Press, New York.

3 Networks of Small–Medium Enterprises Operating in Forestry: Some Theoretical Concepts and Empirical Evidence

Davide Pettenella* and Daria Maso

Abstract

The aim of this chapter is to present and discuss some theoretical concepts and empirical evidence in network analysis in relation to the innovation process in the forestry sector. Attention is given to the role of networks among small and medium enterprises (SMEs), the most common ones in both developed and developing countries. The chapter is organized in three main sections. First, some definitions of networks and their main organizational aspects are considered. Secondly, the types of networks are described, making reference to the work by Varamäki and Vesalainen (2003). The role of networks in the innovation process is discussed in the last section. The chapter highlights the strong link between absolute innovation through patentable research results and single, generally large, research units; conversely, SMEs, also through their networks, can more easily access and share other types of innovation thanks to the role played by a network in scope and scale economies. Innovation not only depends on personal characteristics of the single entrepreneur but also on the context in which he or she lives and operates, the social 'milieu'. A well structured and developed social milieu is the result of the accumulation of a social capital. Networks can be the logical and direct output of a milieu characterized by an accumulated social capital.

3.1 Introduction

This chapter focuses on some theoretical concepts of network analysis in relation to the innovation process in the forestry sector. Attention is given to the role of networks among small and medium enterprises (SMEs), the most common ones in both developed and developing countries. The role of SMEs is increasing with the shift from timber-based forest economies to those where non-wood forest products and services (NWFP&S) are becoming a more important part of income generation. It is clear that networks can play an important role in this process: the creation of networks among NWFP&S producers, traders or sellers (or, in a broader sense, nature-based SMEs) and also among these and other, even larger, enterprises from different sectors, can lead to the supply of a critical mass of products and services. When networks are able to coordinate and harmonize tourist, sport, recreational and cultural services, diversity can result in a successful marketing strategy and not a weakness, as can happen with wood products for a mass market.

In the following, the concept of network is explored from the definitions to the main

* Corresponding author.

types identified in the literature and the role played in the implementation of innovative processes. This work thus provides a theoretical framework where the following contributions of this publication (and especially that by Maso, Matilainen and Pettenella including the case studies) can be analysed.

3.2 Theoretical Instruments to Analyse the Networks

Many different definitions of the term network can be found in the literature (Box 3.1). They are usually quite similar and differences are mainly related to the sector for which they have been developed and to which they are applied. Moreover, the terms 'network' and 'alliance' are quite frequently used as synonyms.

The fundamental concepts that should be included in an exhaustive definition of network, according to what emerges from the definitions in Box 3.1, are:

- a structure made of nodes (firms) and links (relationships between firms);
- firms involved may also have relationships with firms outside this structure;
- firms can be similar but also complementary;
- relationships can be of different types;
- relationships can be both vertical and horizontal; and
- relationships can vary over time.

Box 3.1. Definitions of 'network'.

Social network: a 'set of nodes' (persons, organizations, etc.) linked by a set of social relationships (friendship, transfer of funds, etc.) (Gulati, 1998).

Business network: structure of exchange relationships among business actors (firms as well as individuals), structure that emerges, evolves and dissolves over time in a continuous and interactive process (Halinen and Törnroos, 1998).

Network: a mode of organization that can be used by managers or entrepreneurs to position their firms in a stronger competitive stance (Jarillo, 1988).

Strategic network: long-term, purposeful arrangement among distinct but related for-profit organizations that allows the firms in it to gain or sustain competitive advantage *vis-à-vis* their competitors outside the network. Essential to this concept of strategic network is that of 'hub firm': the firm that sets up the network and takes a pro-active attitude in the care of it (Jarillo, 1988).

Jarillo's definition was used as a basis by Provan and Milward (1995) who proposed the following narrower definition:

Network: an intentionally formed group of small- and medium-sized profit-oriented firms in which the firms: (i) are geographically proximate; (ii) operate within the same industry, potentially sharing inputs and outputs; and (iii) undertake direct interactions with each other for specific business outcomes. The interactions may include joint production, new product development, collective marketing and employee training.

Network: two or more organizations involved in long-term relationships. A network may be viewed as consisting of 'nodes' or 'positions' (firms, trade associations, other types of organizations, etc.) and links (interaction between the nodes). The **links** constitute a reflection and recognition of interdependence. They are based on relationships over time (Thorelli, 1986).

Network: a close yet non-exclusive relationship with other members (Dennis, 2000).

Network: an organizational form logically intermediate between the pure market and vertically integrated firm (Nohria and Eccles, 1992, in Dennis, 2000).

Networks: value-adding partnerships that facilitate the exchange of experience and knowledge between member companies (Johnston and Lawrence, 1988, in Dennis, 2000).

Strategic alliances: the pooling of specific resources and skills by the cooperating organizations in order to achieve common goals, as well as goals specific to the individual partners (Varadarajan and Cunningham, 1995).

Robinson and Clarke-Hill (1994) provide a broader definition of **strategic alliance** or **strategic business relationship**: 'a coalition of two or more organizations intended to achieve mutually beneficial goals' (Varadarajan and Cunningham, 1995).

Based on this analysis, a possible exhaustive definition of network could be: a structure formed by organizations of different types and linked (among themselves but also with organizations outside this structure) by multiple kinds of relationships. The organizations involved are located at different levels, depending on their own characteristics and the types of relationships they form, and can change their position and relationships over time.

When dealing with the business field, the term 'organization' normally refers to a firm, but it could also represent a non-profit organization or a public agency.

3.3 The Main Types of Networks and their Characteristics

Based on their research, Varamäki and Vesalainen (2003) proposed a classification of different types of networks differentiated on the basis of the intensity and tightness of the relationships among the participating firms.

Table 3.1 provides a short description and a synthesis of the main characteristics of the different types of networks identified by Varamäki and Vesalainen (2003), in relation to the multilateral cooperation among network participants, from the loosest to the tightest one. The tightness of cooperation is measured by the degree of formalization of the established relationship and the amount of joint investments that the network objectives require.

A Development Circle is a very informal type of network, based on personal bonds (very frequently friendship between firms' owners) and does not require financial investments from the participating firms. On the contrary, a Joint Unit requires very high levels of formal agreements and financial investments. Indeed, it basically consists of the establishment of a new firm that participating firms must join, thereby losing their previous autonomous identities.

In any case, networks are not fixed structures. Firms can change their position within the network, plus existing networks can evolve and change their original form by transforming themselves into a tighter (or, more rarely, a less tight) structure (Lamming *et al.*, 2000). Network position can be defined as 'how individual business actors in the network are related to one another in terms of their function, role and identity in the network structure' (Halinen and Törnroos, 1998).

Each actor is engaged in many exchange relationships with other actors. These relationships define the position of the actor in the network (Halinen and Törnroos, 1998). The position that a company occupies in a network depends on at least three main factors: the domain of the company (i.e. its role in the division of labour), the position of the company in other networks and the power of the company relative to other participants in the focal network (where the power is the company's ability to influence decisions and the actions of others) (Thorelli, 1986).

Firms modify their position in a network if something changes in their own structure and therefore in their power over other network members. But their position can also be modified if and when a new member joins the network. In this situation, the established members may have to reposition themselves to accommodate the new entrant and their new position will depend on the power of the new entrant.

In the forestry sector, all these issues have been broadly analysed in relation to the development of the wood-based industry, more than of forest enterprises. A large number of studies have been published on the wood-based industrial districts or clusters investigating the theoretical framework, paradigms, policies, impacts on regional economies and employment (see Hazley, 2000, for an extensive literature review). Merlo and Fodde (1996) summarize the most relevant approaches developed in the analysis of industrial district development (Table 3.2), whereas Viitamo (2001) provides a synthesis of the empirical research on industrial competitiveness of wood-based clusters.

Regarding network evolution, Fig. 3.1 shows some of the possible steps through

Table 3.1. Characteristics of the different types of networks identified by Varamäki and Vesalainen (2003) (modified by the authors).

Main characteristics	Cooperative arrangements: Development Circle	Loose Cooperative Circle	Project Group	Joint Venture	Joint Unit
Description	A group of entrepreneurs who meet regularly to discuss problems and experiences. The Circle can jointly acquire education, participate in seminars, etc.	A group of firms that share some kind of common resource (e.g. transport or production equipment, etc.), so that each firm can have that particular resource at its disposal and costs can be shared. Participating firms do not run a joint business	A group of firms that combine the resource or products of the partner firms into a joint business	A jointly owned company, set up by participating firms with the central goal being to start, develop and run new joint businesses	Participating firms set up a new firm and join it. The original firms remain as legal entities, but function for internal purposes. It is the most formal and tightest type of cooperative management
Aims	To learn from each other; transfer knowledge	To share resources and so save costs	To develop a joint business by combining the complementary resources and skills of partners; new market channels through each participating firm	To start, develop and run a new joint business	To set up a new firm that participants join
Investments	Not needed	Usually not needed	Non-standard conditions	Considerable	Considerable
Type of cooperation	Very informal, based on personal bonds	Democratic decision making; loose contract bonds; oral or written agreements	Consensus-based decision making (more than democratic); written agreements	Consensus-based decision making; very formal corporate arrangements	Consensus-based decision making (more than democratic)
Trust	Strong	A certain level, at least regarding quality and reputation of partners	High	High	Very high
Critical factors for success	Entrepreneurs must: actively participate in meetings; discuss openly and on equal terms	Firms must agree on the principles of using and sharing common resources	Careful company selection	Careful partner selection; need for a separate company	Careful company selection
Partner selection	Avoid competitors or potential competitors	Limit the number to the one that permits benefits to be gained from the common resource	Resources and skills of partners have to be different but complementary; quality, capacity and economy of each partner must be good	Very important, resources and skills have to be different enough but complementary; partners must have enough capital to invest in the development of new businesses	

Table 3.2. Approaches in industrial districts analysis.

Author	Theoretical framework
Marshall (1890)	Marshall coined the term 'industrial districts' with reference to the 19th-century steel and textile industries in Sheffield and Lancashire; the key point of these districts was external economies but also the concept 'industrial atmosphere' was employed
GeorgescuRoegen (1976 and 1990)	An alternative, and perhaps more realistic theory of production based on time, stocks and flows; in this approach production processes are like a sort of puzzle or interlocking system. It was demonstrated by Tani (1986) and Zamagni (1988) that industrial districts, thanks to network economies (decentralization of production processes into small units), make it possible to employ labour and industrial plants more efficiently, while adopting more modern technologies
Colin Clark (1967)	A theory based on observation of land-use patterns and regional development explaining the sprawl of activities that in recent decades were brought about by lower transportation and communication costs. This approach refers to von Thünen theory and its development by Isard (1956)
Coase (1937)	The approach based on transaction costs led to the emergence of a firm that supersedes the market (Cheung, 1983). In other cases, industrial districts work the other way round, being based on market transactions of commodities rather than on the organization of production factors within firms
Piore and Sobel (1982)	In Piore and Sobel's approach to industrial development, theory flexibility is the key factor of modern production processes as an alternative scale economies towards mass production (see Taylor and Ford's approach)
Porter (1989)	Porter's competitive advantage paradigm refers to nations and areas, i.e. the 'diamond model' based on factor and demand conditions, strategies structure and competitiveness of enterprises, links and support with other industrial sectors, the so called 'clusters'

Note: from Merlo and Fodde (1996), modified; refer to this paper for complete bibliographic data.

which a network can pass while increasing its degree of formalization, strategic intensity, or both.

3.4 Economic Advantages Associated with Network Development

To be considered useful and viable structures and to justify their own existence, networks must provide advantages to participating firms. Networks creation usually implies costs, especially when they are very formal structures and have high levels of strategic intensity (Table 3.1). Networks therefore materialize when expected costs are lower than expected economic and indirect benefits.

According to Jarillo (1988), a network exists only when there is a lowering of transaction costs. But existence is not the only condition a network must satisfy. It also has to be efficient. A network can be said to be efficient if the gains a firm achieves by being part of it are higher (over the long-term) than those the firm would obtain by acting alone (Jarillo, 1988).

One of the main purposes of network creation or of involvement in an already existing network is to try to take advantage of lower investment costs and reduced future capital expenses (Dennis, 2000) and

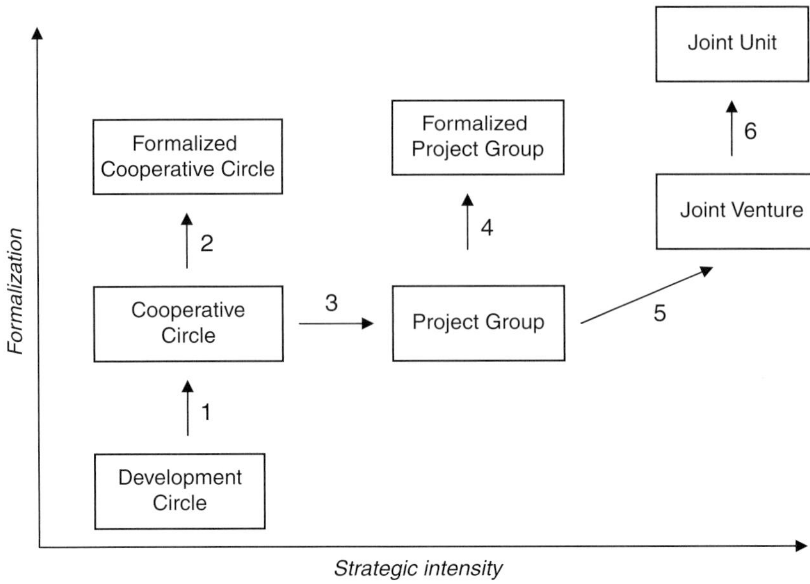

Fig. 3.1. Evolution of cooperative arrangements according to group formalization and strategic intensity (source: Varamäki and Vesalainen, 2003).

risks. In more detail, Dennis (2000) lists the following as the main advantages many SMEs could achieve by network creation or participation:

- they have the opportunity to compete effectively in diverse and often larger markets;
- they can compete on national or international level through the coordination of factors such as research and development, information technology or marketing (without alliances, SMEs would be confined to their local markets);
- links with companies owning complementary expertise and assets allow members to access resources and skills not owned by the company itself; and
- they are encouraged to remain small and to specialize to gain product-specific knowledge, and at the same time they remain flexible and adaptable because of very few resources stored internally.

All these cited probable advantages lead to some economic outcome or results. However, outcomes can be of different types, not only limited to economic aspects. A good outline of the possible outcomes is proposed by Human and Provan (1997). They defined four main general types of outcomes:

1. Inter-organizational exchange: direct transactions or exchanges among network firms, such as buying and selling, jointly producing and marketing a product and exchanging friendship and information.
2. Organizational credibility: firms perceive that their external legitimacy can be enhanced through association with the network. For instance, the network administrative organization (possibly a non-profit organization) can establish relationships with local universities, large suppliers and state agencies that smaller, individual firms would not be able to establish if acting alone. Participation can thus increase the visibility and credibility of member firms.
3. Access to resources: network participation can play an instrumental role in accessing new markets, new product ideas and other valued resources for its companies.
4. Financial performance: economic benefits could occur within a short time after joining the network, but also in a long-term perspective.

These types of advantages have been divided by Human and Provan (1997) into two broad categories:

- transactional outcomes, i.e. enhanced resource acquisition or gains in performance; and
- transformational outcomes, i.e. changes in the way the managers of network firms think, act, or both.

All the various outcomes listed above are very important for SMEs operating in the forestry sector. However, the most important outcomes a profit-oriented entrepreneur will look for are probably the economic ones. Small firms linked by networks face other problems too, like benefit sharing or income distribution derived from the network activities. This is a very complex issue to deal with and, at least until now, no general rules have been defined in order to determine the distribution of benefits among the different organizations participating in the network. It is also necessary to take into account that outcomes can be asymmetrically distributed among firms included in the network: one firm can achieve its objective while another can fail (Gulati, 1998).

Another issue to take into account when discussing the advantages generated by networks is time. According to Jarrat (1998), the objectives the firms want to achieve through network participation change over time. In any case, benefits will have to exceed costs, but strategic intents will vary in the three fundamental periods of the firm's existence:

- beginning of the activity of the firm, when the aim is 'enhancing current business performance' – network participation should help building firm's current capabilities;
- growth and evolution of the firm, when the aim is 'creating new value' – the network should contribute to develop future business potential; and
- mature life of the firm, when the aim is 'defending market position' – the network should help the firm in establishing links that can protect past business development against market or environmental forces.

With respect to firms operating in the forestry sector, we can thus conclude that in order to exist, networks must provide economic or other advantages to participating firms. However, it is not easy to measure these advantages in terms of financial outcomes alone. Over time, and with the changing position of the enterprise within the network, the enterprise will look for different advantages.

3.5 The Role of Networks in the Innovation Process

The classical distinction between process and product innovations can help to explore the role of networks in the innovation process. A process innovation (or organizational innovation) is defined, according to the Oslo Manual (OECD, 1997), as the introduction of significantly changed organizational structures, the implementation of advanced management technique or of substantially improved corporate strategic orientations. All these changes can be considered process innovations only if there is a measurable increased productivity or sales. Product innovation, instead, involves the introduction of a good or a service that is new or substantially improved; it may be useful to make a distinction among three types of product innovation (Dewar and Dutton, 1986):

- absolute new products/services;
- incremental innovation (consisting of the improvement of existing products and services); and
- subjective differentiation (based on communication strategy, branding, etc.).

There is empirical evidence that in the forestry sector totally new products are quite rare, especially when we consider the first steps of the value chain (i.e. forest management activities). Moreover, it is not always possible to protect a new product or service with some kind of patent, the main motivation for investing in research and development (R&D) activities by a private, profit-oriented company. Absolute innovations can still occur in wood harvesting and processing thanks to large and continuous

R&D investments, while – when dealing with process innovation, incremental product innovations or subjective differentiation – innovative products or activities can quickly be copied by other enterprises once launched on the market, rapidly losing their novelty advantage. Thus, as represented in Fig. 3.2, there is a strong link between absolute innovation through patentable research results and single, generally large, research units; conversely, SMEs, also through their networks, can more easily access and share other types of innovation thanks to the role played by a network in scope and scale economies.

Networks and innovations have a mutual and synergic relationship: innovation can support a more efficient, effective and more equitable functioning of a network, fostering its income generation potential and, vice versa, network creation can allow the introduction of innovation.

There are several common examples of this mutual beneficial link between networks and innovations: the creation of wood biomass heating plants to serve a network of residential buildings, the organization of e-marketing services for the tourist and recreational services to benefit a network of forest land owners, the implementation of an infra-red fire monitoring system to protect a network of forest properties, to mention just a few.

Another approach in innovation classification developed by Kubeczko and Rametsteiner (2002) can give an insight into the relation between innovation and networks. These authors describe four levels of innovation:

- institutional level;
- business-to-business level;
- enterprise level; and
- personal level.

SMEs dealing with forest products and services are involved, in various ways and with different intensity, in all these levels of innovation; for what concerns a business-to-business innovation level, SME networks play an especially crucial role. According to Kubeczko and Rametsteiner (2002), especially for SMEs such as forest holdings, innovation or its

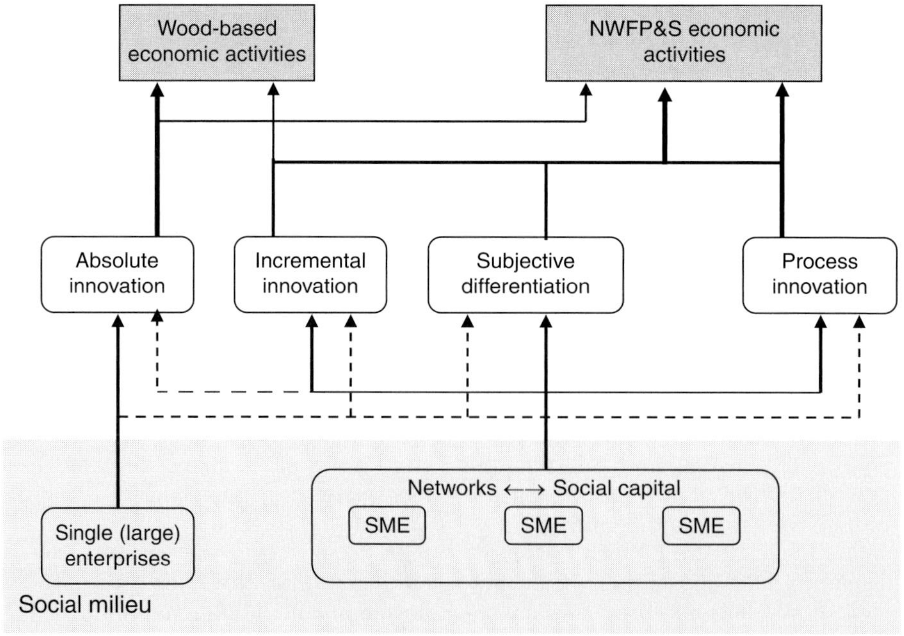

Fig. 3.2. Innovations and networks: a simplified conceptual scheme. NWFP&S, non-wood forest products and services; SME, small and medium enterprise.

diffusion at firm level depends on the attitudes of the individual agents.

Most research done in the field of innovation has been based on single enterprises analysed as autonomous entities, striving to improve their competitive advantages. However, as observed by Gulati *et al.* (2000, p. 203):

> the image of atomistic actors competing for profits against each other in an impersonal marketplace is increasingly inadequate in a world in which firms are embedded in networks of social, professional, and exchange relationships with other organizational actors ... These strategic networks are composed of interorganizational ties that are enduring, are of strategic significance for the firms entering them, and include strategic alliances, joint ventures, long-term buyer-supplier partnerships, and a host of similar ties.

In other words, the forest owner or manager is in many cases the individual agent proposing and deciding about innovation, but innovation not only depends on personal characteristics but also on the context in which the entrepreneur lives and operates, sometimes described as the social 'milieu'.

A well structured and developed social milieu is the result of the accumulation of a social capital. With this term we refer not to a commodity that can be traded, but to (sometimes unwritten) norms, codes, mutual trust, shared common vision and solidarity, which are normally accumulated as an unintended and even unanticipated consequence of some economic activity carried out by the economic actors working in the same milieu. 'Social capital enables firms to improve their innovative capability and conduct business transactions without much fuss and has, therefore, substantial implication for economic performances' (Maskell, 2000, p. 111). Networks can be the logical and direct output of a milieu characterized by an accumulated social capital. The term social capital is commonly used in sociology, political science, economics and other social sciences, with many different definitions (Table 3.3). As a consequence of a broad use of the concept, 'social capital has come to mean so many different things to different researchers that it may border on the meaningless', as critically noted by Bjørnskov and Sønderskov (2010). In the following, we try to go deeper into the relationship between innovation and the social capital.

Quite a number of recent studies have shown that networks relationships create good opportunities for knowledge acquisition, knowledge assimilation and knowledge exploitation, in what can be defined a learning process (Dyer and Singh 1998; Larsson *et al.*, 1998; Yli-Renko *et al.*, 2001). This is true both when we deal with external 'business' knowledge (i.e. knowledge of products, markets and technology) and 'organizing' knowledge (i.e. knowledge of structure and systems) (Eriksson *et al.*, 1997), as in many NWFP&S (see the case study in Box 3.2). Not only competitive but also cooperative interactions among enterprises are important elements for innovation, through the imitation process and the stimulus to innovation that a close interaction creates. This is why innovation is a sort of spillover of networks development and at the same time a key element for their success (Kubeczko and Rametsteiner, 2002).

There is thus a direct and strong link among networks, social capital, knowledge acquisition and innovation as demonstrated by Yli-Renko *et al.* (2001, pp. 1–2):

> the degree to which firms can use external relationships for knowledge acquisition and exploitation is regulated by the amount of social capital embedded in such a relationship. Social capital in a relationship enables the firm to tap into the knowledge resources of its exchange partner. Through close interaction, firms are able to increase the depth, breadth, and efficiency of mutual knowledge exchange.

In general, the smaller the business and the smaller the customer group, the more important the organizational aspects for production, distribution and also market research and promotion become (Pettenella *et al.*, 2006). Networks can improve the business activities of forest enterprises in the production phase, e.g. through the implementation of joint production chains or sub-contracting in order to produce bigger volumes or provide more complete service packages. Moreover,

Table 3.3. Definitions of social capital.

Study	Definition
Coleman (1988, p. 95)	'obligations and expectations, information channels, and social norms'
Coleman (1990, p. 304)	'social organization constitutes social capital, facilitating the achievement of goals that could not be achieved in its absence or could only be achieved at a higher cost'
Putnam (1993, p. 167)	'features of social organization, such as trust, norms and networks that can improve the efficiency of society'
Fukuyama (1997, pp. 378–379)	'the existence of a certain set of informal rules or norms shared among members of a group that permits co-operation among them. The sharing of values and norms does not on itself produce social capital, because the norms may be wrong ones [...] The norms that produce social capital [...] must substantively include virtues like trust telling, the meeting of obligations and reciprocity'
Knack and Keefer (1997, p. 1251)	'trust, co-operative norms, and associations within groups'
Narayan and Pritchett (1999, p. 872)	'the quantity and quality of associational life and the related social norms'
Putnam (2000, p. 19)	'connections among individuals – social networks and norms of reciprocity and trustworthiness that arise from them'
Ostrom (2000, p. 176)	'the shared knowledge, understandings, norms, rules and expectations about patterns of interactions that groups of individuals bring to a recurrent activity'
Paldam (2000, p. 635)	Three families: (i) 'the ability of [an individual) to work voluntarily together with others of [a population]'; (ii) 'the quantity of trust [an individual] has in other members of [a population]; (iii) 'the amount of benefits the individual can draw on his goodwill'
Whiteley (2000, p. 450)	'the willingness of citizens to trust others including members of their own family, fellow citizens, and people in general'
Woolcock (2001, p. 13)	'the norms and networks that facilitate collective action [...] it is important that any definition of social capital focus on the sources rather than consequences [...] this approach eliminates an entity such as 'trust' from the definition of social capital'
Lin (2001, pp. 24–25)	'resources embedded in social networks and accessed and used by actors for actions. Thus the concept has two important components: 1) it represents resources embedded in social relations rather than individuals, and 2) access and use of such resources reside with the actors'
Bowles and Gintis (2002, p. 2)	'trust, concern for one's associates, a willingness to live by the norms of one's community and to punish those who do not'
Knack (2002, p. 42)	'common values, norms, informal networks, and associational memberships that affect the ability of individuals to work together to achieve common goals'
Sobel (2002, p. 139)	'circumstances in which individuals can use membership in groups and networks to achieve secure benefits'
Durlauf and Fafchamps (2004, p. 5)	A feature that 'generates positive externalities for member of a group [...that] are achieved through share trust, norms and values and their consequent effects on expectations and behavior [...] shared trust, norms and values arise from informal forms of organizations based on social networks and associations'
World Bank (2005)	'norms and networks that enable collective action'
Groot *et al.* (2006, p. 1)	'social capital includes all factors that foster social relations and social cohesion'

Note: from Bjørnskov and Sønderskov (2010), adapted from Knowles (2006); refer to these papers for complete bibliographic data.

Box 3.2. Innovation in chestnut marketing: an example of the effects of networks.

Chestnuts are normally sold by forest owners to middlemen at relatively low cost (€1–2/kg, whereas the price for the final consumer is no less than €6/kg); profitability level of chestnut orchard cultivation cannot always cover all the forest maintenance costs (pruning, grafting, sanitary fellings, land cleaning, etc.) and the high harvesting costs. Moreover, on late September–October weekends, day-trippers from urban areas enjoy (illegally) picking chestnuts on private properties. Landowners' reaction to these trespassers is very negative, not to mention the time (and costs) spent on surveillance. This is inducing land abandonment, with the loss of a traditional element of the cultural landscape.

In Tuscany (Italy), in the Garfagnana area, this problem has been transformed into a resource through the creation of a chestnut growers' association that has launched the initiative 'Adotta un Castagno' ('chestnut tree adoption' – www.associazionecastanicoltori.it): a set of services is offered at a price of €75 including the right to picnic under a monumental chestnut tree (clearly identified on a map) for 1 year, 10 kg of fresh nuts, 2 kg of dried nuts, 2 kg of chestnut flour of Certified Origin, a 15% discount for 14 nights at the local agritourist farms, a 20% discount rate for all the local restaurants and food products and one free guide to visit the forest area, a chestnut drying kiln and other historical rural infrastructures.

It was not easy to reach an agreement among the different actors of this innovation: forest owners have to change their role from sellers of a commodity to sellers of a tourist service. Local social capital helped a lot to reach a positive solution: an association was created in 1998; a group of chestnut growers became organic, the EC mark of origin was obtained, and, with the strong support of the local Municipality, they became the initiator of the Associazione Nazionale Città del Castagno (National Association of the Chestnut Municipalities – www.cittadelcastagno.it). The initiative is the result of an effective networking, with quite positive results for the forest owners' profitability and the maintenance of public goods like landscape quality.

the networking activity with other companies and the cooperation during the product development phase is often extremely important in order to generate and spread the needed information related both to the production as well as to the markets. In addition, cooperation in the marketing phase and in the supply of the products is often needed in order to achieve the necessary publicity (Aarne *et al.*, 2005).

The relevant role of networks and networking activity in innovation development in the forestry sector analysed here on more a theoretical level, seems also to be confirmed by the recent field research conducted by Nybakk *et al.* (2009), who affirm that:

> networking can contribute to innovative capacity and innovativeness among local forest owners utilizing NWFP&S by giving them novel ideas and access to resources, and by transferring knowledge. Accordingly, forest owners who invest in networking with local actors will obtain advantage by gaining new ideas, concentrating on core expertise and finding new and better ways to run their businesses.

Innovations in the more dynamic and open NWFP&S market segments may therefore also have positive effects in the innovative attitude of the traditional wood-based value chain, both activities having the same landowners involved and the same social mileu.

3.6 Conclusions

SMEs are still playing a major role in the forestry sector, thanks also to the creation of different types of networks. However, SMEs often face substantial difficulties in innovation development. In this respect, networks play an important role in the innovation process of the SMEs operating in the forest areas, favouring innovation both at the entrepreneur level and at the business to business level.

In order to implement rural development policies based on income generation from nature-based activities, policy makers should pay attention to network creation among SMEs. Properly functioning networks are a pre-condition for the innovation practices; thus, being networks strongly dependent on the accumulation of social capital, stable local communities with a shared strategic vision and mutual trust are the key factors for successful forest development policies.

References

Aarne, M., Hänninen, R., Kallio, M., Kärnä, J., Karppinen, H., Ollonqvist, P., Packalen, K., Rimmler, T., Toppinen, A., Kajanus, M., Matilainen, A., Rutanen, J., Kurki, S., Peltoniemi, J. and Saarinen, J. (2005) Country report: Finland. In: Jáger, L. (ed.) Economic integration of urban consumers' demand and rural forestry production. Forest sector entrepreneurship in Europe: Country studies. *Acta Silvatica & Lignaria Hungarica.* Special Edition.

Bjørnskov, C. and Sønderskov, K.M. (2010) Is social capital a good concept? Paper prepared for presentation at the annual meeting of the Midwest Political Science Association 67th Annual National Conference, Chicago, Illinois 22–25 April.

Dennis, C. (2000) Networking for marketing advantage. *Management Decision* 38, 287–292.

Dewar, R. and Dutton, J. (1986) The adoption of radical and incremental innovations: an empirical analysis. *Management Science* 32, 1422–1433.

Dyer, J.H. and Singh, H. (1998) The relational view: cooperative strategy and sources of interorganizational competitive advantage. *Academy of Management Review* 23, 660–679.

Eriksson, K., Johanson, J., Majkgard, A. and Sharma, D. (1997) Experiential knowledge and cost in the internationalization process. *Journal of International Business Studies* 28, 337–360.

Gulati, R. (1998) Alliances and networks. *Strategic Management Journal* 19, 293–317.

Gulati, R., Nohria, N. and Zaheer, A. (2000) Strategic networks. *Strategic Management Journal* 21, 203–215.

Halinen, A. and Törnroos, J. (1998) The role of embeddedness in the evolution of business networks. *Scandinavian Journal of Management* 14, 187–205.

Hazley, C.J. (2000) *Forest-based and Related Industries of the European Union. Industrial Districts, Clusters and Agglomerations.* The Research Institute of the Finnish Economy (ETLA), Helsinki.

Human, S. and Provan, K. (1997) An emergent theory of structure and outcomes in small-firm strategic manufacturing networks. *Academy of Management Journal* 40, 368–483.

Jarillo, J.C. (1988) On strategic networks. *Strategic Management Journal* 9, 31–41.

Jarrat, D.G. (1998) A strategic classification of business alliances, a qualitative perspective built from a study of small and medium-sized enterprises. *Qualitative Market Research* 1, 39–49.

Johnston, R. and Lawrence, P. (1988) Beyond vertical integration. *Harvard Business Review* July, 481–510 (cited in: Dennis, C. (2000) Networking for marketing advantage. *Management Decision* 38, 287–292).

Knowles, S. (2006) Is social capital part of the institutions continuum and is it a deep determinant of development? World Institute for Development Economics Research, Research Paper 25.

Kubeczko, K. and Rametsteiner, E. (2002) Innovation and entrepreneurship. A new topic for forest related research? IFSPE-Discussion Paper, Vienna.

Lamming, R., Johnsen, T., Zheng, J. and Harland, C. (2000) An initial classification of supply networks. *International Journal of Operations and Production Management* 20, 675–691.

Larsson, R., Bengtsso, L., Henriksson, K. and Spark, J. (1998) The interorganizational learning dilemma: collective knowledge development in strategic alliances. *Organizational Science* 9, 285–305.

Maskell, P. (2000) Social capital, innovation and competitiveness. In: Baron, S., Field, J. and Schuller, T. (eds) *Social Capital: Critical Perspectives.* Oxford University Press, Oxford, pp. 111–123.

Merlo, M. and Fodde, F. (1996) Some annotations on the role of forest based production chains in Italian regional economies. In: Hyttinen, P., Mononen, A. and Pelli, P. (eds) *Regional Development Based on Forest Resources – Theories and Practices. EFI Proceedings* 9, 101–116.

Nohria, N. and Eccles, R.G. (eds) (1992) *Networks and Organisations: Structure, Form and Action.* Harvard Business School Press, Boston, Massachusetts.

Nybakk, E., Crespell, P., Hansen, E. and Lunnan, A. (2009) Antecedents to forest owner innovativeness: an investigation of the non-timber forest products and services sector. *Forest Ecology and Management* 257, 608–618.

OECD (1997) *The Measurement of Scientific and Technological Activities Proposed Guidelines for Collecting and Interpreting Technological Innovation Data.* Oslo manual. Organisation for Economic Co-operation and Development, European Commission, Eurostat.

Pettenella, D., Ciccarese, L., Dragoi, S., Hedegus, A., Hingston, A., Klohn, S., Matilainen, A., Posavec, S. and Thorfinsson, T. (2006) NWFP&S marketing: lessons learned from case studies in Europe. In: Niskanen, A. (ed.) *Issues Affecting Enterprise Development in the Forest Sector in Europe.* University of Joensuu, Faculty of Forestry, Research Notes 169.

Provan, K.G. and Milward, H.B. (1995) A preliminary theory of interorganizational network effectiveness: a comparative study of four community mental health systems. *Administrative Science Quarterly* 40, 1–33.

Robinson, T.M. and Clarke-Hill, C.M. (1994) Competitive advantage through strategic retailing alliances – a European perspective. Recent Advantages in Retailing and Services Science Conference, Banff, Alberta, Canada (unpublished).

Thorelli, H.B. (1986) Networks: between markets and hierarchies. *Strategic Management Journal* 7, 37–51.

Varadarajan, P.R. and Cunningham, M. (1995) Strategic alliances: a synthesis of conceptual foundations. *Journal of the Academy of Marketing Science* 23(4), 282–296.

Varamäki, E. and Vesalainen, J. (2003) Modelling different types of multilateral co-operation between SMEs. *Entrepreneurship and Regional Development* 15, 27–47.

Viitamo, E. (2001) Cluster Analysis and the Forest Sector. Where are we now? International Institute for Applied Systems Analysis, Interim Report IR-01-016, Vienna.

Yli-Renko, H., Autio, E. and Sapienza, H. (2001) Social capital, knowledge acquisition, and knowledge exploitation in technology-based firms. *Strategic Management Journal* 22, 587–613.

4 Innovation-related Knowledge Flows: Comparative Analysis of Finnish and Estonian Wood Sectors

Kadri Ukrainski* and Miika Kajanus

Abstract

Industrial firms are gaining ideas for innovation from various sources and their innovative performance depends, besides their internal knowledge resources, on how successful they are at appropriating knowledge from external sources. This seems to be true also with smaller, low-tech and remote firms as for example wood sector companies often are. By studying the role of different knowledge sources for innovations and comparing Estonian wood sector companies with their Finnish counterparts, it appears that all knowledge sources are exploited slightly less intensively in Estonian wood industries. However, the largest backlog appears to be in utilizing universities and public research organizations. The choices of knowledge sources can be attributed to different capabilities of firms and different capabilities of network partners (consumers, universities, etc.) in creating and utilizing (exploitation and exploration) respective innovation-enhancing knowledge, but also to differences in innovation policy.

4.1 Introduction

Innovation requires creating new knowledge or combining existing knowledge in a novel way. Therefore, it rests on learning, which is largely a social process, especially when transfer and accumulation of tacit knowledge is concerned (Polanyi, 1966; Howells, 1995). The capability of a firm to innovate is expanded by a broader knowledge base, and cost and risk sharing through cooperation with other subjects (suppliers, clients, competitors, universities, etc.) (Rothwell, 1991; Freel, 2000). Firms need considerable in-house capabilities to recognize and evaluate the new knowledge and technology and thereafter to negotiate and adopt this technology (Cohen and Levinthal, 1990).

The spatial or regional aspect in innovation cooperation is discussed in the literature on industrial districts and clusters, where these relationships are defined within the territorial sample of firms connected by interpersonal relationships and similar socio-cultural background of employees, entrepreneurs and politicians (Bianchi, 1998). The strength of the relationships outside the region is treated as a decreasing function of the distance (Howells, 1999). Modern technologies have lowered the distance-related cost and, instead, the role of regulatory barriers of different countries has gained in importance by discouraging the

*Corresponding author.

free movement of goods, services and information (Brown and Butler, 1993). As noticed by Kirat and Lung (1999), instead of spatial proximity, social proximity should rather be considered between firms in the sense of technological, organizational or institutional framework implying that firms in a similar environment are more likely to be ready for mutual cooperation. By analysing the size and products of a firm, it is found that smaller firms should rely more on local cooperation because of their restricted resources for cooperation (Arndt and Sternberg, 2000).

The understanding of the innovation processes has advanced from linear innovation models to more systemic ones. There has also been a transformation in the respective policy approaches from science and technology policies to innovation policies (IPs), that would aim to remedy specific failures in the respective national, regional or sectoral setting (Lundvall and Borrás, 2007). Many countries are faced with the challenge of improving their innovation efficiency, i.e. how innovation assets (education, research and development (R&D) and innovation expenditures) are transformed into innovation results (sales from new products and services, and employment). The lack of innovative growth-oriented small and medium-sized enterprises (SMEs) is one of the identified weaknesses in this context (European Innovation Scoreboard, 2008). Thus, broadening the base of innovative, internationally competitive growth-oriented companies is one of the key focus areas of IP.

The systemic principles for IP are rather general and therefore difficult to apply in practice (Teubal *et al.*, 1991). Rametsteiner and Weiss (2006) find, in exploring respective policies in countries of central Europe, that explicit policies and strategies for providing systematic innovation support for forest and wood sectors are rare and those innovation support measures that do exist are fractioned and not coordinated.

The aim of this chapter is to compare the innovation-related knowledge flows and underlying IPs in Finnish and Estonian wood sectors. There are several reasons for comparing Estonian and Finnish wood and forest industries. Forestry provides one of the few renewable resources in both countries. Therefore, the forest, wood and paper industries have always played an important role in both economies. The Finnish forest cluster is considered to be one of the leading forest clusters in the world (Blomström and Kokko, 2002), and could therefore be taken as a benchmark for Estonian industries. Additionally, wood sectors in both countries have strong connections through foreign direct investments (FDI), supplier relationships (machinery) and customer relationships (raw materials, but also finished products). Finally, the cultural similarities in terms of core values and cooperative styles between Estonian and Finnish managers are found to be rather considerable (Nurmi and Üksvärav, 1994). The differences between the countries stem from their different developmental levels, from the relative importance of the subsectors (in the Finnish wood cluster, pulp and paper industry is a driver, whereas in the Estonian case it plays a minor role) and from the alignment and efficiency of the national innovation system infrastructure in both countries.

4.2 Sources and Characteristics of Innovation-related Knowledge and IP Tools for the Wood Sectors

The empirical evidence shows that sectoral innovation systems differ in the patterns of innovative activities, but for each sectoral system, similarities across countries do exist (Malerba and Orsenigo, 1996). These similarities are stemming from the features of technological regimes, knowledge base (appropriability conditions) and learning processes (cumulativeness) that are relatively invariant across countries (Breschi *et al.*, 2000). The creation and exploitation of the technological opportunities are less similar in sectoral systems across countries, because they are related to the level and specifics of the university research, science-industry bridging mechanisms, inter-firm networks and the types and level of innovation efforts of firms (Nelson, 1993).

Wood industries under discussion belong to supplier-dominated sectors, implying that among the predominant sources of technology and information are suppliers, government-financed research institutions and less frequently large users; often competitor collaboration may be relevant (Pavitt, 1984). In wood-related sectors, constant incremental improvements in technological processes have occurred with close interrelation to resulting product innovations (Laestadius, 2000; Palmberg, 2002). These incremental changes have been occurring by adopting the late achievements of different high-tech sectors requiring no formal in-house R&D. The wood products' markets are indicated as very competitive ones, where the appropriability conditions of technology and innovation are weak, because in many cases the products seem relatively easy to imitate (Palmberg, 2002). The depleted technological opportunities and saturated markets explain to a certain extent the low R&D levels in these sectors. Most of the research is done for these industries by other sectors, like equipment manufacturers or biotechnology firms. Often those firms are highly specialized, concentrated and operate worldwide, which makes the latest technology available everywhere (ILO, 2001). Harvester and forwarder producers or paper mill equipment producers are examples of this. The diffusion of these state-of-the-art technologies to transition countries is also enhanced by FDI.

Knowledge interactions between fields of science and economic sectors are fairly dispersed and do not follow obvious or simple patterns. Firms in traditional sectors (such as wood sector) are mostly not considered to have a close knowledge exchange with a public sector R&D base, but some empirical studies point to the contrary – the industries under discussion revealed higher intensities of interactive learning with universities. Meeus *et al.* (2004) have found that firms with moderate strength of internal knowledge resources but also firms creating innovations with moderate complexity have higher probabilities for interactions with a public R&D base compared with low and high levels of expertise and complexity. Based on an Austrian example, wood- and forestry-related sectors are found to have a well-established division of labour with a public R&D base, where public research provides infrastructure and expertise that is oriented directly to the industry's demand. The science fields with closer interaction include forestry, chemistry, including technical chemistry, other interdisciplinary technical sciences, construction techniques, engineering, biology and hydrology, but the interaction is closest with chemical engineering (Schartinger *et al.*, 2002, pp. 314–315).

The role of different sources of knowledge in the innovation process is summarized in Table 4.1. Most important sources of knowledge relevant in all stages of innovation process are located inside the firm. Not only the basic skills of employees, but also their broader knowledge of modern scientific and technological developments, are relevant for absorbing the knowledge from external sources (Cohen and Levinthal, 1990). Internal innovation capabilities are enhanced by R&D activities in a certain technological field, but there is a kind of technological capability relevant especially to firms in low- and medium-technology (LMT) industries such as wood-related ones. This is the capability of introducing new combinations of existing technologies (including architectural capabilities elaborated by Henderson and Clark, 1990). Such capabilities are not reflected in the industry's R&D intensities or educational indicators, but competitive markets force the companies in the wood sector to be very creative regarding the capabilities of running and readjusting their machinery and equipment. As discussed by Hirsch-Kreinsen *et al.* (2003), this demands from firms more tacit knowledge and captures practical engineering problem-solving by engineers and technicians, but also by shop-floor workers. Laestadius (1995) argues that this kind of knowledge in the firm also supports the innovation stemming from external sources to the firm (through collaboration with universities, R&D institutions or suppliers of machinery and equipment).

Table 4.1. Use of the sources of knowledge, underlying influencing factors and policy measures.

Source of knowledge	Type/stage of innovative activity	Other relevant impact factors	Typical IP measures
Internal	Market finding Design and testing Redesigning and producing Distribution and marketing	Firms with more intensive R&D and personnel development activities can rely more on internal sources while innovating Internal sources are used for absorbing external knowledge	Support for training and mobility (S) Grants for industrial R&D (S)
Suppliers	Design and testing Redesigning and producing Distribution and marketing	Used more by firms with substantial investments in process technologies and for diminishing the barrier of lacking technological knowledge The absence of suppliers with suitable capabilities can be a hindering factor	Networking measures (S) Cluster, supply chain policies (D)
Customers	Design and testing Redesigning and producing Distribution and marketing	Used more for novel product innovations, by exporting firms, for reducing the barrier of lacking market knowledge Non-responsive customers or traders are a discouraging factor	Networking measures (S) Cluster, supply chain policies (D) Public procurement (D) Support for private demand (D)
Competitors and industry associations	Design and testing	Concern members are using less knowledge exchange with competitors Used more by exporters, firms with legal barriers	Networking measures (S) Cluster policies (D) Regulation (D)
Consultants, Knowledge Intensive Business Services (KIBS)	Market finding Redesigning and producing	Firms with financial barriers use fewer consultants	Networking measures (S) Information and brokerage support (S)
Universities, R&D institutes, literature	Design and testing	Used more for novel products, also in an informal way. Used less by SMEs Literature is used by firms having high financing barriers; used to gather mainly technological knowledge	Selective support for public sector research (S) Networking measures (S) Support for training and mobility (S) Cluster policies (D)

Source: Authors' compilation from the literature survey in Ukrainski (2008), taxonomy of IP tools stems from Edler and Georghiou (2007). Note: Supply-side IP measures are denoted by (S), demand-side measures by (D).

Typical IP measures targeted to improving internal knowledge bases of firms are related to training, mobility and subsidization of specific R&D activities. Forsman (2009) emphasizes that enterprises already having appropriate innovation capabilities can best benefit from the provided IP instruments. Highly conceptualized policy instruments meet the needs of large enterprises characterized by efficient internal innovation development processes and clear resource allocation for it (Schumpeter Mark II). A variety of discontinuous public services is focused on accelerating the individual phases of the innovation process. In contrast, the development work in SMEs is fuzzy in nature and goes hand in hand with their daily business without a special development budget. Therefore, Forsman (2009) suggests an alternative path for IP: first, starting to improve capabilities for the development of incremental innovations, then forging ahead via radical innovation development. This process can lead to a success spiral that accelerates internal capability building overlap with innovation development – a continuous and diversified innovation flow as an outcome. This insight is also supported by recent surveys in Europe finding that more than 50% of innovative firms innovate without carrying out R&D (Arundel *et al.*, 2008). They are smaller, active in LMT sectors and located in areas with relatively poor innovation infrastructure. However, they are growing at the same rate as their R&D-performing counterparts. Arundel *et al.* (2008) bring out several different ways to innovate besides R&D: technology adoption, minor modifications or incremental changes, imitation including reverse engineering, combining existing knowledge in new ways and adopting solutions developed by users.

Most external knowledge sources (Table 4.1) are used not only to enhance innovation with complementary knowledge but also to reduce associated risks and costs. Linkage with suppliers is based on either making or buying relations (Teubal *et al.*, 1991). Amara and Landry (2005) suggest that users are exploited as an information source by firms that initiate innovations as the world's first introductions rather than as incremental innovations. Suppliers and customers are sources of foreign knowledge for a firm through exporting goods and services or importing materials and technology, as discussed widely in spillover literature. Both of these partner types have been recognized as a source of innovation knowledge not only for developing processes, but also in developing new products and services and even design activities. Organizational, risk and labour barriers influence knowledge exchange with external partners in a similar way, but the financial constraints have been found to be less relevant in the case of suppliers. In some studies, exporting firms have been found to exert greater pressure for upgrading technologies and cooperating more with suppliers. Estonian wood sector firms are using customer knowledge especially intensively by introducing novel and more complex innovations. The impact of the size of the firm is found to be less relevant for absorbing knowledge from users than from other business sources (Ukrainski, 2008).

Openness of knowledge may speed up the pace of innovation, as competitors are able to build on other innovators' advances rather than being allowed to block the progress of others (Foray, 1997). Amara and Landry (2005) have concluded, on the basis of the existing literature, that the information obtained from competitors is related to the increased complexity and intersectoral nature of new technologies, the reduction of uncertainty and R&D costs associated with market access, or the development of product and process innovations (by acquisition and appropriation of the partner's tacit knowledge, uptake of codified knowledge; by reduction of the period between invention and market introduction). However, collaboration with competitors is different because it is typically more informal (although it can also be formalized via R&D contracts or collaboration activities via industry associations). It is not clear, however, which part of the knowledge is exchanged – technological, design or market knowledge, or all of the aforementioned. In the case of Estonian wood firms, exchanging knowledge with competitors is hindered when firms belong to a larger concern or group because of the greater orientation towards the concern's internal knowledge exchange (Ukrainski, 2008).

The measures promoting the knowledge sourcing from business partners are targeted on networking, but also approach from the demand-side focusing on clusters and supply-chains (Table 4.1). In addition, innovation-demand stimulation by direct support or public procurement is influencing the knowledge interaction with customers. Competitor collaboration is supported for responding to new standards or other regulations.

Consultants and knowledge intensive business services (KIBS) are used as a more formal and costly way of innovating when the firm's own knowledge resources are not sufficient. They are used for several activities (product, process and design innovation, etc.), and the relative importance of these activities probably depends more on the type of innovative activities currently relevant for that specific industry (Ukrainski, 2008). Kajanus and Karhu (2008) find from Finnish cases that when the renewal is important at the firm level only, cooperation with business partners (consultants, suppliers) is emphasized; however, when the renewal is important at market level, then the public sector cooperation with universities, etc. is emphasized.

Universities contribute to local innovation processes in a variety of ways, but most importantly by education and technology transfer. In addition to their own discoveries, universities can help to attract new human, knowledge and financial resources from elsewhere. This is becoming even more important as globalization has moved into third, meta-national phase, meaning for example knowledge 'hunting' from global sources (Doz *et al.*, 2001). The Mode 2 model of knowledge formation emphasizes a multidisciplinary approach and knowledge transfer by informal ways during the ongoing research (Nowotny *et al.*, 2004). There are challenges for universities and the regional innovation environment arising from too little geographical proximity (Boschma, 2005). The availability of expertise in universities is a crucial factor in industry–public sector research linkage. Faulkner *et al.* (1995) show that the low level of use of public research institutes as knowledge sources does not imply a lack of research interest in industry, but rather the paucity of the available specific public knowledge and modern equipment vital for development. The industry–academia knowledge exchange is currently being discussed by academics and politicians. In addition to the technology and science-based 'waterfall' model, other approaches have been introduced. These include, e.g. practice-based and demand-driven models, DUI-model (doing, using, interacting), distributed innovation and non-R&D innovators (Cooke *et al.*, 1997; Arundel *et al.*, 2008).

In the knowledge exchange with universities, it is found that firms' size constraints are even more important than sectoral specificities, because SMEs face considerable resource constraints to both formal and informal linkages and communication problems (Corsten, 1987; Faulkner *et al.*, 1995). The basic issue for an SME is finding a person, enterprise or research organization providing specific complementary knowledge and combining that new knowledge with existing knowledge so that it results in a new approach and innovation. Faulkner *et al.* (1995) found that the recruitment policy is relevant because of the networks new employees can bring to the firm (e.g. employees with doctoral or post-doctoral experience usually have networks within academia, and also have literature sources). The other factors they suggest are the proprietary concerns of the firm, but also the openness of senior R&D staff, which from one side reflects the organizational culture, but from the other side also influences it.

The policies used to enhance this type of knowledge exchange are related to the capability-improvement of research institutions, but also networking promotion, mobility schemes and cluster policies. The basic challenge in this collaboration is to enhance the ability of enterprises and research organizations to find and exploit knowledge and technology produced in the world's best knowledge centres. This in particular is a challenge for the SMEs in LMT sectors located outside of university campuses and technology centres, enterprises that do not have tradition, resources and contacts for distributed innovation.

4.3 Innovation-enhancing Policies in Finland and Estonia

4.3.1 Finland

The Finnish IP structure has a horizontal approach overlapping with sectoral and other horizontal policies. The national IP is described in the proposal for Finland's National Innovation Strategy (June 2008) and the Government's Communication to the Parliament (October 2008), but also included in several different policy documents and programmes (Council of State, 2008). The basis of the IP has been outlined in the reports produced by the Science and Technology Policy Council of Finland (Science, Technology, Innovation, 2006). The documents are related to other government strategies as well as to the Lisbon Strategy for Growth and Jobs – the Finnish National Reform Programme 2005–2008.

The key agency for IP on a national level is Tekes (a former technology centre – now a technology and innovation centre), which aims to coordinate the direction of public subsidies to new entrepreneurship and innovation management. They use the concept of technology programmes as background to the development financing. The technology programme of Tekes has strengthened cooperation and interaction between different parts of the national innovation system; it operates with larger companies even though via different kinds of special projects they have also been supporting SMEs.

The Finnish IP has been characterized by a rapid increase in both public and private research funding. For example, in 1991–2001, R&D expenditures grew from 2.1% to 3.4% of the GDP. Large nationwide corporations have traditionally had a strong influence on the innovation system in Finland (Kautonen, 2006). Also, in recent years, the private sector has covered over 70% of all R&D inputs (Tekes, 2007). This means that a large part of new knowledge and new technologies is produced outside academia or other public research organizations.

The Finnish educational system as well as IP has a strong regional focus, implying that the role of regional key actors and institutions is featured. In a Finnish context, the regional innovation approach on the firm level consists of the four main elements of general innovation environment: market, industry, technology and geography (Kautonen, 2006). The Rural and Regional Development Programme has been among the main supporting policies concerning new entrepreneurship and innovation in forestry, wood products and non-timber SMEs. The implementation of regional IP actions has been intersectorally coordinated through Regional Centres of Employment and Entrepreneurship, which also allocate public support.

The general horizontal policy approach is overlapping with sectoral and other horizontal policies. The Science and Technology Policy Council has identified five subject areas in which concrete measures should be taken: (i) energy and the environment; (ii) metal products and mechanical engineering; (iii) the forest cluster; (iv) health and well-being; and (v) the information and communication industry and services. Innovation promotion was adopted into the original agenda of the National Forest Programme 2010 (FNFP), but it was changed into foresight activity by the time of implementation. The connections between FNFP and rural or regional development policies have not been intensive. The networked advisory services on wood product entrepreneurship (PuuSuomi) have supported rural SME development related to Forest Based Industries Policy and Renewable Energy Policy. Promotion of wood product SMEs has been a separate target in Rural Development Policy activities supporting the strengthening of innovation capacities in related companies and firms.

The recent evaluation of the Finnish National Innovation system welcomes the basic ambition of the broad-based IP balancing the supply and demand sides of innovative activity, including non-technical innovations, as well as emphasizing wider societal considerations (Full Report, 2009). Analysis reveals that the Finnish system is less international than conventionally thought and therefore the IP should aim at tapping deeper into the global knowledge pool. The evaluation takes a strong stance for the university and sectoral research reform and calls for a continuation of the higher education reform recognizing polytechnics as important actors in the system with their strong regional and applied role. The evaluation is cautiously optimistic about the national Strategic Centres for Science, Technology and Innovations (SHOKs), mostly dealing with incrementally renewing larger incumbent companies in traditional industries, but suggests limiting public resources devoted to them. The evaluation calls for a clarification and coordination of the roles and interrelationships of international, national, regional and local innovation and non-innovation policies.

4.3.2 Estonia

The main innovation strategy of Estonia – 'Knowledge-Based Estonia' (KBE) – has been created for the period 2002–2006 and subsequently for 2007–2013. The measures of KBE 2002–2006 concentrated on supporting the R&D projects in firms and universities, strengthening the knowledge base in universities and encouraging the cooperation between industry and academia. The period 2004–2006 is characterized by a low level of R&D funding: about 44–45% from the total R&D activities were publicly funded and the total R&D funding comprised 0.8–1.5% from GDP. The second KBE encompasses a wider variety of measures by adding support for cluster initiatives, funding the diffusion of innovations, but also increased funding for improving innovation awareness of the general public. The second KBE is certainly a step towards a more systemic approach to innovation, but it still omits some crucial activities related to entrepreneurship and networking generally held relevant for systems employing an innovation approach (Edquist, 2001).

Both strategies bring out specific fields that are targeted; however, the wood-related sectors are not among them. It is evident from textual analysis that KBE is well integrated concerning the strategies of higher education and entrepreneurship (as is also evident from the division of funding volumes by the ministries) at the same time containing no links to rural development or forestry (including respective development strategies). Therefore, the conclusion that KBS is not aligned with and integrated specifically towards the forest and wood sector has been brought out by Teder *et al.* (2007).

As in other catching-up economies, Estonian innovation strategy is targeted towards user-friendly information and communication technologies and the development of an information society, biotechnology (in Estonia, biomedicine) and materials' technology. The strategy imitates similar strategies in other countries without aligning it with the structure of the national economy and without seeing the importance of traditional industries (such as the wood sector) as customers of high-tech industries. (The first problem is more deeply discussed in Edquist (2001) and the second in Hirsch-Kreinsen *et al.* (2003) and Von Tunzelmann and Acha (2005).)

Although indirectly, the KBE 2007–2013 can be considered to be targeted more directly to the wood sector as it recognizes the need for supporting innovative attempts of both new and traditional industries. For example, to respond to the needs of traditional industries, starting in 2007, support is being extended for testing and certification procedures, design and productivity management projects, etc. The strategy also aims at strengthening the competences and knowledge base of firms for strategic innovation management and development activities by supporting the engagement, training and counselling of specialists, including development specialists from Estonia and from abroad. By developing industry clusters, priority is established for active use of key technologies that lead to growth in productivity (KBE, 2007). The funding for the above-listed measures is foreseen to comprise 22% of the total funding for KBE for 2007 to 2010.

4.4 Comparison of the Innovation-related Knowledge Flows in Finnish and Estonian Wood Sectors

4.4.1 Data and method

In this study, the EUROSTAT Community Innovation Survey (CIS) aggregated databases covering different time frames in the period of 1994–2004 are used. The first period comprises the years 1994–1996 for Finland (CIS 2) and for 1998–2000 (CIS 3) for Estonia (CIS 2 was not carried out in Estonia). The second period comprises the years 2002–2004 (CIS 4) for both countries. The CIS 5 has been also carried out in both countries for the years 2004–2006; however, Finnish data are not included in the aggregated database of EUROSTAT (no date).

The wood sector comprises in CIS databases the following industries by NACE Rev 2 categories: wood processing and production of wood-related products (16); the pulp and paper industry (17) and the furniture industry (31). It should be noted that in case of older CIS results, the data were aggregated in a way also involving printing and publishing (18) and other manufacturing (32) categories, making the direct comparisons more difficult.

In order to enrich the analysis, but also to discuss the results, secondary sources are used comprising other surveys, interviews with industry managers, etc. These sources include for Estonia an in-depth study of the use of knowledge sources based on CIS micro-data and interviews with managers (Ukrainski, 2008), and a study of regional innovation of manufacturing firms incorporating the analysis of respective public support (Tiits, 2007). For Finland, the study of SME innovation in eastern Finland is included in which implemented innovations during 2003–2005 were investigated. The sample was composed of 370 enterprises altogether and 27 from the wood sector (Kajanus and Karhu, 2008).

There are several aspects that have to be noted when comparing wood sectors in both countries. The differences can be most importantly revealed in the path-dependency issues for the Estonian wood sector. On the other hand, the Finnish pulp and paper industry has been characterized as the most innovative in Europe. The Finnish wood products industry is in second place behind the German wood processing industry (Arundel and Hollanders, 2005). For furniture manufacturing, there is no such comparison available. Discussing the Estonian paper industry, one has to take into account that the industry is relatively small and extremely concentrated with two larger firms and most of the others micro-companies. The Finnish paper industry is characterized by multinational concerns (Table 4.2). In such conditions, the comparability is rather difficult. Considering the size of the firms, CIS surveys in both countries are dominated by SMEs (except for the Finnish paper industry), and the firms in wood processing industry are most comparable across countries; in the furniture industry,

Table 4.2. Average indicators of innovative firms in samples of CIS 4 (2004) in Finland and Estonia.

	Finland		Estonia		Finland	Estonia
Sub-sector	Number of employees	Sales ('000 €)	Number of employees	Sales ('000 €)	Labour productivity ('000 €)	
Wood	111	23,887.33	162	9,150.58	215.22	56.50
Pulp and paper	1,769	475,703.41	34	1,307.10	269.00	38.44
Furniture	61	8,167.24	146	4,832.36	133.89	33.21
Manufacturing	40	6,196.77	25	941.55	155.57	38.43

Source: Authors' calculations based on data from EUROSTAT (no date), Statistics Estonia (no date).

the firms in the Estonian sample are twice as large as those in Finland. Labour productivity differences remain substantial between the two countries.

The use of different sources of knowledge is assessed via two indicators obtained from CIS:

1. The share of innovative firms having innovation cooperation activities with different partners in a certain time period. Here cooperation is understood as active participation with partners on innovation activities (pure contracting out is excluded).
2. Innovative firms' assessment of the relevance of different information sources that provided information on new innovation projects or complemented to the completion of existing innovation projects in the respective time period. The following degrees of relevance were used in the CIS survey: low importance, medium importance, high importance or not used; however, the aggregated database allows the presentation of only the share of innovative firms recognizing respective sources as very important ones.

As the cooperation partners and knowledge sources are structured similarly in CIS, they can be compared as well: the first measure is capturing information exchange within the cooperative framework; however, the success of the knowledge exchange in this cooperation in terms of realization of an innovation project cannot be derived from this indicator. The second measure is more precise in terms of knowledge flows, as it shows the information applied for the specific innovation project and the assessment of its relevance for innovation.

The relevance of public policy is quantitatively assessed from the data available from the CIS dataset and comprises the share of firms that have received public grants for innovation. This measure is rather limited considering only direct financial support forming a part of public policy that was directed to influence the innovation activities. The source of respective grants is identified as local/regional, national, EU or EU Framework Programmes (FP).

As all the micro-data of the CIS are not public information, the published results from EUROSTAT aggregated online database as well as Statistics Finland online Innovation database are used. Estonian CIS micro-data are available to authors and are used where appropriate. There are the following limitations arising from these data sources:

- The intensity of use of different knowledge sources for innovation is not available. The above-described sources allow us comparatively to assess only the more important source of information, but not the total knowledge flows.
- The data across providers are not consistent although resting on the same CIS survey. This problem mainly stems from the different methods for generalizing the data for the population. As a result, in this chapter, we can only compare the patterns or levels of the indicators, avoiding a direct comparison of numerical shares.

4.5 Results and Discussion

Generally, the results reveal that Estonian wood sector firms assess most external information sources with lower relevance for innovation (Fig. 4.1). This is expected, as the lower internal capabilities of firms discourage the absorption of external knowledge. At the same time, the relevance of internal sources has been assessed with similar importance. Therefore, this result could rather reflect some kind of failure in business networks, which can be either partner or institution related. As expected, suppliers and clients as sources of knowledge were the most important external sources. However, clients being more relevant in Finland is a somewhat unexpected result for a supplier-dominated sector, but it stems from the furniture industry, where this source has been extremely relevant for innovations. The only sources assessed more highly by Estonian wood sector firms were fairs and conferences, and other

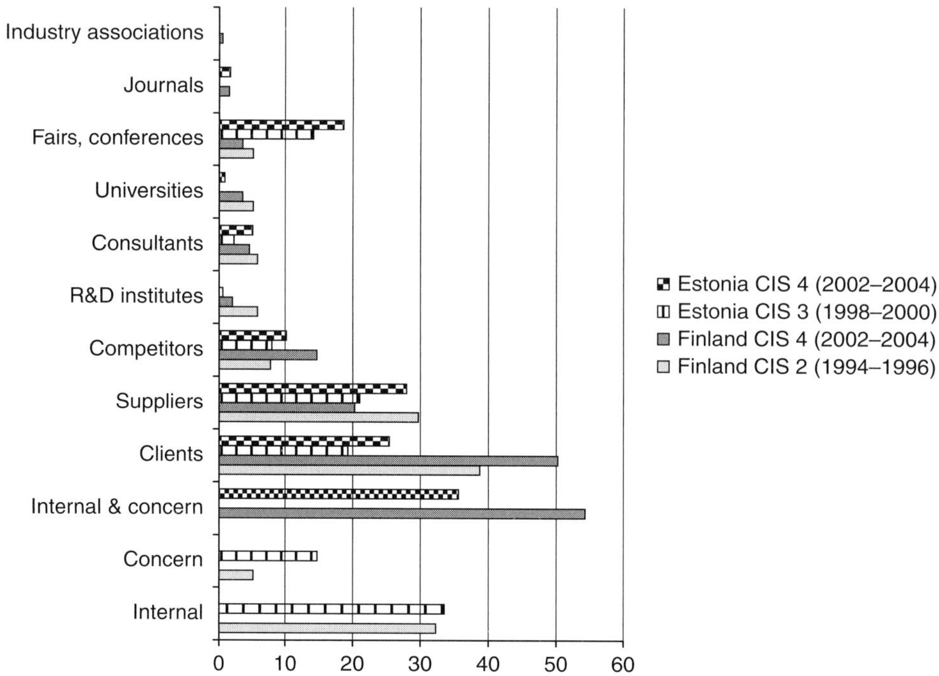

Fig. 4.1. Share of innovative wood sector firms considering the listed sources of information as most important for their innovation projects in Finland and Estonia. Note: CIS 2 and CIS 3 data include NACE 16–18, 31, 32 and CIS 4 NACE 17, 18, 31 categories. The data for conferences in the case of Finland CIS 4 category 17 are not included. Source: authors' calculations based on Finnish data from EUROSTAT (no date), Estonian data from Statistics Estonia (no date).

enterprises within the corporate concern. The last aspect can be explained by the transfer of knowledge and technology from mother firms to Estonia, related to FDI (as foreign-owned firms comprised *c.*70% from all concern member firms in the CIS samples). The use of fairs and exhibitions is higher probably for compensating for the absence of suitable suppliers and customers at the home market. The interviews by Ukrainski (2008) have revealed that the firms are dissatisfied with the knowledge base of local suppliers of machinery, speciality chemicals, etc., as often these are representative offices of the supplier firms without having the expertise needed. With regard to the customers, Estonian firms often do not sell directly to the end-user of the product, and thereby have no direct access to customer knowledge (Ukrainski, 2008).

The suppliers as relevant knowledge sources stems from the importance of process innovations in wood sectors. For example, based on the eastern Finland data on wood sector, SMEs used on average €10,000 for innovation activities in 2005 (average turnover €200,000). Most of the sum was used on machinery (€7700); for the development of marketing activities, €2000 was used (Kajanus and Karhu, 2008). Similar results have been found by Ukrainski (2008): the expenses on machinery and equipment have grown the fastest among innovation expenses (the average innovation expenses in 2004 were €450,000, the median was €77,000; the scale difference from the Finnish data stems from the size difference of the analysed firms).

Universities and public research institutes as knowledge sources were not important for both countries, but almost non-existent for Estonian wood industry companies. One explanation here is that these industries adopt modern technologies created by other sectors. Another is that the SMEs dominating the samples have high barriers to accessing the knowledge of universities in general. An additional explanation is that wood technology research and schooling in Estonia lags behind the actual industry needs. In fact, hardly any specialists in pulp and paper technology, log house technology or wood chemical technology are educated at the tertiary level in Estonia. The public R&D activities of those branches are extremely weakly developed, as recognized also in the developing strategies for the wood and forest sector (Ministry of Environmental Affairs, 2003).

The most important changes in the relevance of knowledge sources are the slight contraction of the role of suppliers in the Finnish case and the increase in the role of clients in both countries. In the Estonian case, all sources have gained slightly in importance, except for the internal and concern sources. The changes in internal as well as concern sources cannot be directly compared, as the method has been changed: in the latter period, the two indicators have been merged in surveys. However, the change can stem rather from the concern side. Earlier, the role of the concern was rather relevant, as was also the role of concern cooperation, which decreased later on in importance.

The above-described results are very similar to the small study done in eastern Finland, finding that the most important sources of knowledge were the internal sources, clients and suppliers. Universities as well as fairs and journals had only minor relevance in this respect. The eastern Finland data also show that wood sector enterprises cooperated not so much with public education and research organizations but rather with the vocational colleges. This shows that vocational education that is outside the CIS methodology can be quite important for LMT firms (Kajanus and Karhu, 2008).

Lower access of the Estonian wood sector firms to several external knowledge sources can be related to the much lower innovation cooperation activity (from innovative Finnish wood sector firms, 51% had formal innovation cooperation arrangements, and from Estonian counterparts, the share was only 28% in the earlier period; in the latter period, the same indicators are 49% and 41%, respectively). By the type of partners, the share of firms with innovation cooperation arrangements is lower in Estonia by each category (Fig. 4.2). It is interesting to

Fig. 4.2. Share of innovative wood sector firms having innovation cooperation with listed partners in Finland and Estonia. Note: CIS 2 and CIS 3 data include NACE 16–18, 31, 32 and CIS 4 NACE 17, 18, 31 categories. Source: authors' calculations based on Finnish data from EUROSTAT (no date), Estonian data from Statistics Estonia (no date).

note that although the cooperation was most prevalent with suppliers, customers were assessed with higher relevance for innovation knowledge. The only category where the cooperation activities and assessments of knowledge flows are at the same level in both countries is the competitor collaboration.

Considering the dynamics, one can see that the share of cooperation arrangements has grown in both countries, but perhaps relatively more in Finland. This can be seen especially concerning the competitor collaboration, which was on the same level initially in both countries, but has grown tremendously in Finland. This result can be policy related, because Finnish IP instruments have supported and encouraged regional and cluster-based collaboration, even between competing firms.

It is interesting to note that in the case of Finland, the share of firms having innovation cooperation with universities has increased from about 27% to 37%. At the same time, the relevance of universities as important sources of knowledge has not increased. It is possible that the role universities are playing in the innovation process (mainly design and testing, Table 4.1) is not assessed as highly relevant, but rather as supportive for conducting an innovation. The second reason can be related to the misalignment of the capabilities of both firms and universities, which is why the barriers are high and knowledge exchange is not very successful despite a cooperation agreement. This is a challenge for universities: to develop their activities to meet the needs of SMEs, because the highly conceptualized development services they typically offer meet the needs of large enterprises only. In large firms, the innovation development projects are characterized by efficient internal processes and a clear resource allocation for innovation projects. These are clearly absent in SMEs. The study on eastern Finland also reveals that the wood sector firms considered short educational courses and informal collaboration as the most beneficial forms of cooperation with universities (Kajanus and Karhu, 2008).

When analysing the location of cooperation partners, one can see that the national partners are prevalent in both countries. In Estonia, the role of historical contacts to Russia and other non-EU countries, and also to Japan, seems more relevant (Fig. 4.3). The EU cooperation has strengthened in both countries, showing deeper integration of this market. When one puts together the results of Figs 4.2 and 4.3, one can see that the share of firms with any type of innovation cooperation in Finland has not increased generally;

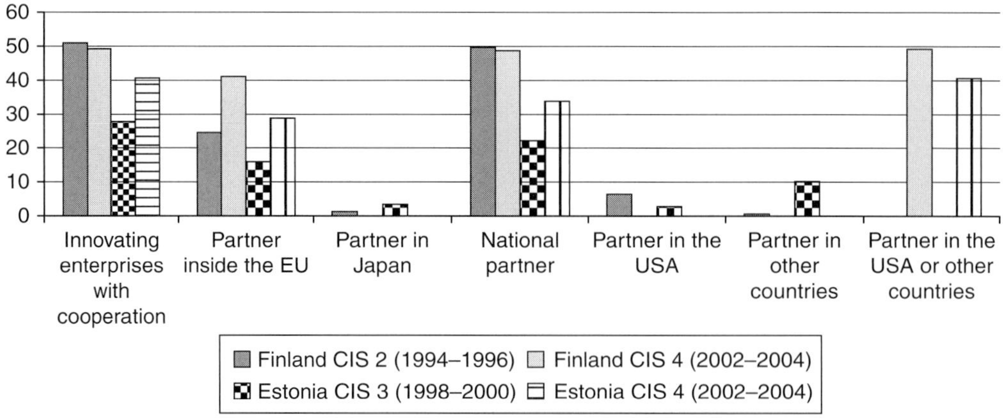

Fig. 4.3. Share of innovative wood sector firms having innovation cooperation by location of partners in Finland and Estonia. Note: CIS 2 and CIS 3 data include NACE 16–18, 31, 32 and CIS 4 NACE 17, 18, 31 categories. Source: authors' calculations based on Finnish data from EUROSTAT (no date), Estonian data from Statistics Estonia (no date).

however, the cooperation arrangements have been diversified and involve now several different partners. In the Estonian case, the share of firms with any type of innovation cooperation has increased quite significantly, by 13 percentage points. In addition, the cooperation with partners outside the EU has widened.

The results above reveal that there is still room for reshaping the international cooperation, where certainly the factors of distance and cultural differences play significant role. Sourcing deeper from the global knowledge pool should become one of the main objectives of IP in order to raise the competitiveness of wood industries in global markets. Distributed innovation means innovating in a globally distributed knowledge environment, where the basic question for an enterprise is where to find special complementing knowledge for its innovation activity. This is a big challenge for the SMEs in wood sector LMT enterprises, which do not have tradition, resources and contacts for distributed innovation. The question is whether universities and higher education institutions (HEIs) can help those enterprises towards a distributed innovation approach. They do, by nature, have international connections, exchange programmes, international education and R&D projects; could those settled assets be adjusted to enhance SMEs' distributed innovation actions? This also means a challenge for the HEIs because this differs very much from the traditional science and technology focus.

Comparing the public innovation support measures in an earlier period, wood sectors in both countries have received relatively less innovation support (33% and 6% from innovative firms in Finland and Estonia, respectively) compared with manufacturing firms in general (48% and 7%, respectively). The difference by country is very large. In both countries, wood processing and pulp and paper industries have received relatively more support compared with the furniture industry, especially in Finland, where more than 30% of innovative wood and paper firms received some public financial support for innovation. For the latter period (Fig. 4.4), the public support has been extended to a larger circle of enterprises. Wood sector firms in both countries have received public support in a larger number of cases than the manufacturing average.

Most of the public financial support has been allocated via mechanisms coordinated by the government. The difference between Estonia and Finland is revealed by absent region-level support for innovation in Estonia stemming from the non-responsiveness of regional policy in the

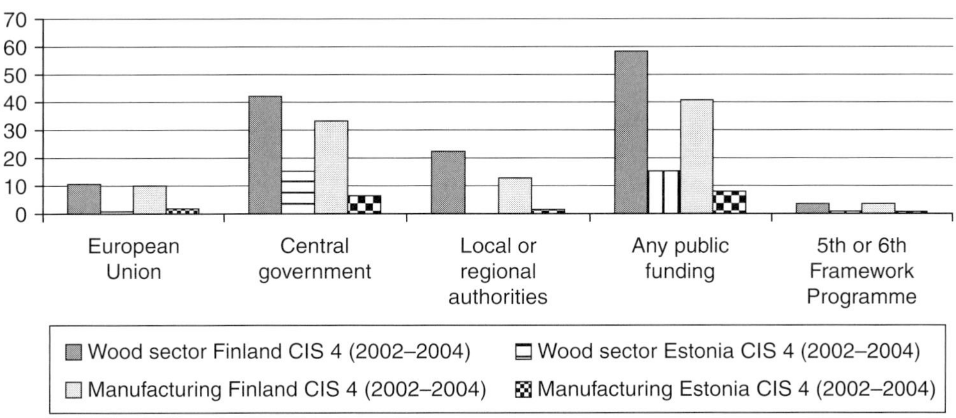

Fig. 4.4. Share of innovative wood sector and manufacturing firms receiving public support in Finland and Estonia. Note: The wood sector includes NACE 17, 18 and 31 categories. Source: authors' calculations based on Finnish data from EUROSTAT (no date), Estonian data from Statistics Estonia (no date).

respective field (Teder *et al.*, 2007). In addition, the receipts of EU support, including fifth and sixth FPs, are considerably lower in Estonia. The study of eastern Finland reveals that the wood sector firms used more development projects (mostly structural fund projects), but also public business advisers as a source of knowledge in their innovation processes (Kajanus and Karhu, 2008). According to the CIS data in the Estonian case, the role of innovation-related public funding has increased mainly in funding training expenses (Ukrainski, 2008).

As specific innovation support is not notable for Estonian wood sector firms, more indirect measures are also discussed here. In Fig. 4.5, one can see that wood and furniture sector firms in Estonia have received considerably more support for infrastructure and exports (including also the support for participation in international fairs and exhibitions), but less for consulting and R&D (only furniture firms had some R&D support).

The ignorance of wood-related sectors of innovation strategy is reflected in the respective public innovation support. In contrast to other sectors, Estonian wood sector firms have received no public support for acquiring external knowledge resources (consultants, universities, etc.) for innovation; however, they have relied upon those resources just as intensely as those industries receiving support. Wood, paper and furniture firms received significantly more public funding supporting export, training, counselling, infrastructure and start-up projects compared with other industries in manufacturing. The importance of export support (including the participation on fairs) can be also connected to the intensive use of fairs and exhibitions as a knowledge source by Estonian wood sector firms. However, as also Forsman (2009) suggested, instead of separate discontinuous public services focused on accelerating the individual phases of the innovation process, IP instruments are needed that accelerate capability building overlapping with innovation development. Thereby the whole variety of different ways to innovate (not only R&D) should be taken into account.

The main results concerning the use of different knowledge sources for innovation, but also cooperation activities and supporting policies, are summarized in Table 4.3. It shows that generally the knowledge sources are used with lower intensity in Estonia. The reasons are associated mainly with capability failures in innovating firms, but also in partner networks. This can also explain why more indirect knowledge sources (such as fairs, exhibitions and conferences) play a relatively more important role in Estonia.

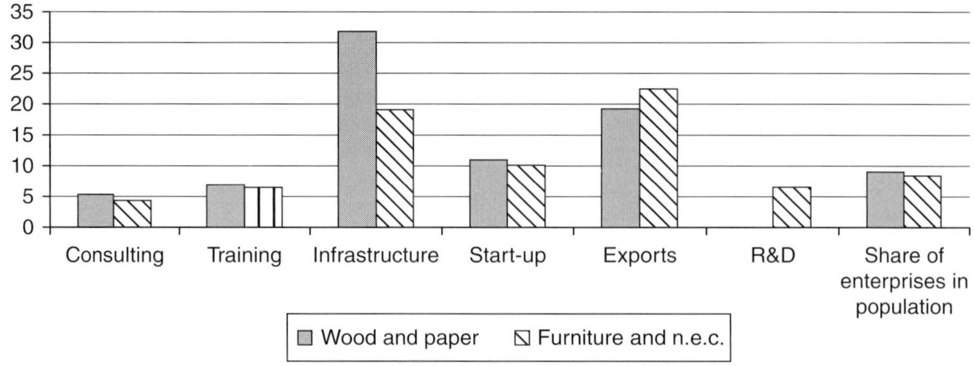

Fig. 4.5. Share of wood sector firms in Enterprise Estonia supported projects in 2004–2005. Note: wood and paper include NACE 17–18 and furniture 31–32 categories. Source: authors' calculations based on Tiits (2007, pp. 138–139).

Table 4.3. Summary of the use of knowledge sources and respective policy measures in Estonia (E) and Finland (F).

Source of knowledge	Level of use, dynamics	Relevant factors	Relevant (innovation) policy measures
Internal (Concern)	Relevant source Initially: Internal: E≈F Concern: E>F In latter period, combined: E<F	E: lower capabilities in general, in earlier period technology transfer from mother company important	E: support for training F: grants for industrial R&D
Suppliers	Relevant source Initially, E<F Later E≈F Relatively stable importance Strengthened in E	Most prevalent cooperation partner Process innovation (new machinery, equipment) very important in both countries E: network failure (local suppliers not competent)	F: networking measures F: cluster, supply chain policies
Customers	Relevant source E<F Strengthened in both countries	F: in latter period most important source stemming from the importance in furniture manufacturing E: lower cooperation, access to the knowledge is restricted while the products are marketed via traders	F: networking measures F: cluster, supply chain policies (D)
Competitors and industry associations	Relatively low importance E≤F Stable in E, strengthened in F	Industry associations not relevant as direct source of knowledge F: growth in cooperation arrangements, not reflected in knowledge flows	F: networking measures Cluster policies Regional measures
Consultants, Knowledge Intensive Business Services (KIBS)	Low importance E≈F Stable	Cooperation activities are more important F: cooperation activities have grown	F: networking measures, information and brokerage support E: networking measures
Universities, R&D institutes literature	Low importance E<F Slightly strengthened	F: cooperation activities more important and grown; barriers in SMEs accessing and using the knowledge; vocational education institutions more relevant E: university competence is absent in some areas, firms are absorbing rather than creating new technologies	F: support for public sector research Networking measures (S) Support for training and mobility (S) Cluster policies (D)
Fairs, exhibitions, conferences	E>F Stable in F, strengthened in E	E: relevant because is replacing to a certain extent the networking problems with suppliers and customers	E: export promotion measures

Note: supply-side IP measures are denoted by (S) and demand-side measures by (D).

4.6 Conclusions and Policy Relevance

To summarize, both countries' wood sector firms used internal sources, but also customers and suppliers as the main knowledge sources for innovation (see also summary in Table 4.3). Lower knowledge transfer was found from universities and other research organizations. This indicates that minor incremental innovations are more important at the firm level compared with the market level in wood sector firms of both countries.

Wood sector firms are typically SMEs, which restricts their ability to absorb and use the external knowledge. Our results show that even if the firms had innovation-related cooperation with universities, consultants and competitors, the knowledge obtained seems not to be assessed very highly from the innovation point of view. This points to the need for improvement of innovation-related internal (strategic) and external capabilities of wood sectors firms, universities and consulting firms.

Estonian wood sector firms have remarkably less innovation cooperation arrangements compared with their Finnish counterparts, but similarly, the firms in both countries concentrate on national and EU-wide cooperation. However, the cooperation arrangements are still more important with business partners (consumers, consultants, suppliers), which is characteristic of the incremental innovations at the firm level. When the policy goal is to enhance also innovations important at market level, public sector cooperation with universities, etc. should be encouraged more. This means an important challenge for universities to develop their services and capabilities to be compatible with SMEs.

The IP impact in terms of direct financial support is much lower in Estonia. Differently from Finland, the wood sector is also in the national IP framework and not considered as important and specifically targeted in Estonia. It has to be noted that wood sector firms in both countries have received public subsidies for innovation in more cases than other manufacturing firms. Most public support stems from the central government, but in the case of Finland, the regional funding is also important. In the case of Estonia, it is found that some measures of export promotion have been encouraging the use of fairs and exhibitions as knowledge sources for innovation.

The above results point to the need to change separate discontinuous public services focused on accelerating the individual phases, but also selected cooperation partners of the innovation process. It is needed rather to remedy specific network or institutional failures. IP instruments are needed that accelerate capability building overlapping with innovation development. It would be necessary to start by improving capabilities for the development of incremental innovations, which is subsequently moving ahead of radical innovation development. The whole variety of different ways to innovate, including non-R&D activities, as technology adoption, minor modifications or incremental changes (learning by doing), and imitation including reverse engineering, combining existing knowledge in new ways and adopting solutions developed by users (Arundel *et al.*, 2008), should be taken into account.

One of the questions is how SMEs search and exploit globally distributed knowledge in their innovation processes. Often they have a lack of tradition, resources or capabilities for that. Special interest should be given to HEIs, like universities and others, and whether their settled international assets (e.g. networks, research and education programmes) could be adjusted and developed so that they will enhance and facilitate SME's distributed innovation processes. Often there are no personal connections between organizations, and a broker or mediator is needed to facilitate the communication. It is still a very difficult challenge, how to integrate the interests and routines of HEIs and SMEs in a mutually beneficial and innovation-enhancing way.

The results of our analysis show that ensuring knowledge creation, e.g. by establishing regional knowledge centres, is not enough. In addition, there is a need for improvement of innovation-related strategic, internal and external capabilities of wood sector firms. Universities and consulting firms also need to exploit new knowledge created in science, in practice and by end-users. This means a challenge for universities and other actors to develop their services and capabilities to meet the needs of SMEs and the manner of SMEs to implement innovation processes hand in hand with daily business. Moreover, there is a need to enhance enterprises' ability to utilize knowledge created and situated elsewhere in the world's knowledge centres. This is done, e.g. by developing higher education institutions' international assets in networks and developing research and education programmes to enhance and facilitate SMEs' distributed innovation processes.

Acknowledgement

The authors acknowledge financial support from the Estonian Ministry of Education and Research target funding SF0180037s08. ESF grants 8311, 8580.

References

Amara, N. and Landry, R. (2005) Sources of information as determinants of novelty of innovation in manufacturing firms: evidence from the 1999 statistics Canada innovation survey. *Technovation* 25, 245–259.

Arndt, O. and Sternberg, R. (2000) Do manufacturing firms profit from intra-regional innovation linkages? An empirical-based answer. *European Planning Studies* 8, 465–485.

Arundel, A. and Hollanders, H. (2005) European Sector Innovation Scoreboards. Discussion paper from the Innovation/SMEs Programme, EC.

Arundel, A., Bordoy, C. and Kanerva, M. (2008) Neglected innovators: how do innovative firms that do not perform R&D innovate? Results of an analysis of the Innobarometer 2007 survey. INNO-Metrics Thematic Paper 125.

Bianchi, G. (1998) Requiem for the third Italy? Rise and fall of a too successful concept. *Entrepreneurship and Regional Development* 10, 93–116.

Blomström, M. and Kokko, A. (2002) From natural resources to high-tech production: the evolution of industrial competitiveness in Sweden and Finland. EIJS Working Paper Series 139.

Boschma, R.A. (2005) Proximity and innovation: a critical assessment. *Regional Studies* 39, 61–74.

Breschi, S., Malerba, F. and Orsenigo, L. (2000) Technological regimes and Schumpeterian patterns of innovation. *Economic Journal* 110, 388–410.

Brown, B. and Butler, J. (1993) Networks and entrepreneurial development: the shadow of borders. *Entrepreneurship and Regional Development* 5, 101–116.

Cohen, W. and Levinthal, D. (1990) Absorptive capacity: a new perspective on learning and innovation. *Administrative Science Quarterly* 35, 128–152.

Cooke, P., Uranga, M.G. and Etxebarria, G. (1997) Regional innovation systems: institutional and organisational dimensions. *Research Policy* 26, 475–491.

Corsten, H. (1987) Problems with cooperation between universities and enterprises: a comparative study on size of enterprise. *Technovation* 6, 295–301.

Council of State (2008) Government's Communication on Finland's National Innovation Strategy to The Parliament. 9 October.

Doz, Y., Santos, J. and Williamson, P. (2001) *From Global to Metanational. How Companies Win in the Metanational.* Harvard Business School Press, Boston, Massachusetts.

Edler, J. and Georghiou, L. (2007) Public procurement and innovation – resurrecting the demand side. *Research Policy* 36, 949–963.

Edquist, C. (2001) Innovation policy – a systemic approach. In: Archibugi, D. and Lundvall, B.-Å. (eds) *The Globalizing Learning Economy.* Oxford University Press, Oxford, pp. 219–239.

European Innovation Scoreboard (2008) Comparative analysis of innovation performance. Pro Inno Europe Inno Metrics.

EUROSTAT Database on Science, Technology and Innovation (no date) Available at http://epp.eurostat.ec.europa.eu/portal/page/portal/science_technology_innovation/data/database (accessed 11 November 2009).

Faulkner, W., Senker, J. and Velho, L. (1995) *Knowledge Frontiers: Public Sector Research and Industrial Innovation in Biotechnology, Engineering Ceramics and Parallel Computing.* Clarendon Press, Oxford.

Foray, D. (1997) Generation and distribution of technological knowledge: incentives, norms and institutions. In: Edquist, C. (ed.) *Systems of Innovation: Technologies, Institutions and Organisations.* Frances Pinter, London, pp. 64–85.

Forsman, H. (2009) Innovation capabilities of small enterprises. A cluster strategy as a tool. *International Journal of Innovation Management* 13, 221–243.

Freel, M.S. (2000) External linkages and product innovation in small manufacturing firms. *Entrepreneurship and Regional Development* 12, 245–266.

Full Report (2009) Evaluation of the Finnish National Innovation System – Full Report. 28 October.

Henderson, R. and Clark, K. (1990) Architectural innovation: the reconfiguration of existing product technologies and the failure of established firms. *Administrative Science Quarterly* 35, 9–30.

Hirsch-Kreinsen, H., Jacobson, D., Laestadius, S. and Smith, K. (2003) Low-tech industries and the knowledge economy: state of the art and research challenges. Paper written within the context of research project 'PILOT: Policy and innovation in Low-Tech'.

Howells, J. (1995) Tacit knowledge and technology transfer, WP 16. Centre for Business Research, University of Cambridge, Cambridge.

Howells, J. (1999) Regional systems of innovation? In: Archibugi, D., Howells, J. and Michie, J. (eds) *Innovation Policy in a Global Economy.* Cambridge University Press, Cambridge, pp. 67–93.

ILO (2001) *The Globalization and Sustainability: The Forest and Wood Industries on the Move.* ILO, Geneva.

Kajanus, M. and Karhu, S. (2008) Innostuva ja oppiva Itä-Suomi (INNO-STU) -hanke. Loppuraportti. Savoniaammattikorkeakoulun julkaisusarja D 9/2007 (in Finnish).

Kautonen, M. (2006) The regional innovation system bottom-up: a Finnish perspective. A firm-level study with theoretical and methodological reflections. Electronic dissertation. *Acta Electronica Universitatis Tamperensis* 545.

Kirat, T. and Lung, J. (1999) Innovation and proximity: territories as loci of collective learning. *European Urban and Regional Studies* 6, 27–38.

Knowledge-based Estonia (KBE) (2002) Estonian Strategy for Research and Development 2002–2006. Tallinn.

Knowledge-based Estonia (KBE) (2007) Estonian Research and Development and Innovation Strategy 2007–2013. Tartu.

Laestadius, S. (1995) Tacit knowledge in a low-tech firm. *European Journal of Vocational Training* 6, 27–33.

Laestadius, S. (2000) Biotechnology and the potential for radical shift of technology in forest industry. *Technology Analysis & Strategic Management* 12, 193–212.

Lundvall, B.-Å. and Borrás, S. (2007) Science, technology and innovation policy. In: Fagerberg, J., Mowery, D.C. and Nelson, R.R. (eds) *The Oxford Handbook of Innovation.* Oxford University Press, Oxford, pp. 599–631.

Malerba, F. and Orsenigo, L. (1996) Schumpeterian patterns of innovation. *Cambridge Journal of Economics* 19, 47–65.

Meeus, M.T.H., Oerlemans, L.A.G. and Hage, J. (2004) Industry–public knowledge infrastructure interaction: intra- and inter-organizational explanations of interactive learning. *Industry and Innovation* 11, 327–352.

Ministry of Environmental Affairs (2003) The Strategy of Estonian Forestry to 2010. Available at: http://www.envir.ee/metsandus/arengukava/arengukava.html (in Estonian).

Nelson, R.R. (ed.) (1993) *National Innovation Systems: A Comparative Analysis.* Oxford University Press, Oxford.

Nowotny, H., Scott, P. and Gibbons, M. (2004) *Re-Thinking Science. Knowledge and the Public in an Age of Uncertainty.* Polity Press, Cambridge.

Nurmi, R. and Üksvärav, R. (1994) Fenno-Ugric Core Values? The case of Estonia and Finland. In: Papers of the Workshop on Cross-Cultural Perspectives: Comparative Management and Organization, European Institute for Advanced Studies in Management, 11–13 November. Available at http://www.uku.fi/~paganuzz/xcult/heroes/Fenno-Ugric.html (accessed 6 January 2011).

Palmberg, C. (2002) The many faces of absorptive capability in low-tech industries–the case of glue-lam timber and foodstuffs. Paper presented at the DRUID Summer Conference on Industrial Dynamics of the New and Old Economy – who embracing whom? Copenhagen/Elsinore, 6–8 June.

Pavitt, K. (1984) Sectoral patterns of technical change: towards a taxonomy and a theory. *Research Policy* 13, 343–374.

Polanyi, M. (1966) *The Tacit Dimension*. Routledge and Kegan Paul, London.

Rametsteiner, E. and Weiss, G. (2006) Innovation and innovation policy in forestry: linking innovation process with systems models. *Forest Policy and Economics* 8, 691–703.

Rothwell, R. (1991) External networking and innovation in small and medium-sized manufacturing firms. *Technovation* 11, 93–112.

Schartinger, D., Rammer, C., Fischer, M.M. and Fröhlich, J. (2002) Knowledge interactions between universities and industry in Austria: sectoral patterns and determinants. *Research Policy* 31, 303–328.

Science, Technology, Innovation (2006) Science and Technology Policy Council of Finland. Available at: http://www.minedu.fi/export/sites/default/OPM/Tiede/tiede-_ja_teknologianeuvosto/julkaisut/liitteet/Review_2006.pdf?lang=en

Statistics Estonia (no date) CIS3 and CIS4 databases, CD-ROM.

Teder, M., Ukrainski, K., Prede, M. and Kaimre, P. (2007) Assessing the alignment and integration of innovation and development policies for the forest sector in Estonia. *Forestry Studies* 46, 102–117.

Tekes (2007) The Web pages of Tekes. Available at: http://www.tekes.fi/eng/innovation/statistics/statistics.htm#Total%20R&D%20input%20in%20Finland%202001 (accessed 6 January 2011).

Teubal, M., Yinnon, T. and Zuscovitch, E. (1991) Networks and market creation. *Research Policy* 20, 381–392.

Tiits, M. (ed.) (2007) *The Nation of Merchants*. Estonian Academy of Sciences, Tartu.

Ukrainski, K. (2008) *Sources of Knowledge Used in Innovation: An Example of Estonian Wood Industries*. Tartu University Press, Tartu.

Von Tunzelmann, N. and Acha, V. (2005) Innovation in 'low-tech' industries. In: Fagerberg, J., Mowery, D.C. and Nelson, R.R. (eds) *The Oxford Handbook of Innovation*. Oxford University Press, Oxford, pp. 407–432.

5 Institutional Innovation in European Private Forestry: the Emergence of Forest Owners' Organizations

Américo M.S. Carvalho Mendes,* Blaž Štefanek, Diana Feliciano, Diana Mizaraite, Dragan Nonić, Emil Kitchoukov, Erlend Nybakk, Gabriel Duduman, Gerhard Weiss, Liviu Nichiforel, Maria Stoyanova, Pekka Mäkinen, Rosário Alves, Vojislav Milijić and Zuzana Sarvašová

Abstract

The emergence and development of organizations of private forest owners in situations where they were not previously collectively organized is a relevant institutional innovation in forestry. This chapter looks at the factors that may have contributed to this institutional change in the following countries: Austria, Bulgaria, Croatia, Finland, Lithuania, Norway, Portugal, Romania, Serbia and Slovakia. The conceptual framework used to present and discuss these country cases considers the following types of factors: (i) the structural changes in the social and economic environment of private forestry when forest owners' organizations emerged, and the needs for collective action of private forest owners triggered by those changes; (ii) the factors contributing to cope with the 'free riding' problems involved in collective action; (iii) the mechanisms leveraging the capacities of forest owners' associations beyond the initial domain where they emerged and contributing to give them the 'critical mass' needed for having substantial impact on forestry economic conditions; and (iv) the possible existence of 'path dependence' phenomena, where the conditions prevailing when forest owners' organizations emerged have a lasting influence throughout their lifetime. With different specifications according to the characteristics of each country, these four sets of factors appear to be useful as a common framework for organizing the explanation of how forest owners' associations emerged and developed in the countries considered here.

5.1 Introduction

According to the Global Forest Resource Assessment 2005 (FAO, 2006), 58% of forest area in Europe was privately owned in the year 2000. This excludes Russia, Belarus and Ukraine, where almost 100% of forests are publicly owned.

Based on this source and on complementary data, Schmithüsen and Hirsch (2009) report a 'noticeable' increase in private forest ownership in Europe during the period 2000 to 2005. According to the same authors, this was due to reforestation of marginal agricultural and pasture land in countries such as Ireland and Norway, and to land restitution and privatization processes in the Central and Eastern European countries. Since these two types of processes are still going on, private forest ownership in Central and Western Europe may be still increasing.

* Corresponding author.

In the Central and Western European countries covered by Schmithüsen and Hirsch (2009) for which there are data on the distribution of forest holdings' size classes, small-scale holdings prevail in number and have a non-negligible share in area: size classes up to 5 ha represent 86% of the number of holdings and 19% of the forest land.

With this structure of forest land ownership, private forest owners should play a major role in any policy aiming at successfully promoting sustainable forest management. This kind of policy has to include instruments to increase the production of some forest public goods and to reduce some negative externalities (e.g. forest fires), which requires cooperation and collective action by private forest owners, especially when private forest ownership is fragmented.

At least for these reasons, it is a major step towards sustainable forest management when private forest owners move from a situation where they are not collectively organized to a situation where they set up associations or other forms of collective organization to cooperate among themselves and to represent and protect their common interests in their interactions with other stakeholders. The emergence of these special-purpose organizations changes the way private forest owners interact among themselves and with other stakeholders. It can be said to generate new institutions in forestry if, by 'institution', we mean: 'a set of rules that structure social interactions in particular ways… shared by the members of the relevant community or society' (Knight, 1992: 2–3).

This chapter will look at how this institutional change emerged and developed in ten countries from Central and Western Europe, building mainly on the work done by Mendes *et al.* (2006), complemented with contributions drawn from other authors mentioned in the next section.

5.2 Research Approach and Conceptual Framework

The cases presented here are not meant to be an empirical test of a previously deduced theory encompassing the explanation of the trajectory of forest owners' organizations from their birth and throughout their lifetime. Even though the discussion of the cases will be organized by making use of concepts coming from different theories, there is no claim here that they are already all well fitted together in a fully fledged model. These cases more serve the purpose of an exploratory approach having some similarities with what Smith (no date) calls a 'nomoempirical' approach, in which, by describing how forest owners' organizations emerged and evolved in different environments, one may identify possible explanatory factors of their emergence and evolution that may be useful for future theoretical developments.

Following the literature survey proposed by Knight (1992), there are two main schools of thought in the classical accounts of the emergence and development of institutions. One emphasizes the role of institutions as forms of cooperation and coordination of individual actions to achieve collective benefits. The other emphasizes the conflicts of interests inherent in distributional problems and the role of institutions in securing advantages for some parties involved in those conflicts.

The first school of thought has several variants deriving from the works of Thomas Hobbes (1963), David Hume (1978), Adam Smith (1969, 1976) and Herbert Spencer (1969). For Thomas Hobbes, new institutions emerged as forms of a social contract intentionally designed and accepted by members of a society who see in this kind of social state more benefits than costs. Costs come from additional constraints on individual behaviours when abiding by the social contract. Benefits are related to the avoidance of conflicts that would arise if individual behaviours were less constrained.

For David Hume and Adam Smith, the institutions reducing this risk of detrimental social conflict is not the result of intentional design, but the outcome of spontaneous and gradual emergence in social intercourse. Through a process of trial and error, humans come to the realization that some patterns of social rules are more beneficial than unregulated individual behaviours.

Herbert Spencer emphasized the role of fitness in institutional emergence and development. New institutions are created and persist when they contribute to the survival of the society.

The second school of thought in the classical tradition derives from the works of Karl Marx (1971) and Max Weber (1978, 1989). Marx emphasized the role of social class conflicts in institutional change. For Max Weber, new institutions come as a necessity for some social groups to secure their interests.

Contemporary theories of institutional change with microeconomic foundations are related to the first school of thought in the classical tradition, where institutional change is explained as the emergence of new forms of coordination or cooperation to achieve collective benefits. Here institutions are viewed as an equilibrium configuration of strategies of the social actors faced with recurring collective action problems, subject to some competitive selection process that led to that configuration rather than other possible ones. Often these are problems related to the provision of public goods, as those dealt with in the influential book by Olson (1971).

An important school of thought in this large body of literature is the theory initiated by Coase (1960) and Williamson (1975, 1985), in which organizations emerge to economize on transaction costs. Economic efficiency factors are also the basis of North's (1990) explanations of institutional change. Another influential contemporary school of thought appeals to the tools of game theory to renew and develop the classical social contract tradition. Binmore (1994, 1998) is a good representative of this kind of literature.

The approach adopted here also looks at the emergence of forest owners' associations as related to the appearance of new collective action problems private forest owners had to face at a certain point in time. These problems can take different forms according to the characteristics of each country and the moment in time when they arise. In spite of these differences, the general hypothesis here is that they are related to structural changes in the social and economic environment in which private forest owners operate. So if one wants to explain why forest owners' associations emerged in a country at a particular moment in time with a specific form, one should look at those structural changes and their characteristics. One should also look at the specific forms of collective action problems forest owners had to face as a consequence of those changes.

As stressed in contemporary theories of institutional change, those collective action problems are often related to the provision of public goods, or to other situations in which individual action alone may be incapable of coping with economic inefficiencies. For some of those public goods, forest owners are the main and most direct beneficiaries. Such is the case, for example, of the effects of forest owners' organizations in protecting the interests of private forestry, improving the public support for this kind of activity and reducing some risks affecting forests (e.g. fires).

However, improving the cooperation among forest owners generates public goods not only for the forest owners themselves, but also for the society at large because it may enhance the positive environmental services of forest production.

When the provision of public goods is at stake, 'free riding' problems are likely to occur. So another relevant issue to examine, with the material provided in the country cases, is to identify factors contributing to cope with those problems, such as:

- social identity factors, i.e. forms of common identity of forest owners facilitating voluntary cooperation among themselves;
- joint supply of public and private goods and services by the forest owners' organizations, where these private goods and services are what Olson (1971) calls 'selective incentives', i.e. goods and services sold by the organizations only to their members in better conditions than they could find elsewhere; and

- incentives (financial or in kind) provided by the public administration (national and local).

Coping with 'free riding' is more critical in the initial stages of the lifespan of forest owners' organizations, when the number of members is small and their activity does not have yet noticeable positive effects on the economic conditions of private forestry. Therefore another relevant issue to explore with the materials provided by the country cases is whether, in the evolution of forest owners' organizations, there have been some factors facilitating their development up to a point where they can start having sizeable effects on the economic conditions of private forestry.

Marwell and Oliver (1993) used the term 'critical mass' to describe the level of provision of a public good where it has received enough interest and resources from its contributors that its provision becomes self-sustained by these voluntary contributions. The attainment of this critical mass depends on several factors, three main ones being the shape of the production function of the public (or club) good, the degree of heterogeneity within the potential and actual group members, and the type of correlation between resources and interest for each member. If the production function has increasing marginal returns for lower levels of group membership, there is a start-up problem: each contribution yields a low return. However, if among the potential group members there are some with a strong interest in the public good, they may be willing to be the initial contributors, hoping that others will join later. Since there are increasing marginal returns, there may be a snowball effect, each additional contribution by a new member raising the marginal returns for the next members.

Reaching a critical mass may be achieved by other means than those proposed by Marwell and Oliver (1993), and the material provided in the country cases may help to identify some of those other possible means, such as:

- building partnerships between forest owners' organizations and other organizations of the same nature (e.g. building up federations of forest owners' organizations) or of a complementary nature;
- combining voluntary action with public authority by entrusting forest owners' organizations with some legal authority to collect fees or to extend the organization's influence beyond the limits of its members' holdings;
- combining voluntary action by forest owners' organizations with public collective investments protecting and enhancing forest resources;
- expanding the scope of forest owners' organizations by exploring complementarities between existing activities and new ones that can bring additional benefits to the members (e.g. a forest owners' organization already established in a region, mostly with the mission of providing technical advice to the owners, can set up a regional forest certification scheme); and
- vertically integrating forestry with forest product marketing and industrial activities.

In spite of the diversity of these examples, what is common to all of them is a sort of 'leverage mechanism': the position the forest owners' organization already has in its initial domain of influence is amplified by reaching out beyond that initial domain; this has a positive feedback effect that contributes to reinforcing the influence of the organization.

Finally, since we are dealing with organization dynamics, it is relevant to look for the existence of 'path dependence' phenomena (David, 1985; Arthur, 1994). These are situations where the conditions (economic, social, cultural, political) prevailing in the initial stages of development of forest owners' organizations have a long-lasting influence in the later stages; that is, initial conditions matter for the evolution of the organizations even though they no longer apply.

5.3 Country Cases

5.3.1 Finland: cooperative tradition, critical mass and vertical integration

Factors contributing to the emergence of forest owners' associations

The first Forest Management Associations were founded in 1907, when the market power of timber buyers was increasing. This change in market structure triggered a need for cooperation among forest owners to counteract the decrease in their market power.

Supporting factors for coping with 'free riding' problems

This associative movement benefited from an old tradition of cooperative arrangements in farming, still very vivid at a time when most forest owners were also farmers. In the beginning it also benefited from public support for collective investments to improve timber supply, as public authorities and other stakeholders feared that a weakening of the economic situation of forest owners could lead to timber shortages.

An important next step in the public support for this associative movement was taken in 1928, when the Private Forest Law was enacted and the Forest Centres were created to help in the implementation of this law.

In 1950 the Act on Forest Management Associations enabled these organizations to collect a fee from forest owners. Every forest owner pays the fee and thus is automatically a member of the Forest Management Association in the area where his or her forest is located. These fees account for approximately 20% of the associations' turnover. The rest is generated by the services provided.

Nowadays there are 135 Forest Management Associations. They assist forest owners in forest management, wood sales and certification. They also carry out silvicultural and forest improvement works and increase the professional knowledge and skills of forest owners.

These associations have representatives in the Forestry Council of the Central Union of Agricultural Producers and Forest Owners (MTK). This council looks after private forest owners' interests by providing information on wood markets and prices, influencing forest policy and developing the operations of the local Forest Management Associations and their ten regional unions.

Leverage mechanisms

In 1934 forest owners' organizations made a step forward in strengthening their position in the timber market when they established Metsäliitto Oy, a joint stock company, to coordinate their sales of timber to customers outside of Finland. The company gathered valuable experience in selling timber and built solid business relationships with customers in both the domestic and export markets.

In 1947 Metsäliitto Cooperative (Metsäliitto Osuuskunta) was founded to continue the work of Metsäliitto Oy. The intention was to invite a large number of forest owners to become members of the cooperative and sell their timber through it.

Having built a strong basis in timber marketing was instrumental for the forest owners' organizations to engage in a strategy of vertical integration. Nowadays Metsäliitto Group, the parent company of Metsäliitto Cooperative and its subsidiaries Metsä-Botnia, M-real and Metsä Tissue, operates in 30 countries and is the eighth largest forest industry group in the world. Its five business areas are wood supply, wood products, pulp, board and paper, and tissue and cooking paper. With 20,000 employees, its sales are about €8 billion.

5.3.2 Norway: critical mass and vertical integration

Factors contributing to the emergence of forest owners' associations

In Norway, like in Finland, forest owners' organizations also emerged at the beginning of the 20th century and for similar reasons: counteracting the increasing market power of timber buyers. The first forest owners' association was established in 1903. During the following 10 years, many other associations started up as well. (This analysis is also based on contributions from Ståle Størdal, co-author of the paper by Mendes *et al.* (2006)).

Supporting factors for coping with 'free riding' problems

Timber was transported by floating it along waterways. This fact contributed to neighbouring forest owners along the same river joining efforts and organizing their associations in geographic areas corresponding to the territories driven by the same waterway.

Leverage mechanisms

In 1913 the forest owners' associations managed to create an organization to represent them at the national level, the Norwegian Forest Owners' Association. This association gave them some influence in the public policy-making process. This growing influence contributed to increasing their membership and their market share in timber sales, especially during the 1930s.

Having built this strong basis before the timber market was regulated put the forest owners' organizations in a good position to influence forest policy in the post-World War II period when timber prices were under public regulation.

Like in Finland, for these organizations their strong position in timber markets was also instrumental to engage in a strategy of vertical integration by becoming large shareholders in pulp and paper companies and several lumber mills.

Today, all but one of the forest owners' associations (Norskog, the organization of big forest owners, including industrial forest owners) are federated in a national umbrella organization, the Norwegian Forest Owners' Association, representing 85% of the timber production in Norway. They have also significant shares in Norwegian and international forest industry.

5.3.3 Austria: 'social partnership'

Supporting factors for coping with 'free riding' problems

Cultural and political factors are very important in understanding the forms of collective organization of private forest owners in Austria. The representation of group interests is transferred from the state to self-governing bodies called 'chambers' (*Kammern*). Chambers are statutory interest organizations established by public law and with obligatory membership. As central pillars of the so-called 'social partnership' (*Sozialpartnerschaft*), chambers are an omnipresent and powerful political player typical of the Austrian political system. The Austrian social partnership is mainly rooted in the cooperation of five large economic interest groups, which, in general, represent all relevant entities in their specific economic sector:

- the Chambers of Commerce (*Kammer der gewerblichen Wirtschaft*);
- the Chambers of Labour (*Arbeiterkammer*);
- the Chambers of Agriculture (*Landwirtschaftskammer*);
- the Austrian Trade Union Federation (*Österreichischer Gewerkschaftsbund*); and
- the Association of Austrian Industrialists (*Industriellenvereinigung*).

Agricultural and forestry interests are looked after by the Chambers of Agriculture (*Landwirtschaftskammern*). Statutorily the chambers engage in three different fields: (i) representation of group interests; (ii) provision of consulting services to foresters and forest land owners; and (iii) access to public subsidies.

This makes the character of the chambers ambivalent to some extent: on the one hand, they act as powerful and effective lobbyists; on the other hand, they behave as semi-public institutions carrying out state functions.

Further analyses of Austrian private forestry can be found in Hogl *et al.* (2005), Pregernig and Weiss (1998), Rametsteiner *et al.* (2005) and Weiss *et al.* (2007).

Factors contributing to the emergence of forest owners' associations

The Chambers of Agriculture played an important role in initiating forest owners' organizations at the local and regional levels. These organizations are quite active in supporting forest management of small-scale owners. They are voluntary membership groups, the main ones making up the Austrian Federation of Forest Owners' Associations (*Hauptverband der Landund Forstwirtschaftsbetriebe Österreichs*) and the Austrian Forest Association (*Österreichischer Forstverein*). As a voluntary interest organization, the Austrian Federation of Forest Owners' Associations looks after the interests of private farm and forest land owners. Due to a high degree of organization, approximately 80% of larger estates actually join the association, and the Federation is a powerful player in Austrian forest politics. The Federation mainly tries to safeguard the rights of private ownership and to fight any restrictions on the right of free disposal of private forest property. Besides this political aim, forest owners' associations also have important economic functions such as:

- provision of forestry-related services;
- sharing of common logistic projects and infrastructures;
- implementation of forest certification;
- joint timber sales on behalf of their members; and
- joint supply of biomass for wood heating projects.

Their role in timber and biomass supply has been increasing in the last decade.

Leverage mechanisms

At the federal level, the President's Conference of Chambers of Agriculture (*Präsidentenkonferenz der Landwirtschaftskammern*) represents agrarian and forestry interests with respect to other social interests within the social partnership. In Austrian forest politics, the President's Conference is a rather influential institution. As a constituent part of the social partnership, the Chambers of Agriculture, or rather the President's Conference, is granted institutionalized influence on policy formulation. Already in the preparatory stage, they get informed about draft legislation and are allowed to comment on it. This applies to laws as well as to ordinances. Usually the comments are incorporated into the final draft before the bill is sent to parliament. In addition, chambers often get the opportunity to send 'their' experts into parliamentary subcommittees where draft bills are formulated and finally voted on.

The second powerful voluntary interest organization is the Austrian Forest Association. The Association is open to forest owners as well as to forest professionals working in private enterprises, chambers and public administration. Roughly two-thirds of the potential members belong to the Association. For most forest professionals, membership is taken for granted; it results from tradition. Public relations have always been an important task for this association: inwards, to find the 'lowest common denominator' among the members; and outwards, to represent the 'common position of forestry'.

5.3.4 Lithuania: cooperativism and economic networking

Factors contributing to the emergence of forest owners' associations

After the declaration of independence there have been several structural changes in the Lithuanian forest sector:

- new modes of ownership (private forests) and enterprise (private business logging companies);
- privatization of forest industry;
- formation of a free timber market; and
- increasing levels of timber exports.

The main problems of the private forestry sector are:

- small-sized forest holdings;
- weak cooperation among forest owners;
- owners with poor forest management knowledge; and
- no public financial incentives for improving private forest management.

In the 'Policy of Lithuanian Forestry and its Implementation Strategy' approved in September 2002, some policy instruments were identified to cope with those problems, namely:

- training of private forest owners;
- correction of the boundaries between state and private forests through equivalent land exchanges in order to improve the efficiency of forest management operations;
- introduction of a compensation scheme for the restrictions on forest utilization in newly established protected areas;
- integration of private forestry development into the general rural development programmes co-funded by the EU; and
- public provision of technical advisory services to private forest owners.

In the context of these structural changes forest owners' associations appeared. There are two officially registered associations of this kind: (i) the Forest Owners' Association of Lithuania (FOAL), a national organization established in 1993, with 38 local units whose main missions are to represent the interests of private forest owners and to promote the networking of private forest owners' cooperatives and forest related companies; and (ii) the Private Forest Owners' Association (PFOA), more oriented to the provision of advisory services to their members.

The non-state organizations that have more impact on the management of private forests are the cooperatives of private forest owners. More than 20 forest owners' cooperatives have been established during the last decade and their capacities are increasing. They cover a broad range of activities such as technical advice and training of private forest owners, timber trade, preparation of forest management plans, forest cuttings, reforestation and improvement of recreational areas, among others.

Leverage mechanisms

The strategy followed by some forest owners' cooperatives to build up increasing power from the initial positions they developed in forest activities is to engage in networking among themselves and with private companies. There are three networks of forest owners' cooperatives and other forest-related companies: (i) Jungtiniai miškai (United Forests), including one wood marketing company and four forest-related companies providing a broad range of services to private forest owners; (ii) the network promoted by FOAL where the biggest company, Ekomediena Ltd, coordinates timber sales from private forests on a large scale; and (iii) a network collaborating with FOAL, which includes around 30 small- to medium-sized companies and cooperatives, providing all forestry-related services and employing more than 130 skilled specialists.

5.3.5 Slovakia: advocacy coalition

Factors contributing to the emergence of forest owners' associations

Forest land private ownership rights were violated, but not abolished, by the collectivization process started in 1948. This private ownership and use lasted until 1977, when a new Forest Act came into force. This law abolished the private use of forests in reality, but private ownership was preserved *de jure*. It was only in 1990, after the fall of socialism, that full ownership was restored to private forest owners. To get their forests back, owners had to cope with the procedures to claim the land, and did not always have sufficient resources (e.g. equipment) for forest management. In response to this situation, they created associations that could advocate their interests. The first association of non-state forest owners, called the Association of Private, Associated and Communal Forests, was set up in 1991. Its ambition was to cover the whole non-state forest sector in Slovakia with smaller regional units, but it did not succeed (Konôpka *et al.*, 1999). Only some regional associations of non-state forest owners were created, starting in 1996.

After that first attempt, a new organization appeared in 1996, called the Union of Regional Associations of Non-state Forests, which now represents ten regional associations. This union is considered a social partner by the Ministry of Agriculture, having the right to participate in the forest policy-making process. Other important regional associations of non-state forests were set up during the 1990s, such as the Association of Owners of Private Forests and Forests in Shared Ownership in Banská Bystrica County (1991), the Association of Municipal Forests in Slovakia (1994) and the Union of Diocesan Forests in Slovakia (1998). The Association of Owners of Private Forests and Forests in Shared Ownership in Banská Bystrica County is a comprehensive forum whose members are not only individual forest owners and communities with forests in shared ownership, but also businessmen working in the forest sector and individuals with a forestry education background providing training for this association.

The membership of these organizations is as follows:

- Association of Municipal Forests in Slovakia: 62 members and 134,054 ha;
- Union of Diocesan Forests in Slovakia: 13 members and 40,000 ha;
- ten regional associations represented by the Union of Regional Associations of Non-state Forests: managed area of 276,200 ha; and
- Association of Owners of Private Forests and Forests in Shared Ownership in Banská Bystrica County: 534 members and 134,011 ha.

Leverage mechanisms

In 2006, these four organizations established the Council of Non-state Forest Owners' Associations to cooperate and coordinate their activities in the advocacy of non-state forests' interests, especially in the public policy-making process. There are already several cases where this organization has participated actively in this arena (e.g. game law, preparation of the National Forest Programme).

5.3.6 Serbia: a forest policy- and international cooperation-driven process

Factors contributing to the emergence of forest owners' associations

After the political changes in 2000, significant reformative steps were taken in public policy for the development of forestry, including private forestry. The share of private and state forests in Serbia is almost equal in terms of area, but private forest management is handicapped by the high fragmentation of holdings, the large number of forest owners (around 900,000) and their insufficient organization for forest management. However, the large area that is covered by private forests, and the significant values of timber volume and annual increment, which are similar to those in state forests, stress their importance as a very valuable natural resource in Serbia.

This importance is recognized in several pieces of legislation, from the Law on Forests of 1991 to the Constitution of the Republic of Serbia of 2006. Section 86 of the Constitution determines that private and state ownership rights are equal and should have equal legislative protection.

Following the law of 1991, forest regions were formed, including both state-owned and private forests. Within these regions all forest management operations should be equally implemented in all forests, disregarding their type of ownership. According to that law, public forest companies and national park public enterprises were entrusted to perform forest management operations not only in state forests, but also in private forests in their region.

This situation is being changed with the new Forest Law. This reform aims at giving private forest owners the option to choose among alternative service providers. Also, there is room for forest owners' associations or their companies to perform tasks of public interest. Forest owners have the responsibility for harvesting and protecting their forests. Advisory and technical tasks must be carried out by registered companies. Forest holdings above 100 ha are required to have a forest management plan and a contract with a company registered for technical and advisory services. The smaller holdings have less stringent requirements and benefit from technical services provided by a registered company appointed and paid by the state. The new legislation also declares the important role of forest owners' organizations. Further analysis of Serbian private forestry can be found in Nonić (2004), Nonić and Herbst (2005), Nonić and Milijić (2008) and Nonić *et al.* (2006, 2008).

Supporting factors for coping with 'free riding' problems

The new Forest Law establishes mechanisms for public financial support to private forestry. More precisely, what is proposed in the National Forest Action Programme (NFAP) is:

- establishment of a Forest Fund;
- an annual policy of financing forestry activities under the orientation set in the NFAP; and
- coordination of the public financing of forestry with the rural development funds.

Support for private forest owners' associations is to be financed by the Forest Fund, which will cover advisory services supporting the creation of regional and national forest owners' organizations and forest management operations carried out by forest owners' organizations.

This new legislative framework is under preparation, but has not yet been implemented. However, an important step towards the development of forest owners' organizations has already been taken within the framework of some joint projects supported by the Food and Agriculture Organization of the United Nations, the Confederation of European Private Forest Owners, the Directorate of Forests and bilateral cooperation (Beguš, 2006; Kadovic *et al.*, 2008; Glück *et al.*, 2009; CEPF-PROFOR, no date; FAO, no date a,b). Numerous workshops for private forest owners were organized, the result being the effective creation of 19 local forest owners' associations between 2006 and April 2009.

The legal status of these organizations is similar to that of non-governmental organizations (NGOs), so their main activities are not yet a direct engagement in the forest management operations of their individual members. Each member is still fully in charge of the management of his or her own forests. What these associations do is: represent the interests of their members; promote and coordinate some collective operations in forest infrastructures; support the marketing of forest products; and organize training courses and cooperation with other associations related to forests.

The average forest holding size of the members of these associations is above the national average, which means that at the initial stage they attracted mostly owners with large holdings, living in regions where residents are traditionally linked to forests and oriented to entrepreneurship in forestry (Milijić, 2007; Milijić *et al.*, 2007).

Leverage mechanisms

The international cooperation projects mentioned above supported the establishment of the Serbian Federation of Private Forest Owners' Associations (SFPFOA) at the end of May 2009. Although this federation is very young, it is expected to lead to further development of the private forest sector, namely by improving the position of private forest owners in the forest policy process.

5.3.7 Croatia: a forest policy-driven process

Factors contributing to the emergence of forest owners' associations

When the first Forest Law was approved in 1990, there was not yet much concern for private forests. This law treated all forests in the same way, regardless of their form of ownership, so private forests were not the object of specific and adequate regulations and support. What happened then was a situation of frequent illegal logging and mismanagement. Hrvatske šume Ltd, the company created for forest management in all kinds of forests, was only interested in public forests.

This situation of abandonment of private forests led to changes in forest policy (Ministry of Agriculture, Forestry and Water Management, 2003, 2005). In 2005 a new Forest Law was approved, which included some specific provisions to support private forests and the associative organization of private forest owners. The main provision was the creation, in 2006, of a public Forest Extension Service, which started its activities in 2007 with the mission of providing services and public financial incentives to improve private forest management. These services are focused on the establishment of a private forestry cadastre and on the preparation and implementation of forest management plans for private forests whose costs are supported by the Forest Extension Services. These plans are mandatory for forest holdings above 1000 ha. The aim is to go from the current low level of coverage of these plans (6% of the area of private forests) to full coverage in about 10 years. Besides preparing and supervising the implementation of forest management plans, the Forest Extension Service is also in charge of forest infrastructures (roads, infrastructures for firefighting, etc.) and fire surveillance operations. The other main mission of this public agency is to promote and finance the creation of forest owners' associations.

Supporting factors for coping with 'free riding' problems

The activities of the Forest Extension Services are funded by a percentage of the so-called 'green tax'. This is a tax paid by all legal individuals who conduct economic activity in Croatia, corresponding to 0.07% of their corporate income. The economic justification for this tax is that all economic activities benefit from the public goods provided by forests. The proceeds of this tax are allocated according to land ownership distribution, that is, 22% for private forests and 78% for state-owned forests. The uses of these funds are specified in annual operating plans prepared by the Forest Extension Service for private forests and by Croatian Forests Ltd for state-owned forests.

With the help of this favourable public policy, Croatia now has about 30 forest owners' associations relying on the public support channelled through the Forest Extension Service. One of the main demands coming from these organizations is the construction and repair of forest roads.

Leverage mechanisms

On 7 June 2008, a union federating 25 of the existing forest owners' associations was created for their representation at national and international levels and for promoting cooperation among the member organizations.

5.3.8 Bulgaria: path dependence

Factors contributing to the emergence of forest owners' associations

In Bulgaria the land restitution process is going on, but in terms of forest ownership it has not yet reached the situation prevailing before nationalization in 1947, when the state forests represented only 26.6% of the total forest land. Nowadays most forests are still in the hands of the state.

The management of existing private forests in Bulgaria is carried out in different ways. Forest owners with large holdings are capable of managing their forests without external support and some of them do it, but the smaller ones cannot. Religious communities (monasteries and churches), which, in several cases, own large areas of forests are also capable of managing them without external support, except hiring the services of forest consultants. Municipalities that are getting back the forests they owned before nationalization are establishing community enterprises to manage their forests. In the region of Smoljan, in Mountain Rodopi, where there is an old tradition of management of private forests by cooperatives, this form of organization was re-established after the land restitution process, so that now there are about 30 cooperatives for management of private forests, including harvesting, timber sales and some wood processing.

Further analyses of private forestry in Bulgaria can be found in Kitchoukov and Stoyanov (2002), Stoyanov and Kitchoukov (2002) and Stoyanov and Stoyanova (2003, 2005).

Supporting factors for coping with 'free riding' problems

Public engagement with private forests is mostly in terms of legislation regulating forest management and the provision of some training and technical support. There are no financial incentives or other policies supporting the establishment of forest owners' associations.

Path dependence

In this context, what emerged after the land restitution process as the main forms of collective organization in non-state forests was the revival of organizations that existed before the nationalization of forests in 1947, namely the municipalities and the forest cooperatives in the Smoljan region. Therefore this is a case of path dependence, as the situation prevailing in the past is still very influential in the current forms of organization of forest management.

5.3.9 Romania: a property rights-driven process

Factors contributing to the emergence of forest owners' associations

In the overall context of a chaotic transition and accession to the EU, land restitution has been a sensitive issue for Romanian politicians. The result is a piecemeal approach to restitution regulated by three main laws (18/1991, 1/2000 and 247/2005). The first regulation targeting the management of private forests appeared only in December 1999, requiring a forest management plan before conducting forest operations in private forests. Problems such as overcutting, timber theft, lack of forest regeneration and absence of forest investment appeared early in 'newly' private forest estates (Bouriaud and Schmithüsen, 2005: 297). Almost half of the first privatized forest land (120,000 ha) was clear-felled or overharvested in a short period of time (Nichiforel, 2007: 32).

In the first decade after the revolution, the creation of associations of local forest owners was driven more by the need to secure private ownership rights on forest land and less by the willingness of forest owners to get organized for coping with those bad management problems.

Supporting factors for coping with 'free riding' problems

After 2002 the main driver for the creation and development of local forest owners' associations was the possibility to constitute Private Forest Districts (PFD), which allowed these organizations to gain more administrative and management rights over all private forests, under some minimal size restrictions. According to the law, an association of forest owners can create a PFD only if it brings together forests covering at least 3000 ha in the plains, 5000 ha in the hilly regions or 7000 ha in mountains. The first PFD was established in 2002. Since then the number of PFDs increased to 104 by the end of 2005, managing about 1 million ha of non-state forests, belonging mostly to towns, villages and communities. In 2005, the managers of the PFDs established the Association of Forest Administrators in Romania, which started to play an important lobbying role for the interests of non-state owned forests (Bouriaud *et al.*, 2005).

Leverage mechanisms

The implementation of the latest land restitution law (247/2005) is moving the country from a situation where state-owned forests prevailed to a situation where non-state ownership will account for two-thirds of forests. This change may bring about more interest in policy makers and in the society at large to support improved management in private forests, including support for capacity building in forest owners' associations. An example is the project called 'Support to the Establishment and Development of Associations for Local Forest Owners', carried out by INDUFOR between February 2006 and July 2007. The purpose of this project was to prepare and implement a methodology supporting the establishment of associations of local forest owners and to achieve responsible forest management by setting up group forest management certification schemes in nine pilot projects.

Path dependence

By following a bottom-up approach, local forest owners' associations are more directly linked to local communities and traditions and to forest owners than another organization called the Association of Private Forest Owners (APPR). APPR was established in 1998 as a national level umbrella organization claiming to represent all categories of private forest owners in Romania. APPR has 30 county branches but associations of local forest owners work better than these APPR branches, at least in terms of logistical support and networking of private forest owners.

The top-down approach at the origin of APPR made this organization vulnerable to conflicts among its leaders. Another sign of weakness is the fact that, having obtained the status of NGO of public interest in 2004, meaning that the organization could apply for financial support from the state budget, this never happened because APPR was never able to produce an official balance sheet, which is the basic condition for using public funds and an obligation for any NGO of public interest (INDUFOR, 2007: 5).

5.3.10 Portugal: a by-product of forest policy changes and negative externalities

Factors contributing to the emergence of forest owners' associations

Forest owners' organizations started to develop in Portugal in the early 1990s. Several factors contributed to this structural change in private forestry:

- the re-establishment of a democratic political regime in 1974, after almost 50 years of dictatorship not sympathetic to free associative movements;
- more openness to international cooperation, including with countries having longer experience of forest owners' associations;
- the crisis of the State Forest Services as an agency with a vocation for direct intervention in forest management after the end, in the 1970s, of the major project of this agency – the afforestation of common lands;
- the intense rural out-migration during the 1960s, 1970s and 1980s, coupled with technological change in farming, and an increasing decoupling between farming and forestry;
- the increasing risk of forest fires, especially in the regions of small-scale forestry, mainly due to the land abandonment resulting from the rural out-migration; and
- the agricultural and forestry programmes co-funded by the EU since 1987 providing financial incentives to private farmers and forest owners for afforestation and stand improvement.

These EU co-funded programmes combined with the crisis in the State Forest Services brought about a structural change in the public financing of forestry: instead of direct public intervention based on the action of State Forest Services mostly in the communal forests this agency has established, public finance to forestry moved to the provision of financial incentives (mostly subsidies for afforestation and stand improvement) to private forest owners. Since there was almost no public support for the advisory services to forest owners, the ones with small or medium sized holdings were handicapped with respect to the larger ones in terms of preparing applications for those incentives. This was a triggering factor for the emergence of forest owners' associations in small-scale forestry (Northern and

Central Portugal). Initially these organizations were focused mostly on supporting their members in preparing the applications for the EU co-funded programmes and helping them in their deals with the forest contractors who carried out the afforestation and the stand improvement works. Further analyses of the forest owners' associative movement in Portugal can be found in Mendes (2006a,b, 2008).

Supporting factors for coping with 'free riding' problems

As the risk of forest fires continued to increase, reaching tragic dimensions in 2003 and 2005, not only forest owners but also policy makers became more aware of the need to change the focus of forest investment from afforestation to fire prevention. So, in 1999, a new public incentive scheme was set up to support forest owners' associations in creating teams of forest sappers dedicated to silvicultural works reducing the risk of forest fires. Also, in 2004, after the big fires of 2003, a Forest Fund created by a Forest Law of 1996, but not implemented since then, was finally set up, providing some support to this and other types of activities in forest owners' associations.

Leverage mechanisms

Another important step in supporting the development of forest owners' associations, also mainly driven by the increasing risk of forest fires, was a decree in 2005 regulating the creation and management of Forest Intervention Zones (ZIFs). This is an interesting case of a combination of voluntary association of private forest owners and public authority to cope with a negative externality. If in a certain area more than 50% of the forests belong to owners who voluntarily agree to abide by some common forest management rules, they can request from the state the establishment of a ZIF. Once this is approved by the Ministry of Agriculture, that group of forest owners has to prepare and implement a Forest Management Plan and a Plan for Protection of the Forests Against Fires for the whole area. These plans are mandatory not only for the voluntary members of the ZIF, but also for the others owning forests within the limits defined for the ZIF.

Forest owners' associations in the regions of small-scale forestry of Northern and Central Portugal have been the most active players in the establishment of ZIFs. Once they are officially approved, they also tend to become the entity to which the ZIF members delegate the responsibility for preparing and implementing the Forest Management and Fire Protection Plans.

Path dependence

Forest owners' associations in Portugal do not yet make any significant intervention in the trade of forest products. These associations emerged at a time when the markets of forest products already had an oligopsonic structure. Since this structure has been in place for several decades and is strong, it is difficult to break up in a short period of time. Being out of the trade of forest products, these organizations cannot get revenues from timber and cork sales. Therefore in the current state of their existence they are dependent on whatever they can get by applying to public financial incentives, complemented by the revenue from the private services they can sell to their members and other stakeholders.

5.4 Discussion

In all cases presented here the emergence of forest owners' organizations is associated with structural changes in the social and economic environment in which private forest owners operate. In the Scandinavian countries, these changes were the increasing market power of timber buyers and public investments in collective forest infrastructures to counteract the weakening of private forest owners' conditions. In the countries with economies in transition, the changes were the fall of communist regimes and the land restitution processes that followed from there. In Portugal, the change was a structural one in forest policy when substantial financial incentives started to be available for private forest owners willing to expand or improve their forests. Another structural change was the increasing risk of forest fires.

Also, in all cases presented here, those structural changes in private forest owners' environments generated the need for cooperation among forest owners to cope with problems of common interest to them. Since the structural changes were different from country to country, the same thing happened with the need for cooperation. In the Scandinavian countries the main motivation for forest owners' associations was the need to move from individual to collective negotiation of timber prices when the market power of timber buyers was increasing. In Portugal it was the need felt mostly by small and medium size forest owners for technical advice to prepare applications for public financial incentives and to carry on the corresponding forest investment projects, in a situation where the provision of those services was not available or was too costly for them. Combined with this need, there is the increasing risk of forest fires, which cannot be reduced on a individual basis. In the countries with economies in transition, forest owners' organizations arose as a response to the fragmentation of land ownership and its devolution to private agents lacking the capacity to manage them efficiently.

Moving now to the factors helping to cope with 'free riding' problems, in the Scandinavian countries we find them in the old traditions of cooperative arrangements in farming and in the use of common watercourses to transport timber. Also, since forest owners' associations started with a substantial engagement in the provision of private services to their members from which they could get a substantial income, this allowed them to cover the costs of other services with a public goods nature. In Austria we find these kinds of factors in cohesion role-played by the 'social partnership' political culture, as well as in the public authority capacities delegated to the Chambers of Agriculture. In Serbia a joint project involving international cooperation provided a common supporting ground for these organizations to take off. In some countries with economies in transition and in Portugal, some public financial support was important for the start-up of forest owners' associations. The availability of financial support in these cases was important where forest owners' organizations did not start with a direct engagement in timber marketing, but in poorly paid activities of technical advice to small-scale forest owners.

Concerning the 'leverage mechanisms' contributing to forest owners' associations to reach out beyond the initial domain where they started, in Scandinavia they achieved this through vertical integration after having built a strong foundation in timber marketing. The Austrian Forest Association and the delegation of public authority to the Chambers of Agriculture play this kind of role in Austria. In Portugal, the ZIFs are a combination of voluntary collective action of forest owners with public authority enabling forest owners' organizations to expand their influence beyond the limits of their members' holdings. In Romania, PFDs play a role similar to the ZIFs in Portugal. In Lithuania, forest owners' associations expand their influence through cooperation and networking among forest cooperatives and between cooperatives and private forest companies.

Finally, concerning the issue of 'path dependence', this is clearly present in the Scandinavian countries where forest owners' organizations started their existence very focused on timber sales, and a century later this is still their core business as far as forestry is concerned.

In most of the other countries forest owners' organizations did not start with such a strong focus on commercial activities. Instead, they started with a focus on the representation of forest owners' interests and the provision of technical advice services that do not generate enough income to pay for their costs. Therefore the development of these organizations has been more reliant on public policies and public financial incentives than in Scandinavia, so that their financial survival remains a major issue of their sustainability.

Bulgaria is an interesting case of path dependence with different contours. In the absence of active public policies to support forest owners' associations, what emerged with the land restitution processes in terms of collective forest management were the old forms of cooperative management. A similar situation is found in Romania, where a local forest owners' association inherited characteristics of the same type of association that existed there before 1948.

5.5 Directions for Future Research

Even though the cases presented here show that the emergence of forest owners' organizations is associated with structural changes in private forest owners' environments, further work is needed to examine the nature of this correlation. Is there linear causality going from these changes to the emergence of forest owners' organizations, or interdependence between the two processes? Since, as the cases also show, different types of structural changes gave rise to the emergence of forest owners' organizations, why in some cases did one type of change have this effect and in other cases this did not happen and the triggering factors were different?

Another interesting direction for further research suggested by these cases is the connections between the type of structural change associated with the emergence of forest owners' associations and the initial and subsequent development in the scope of services they produce.

Even though forest owners' organizations are meant to produce services that, for the most part, are private services to private forestry, they involve the provision of several kinds of public goods benefiting the forest owners themselves and the society at large. So insights coming from the theory of public goods are useful here and the country cases showed the role played in forest owners' organizations' development of factors helping to cope with 'free riding' problems. However, as was stated before, these are not organizations focused on the provision of public goods only, but organizations where there are several forms of joint supply of public goods and private goods and services. Therefore, it is along these lines that further work should be pursued for better understanding 'free riding' problems and their solutions through this kind of organization.

Another direction for future research pointed out by these cases is to examine the role played by the 'leverage mechanisms', defined here as factors that may be combined with existing capacities of forest owners' organizations to enable them to reach beyond the initial domain where they got started.

Finally, the cases presented here recommend that in the dynamics of these organizations attention should be given to path dependence phenomena.

Acknowledgement

This work was supported by the Slovak Research and Development Agency under the contract No. APVV-0692-07.

References

Arthur, W.B. (1994) *Increasing Returns and Path Dependence in the Economy.* The University of Michigan Press, Ann Arbor, Michigan.

Beguš, J. (2006) Report of International Consultant for Capacity Development of Public Forest Service and Private Forest Owners' Associations. FAO/GCP/FRY/003/FIN Project 'Forest Sector development in Serbia', Belgrade. Available at: http://www.forestryprojectserbia.org.rs (accessed 21 October 2010).

Binmore, K. (1994) *Game Theory and the Social Contract.* Volume 1: *Playing Fair.* MIT Press, Cambridge, Massachusetts.

Binmore, K. (1998) *Game Theory and the Social Contract.* Volume II: *Just Playing.* MIT Press, Cambridge, Massachusetts.

Bouriaud, L. and Schmithüsen, F. (2005) Allocation of property rights on forests through ownership reform and forest policies in Central and Eastern European countries. *Swiss Forestry Journal* 156, 297–305.

Bouriaud, L., Nichiforel, L., Padureanu, L., Nastase, C. and Borlea, F. (2005) Romania. Country report for Phase 1 of the COST Action E30. In: Jáger, L. (ed.) COST E30 Economic Integration of Urban Consumers' Demand and Rural Forestry Production. Forest Sector Entrepreneurship in Europe: Country Studies. *Acta Silvatica & Lignaria Hungarica* (Special Edition 2005), 643–694.

CEPF-PROFOR (no date) Private and community forestry – developing livelihoods on the basis of secure property rights in selected countries of South East Europe. Available at: http://www.skog.no/cepf/profor.asp (accessed 21 October 2010).

Coase, R.H. (1960) The problem of social cost. *The Journal of Law and Economics* 3, 1–44.

David, P.A. (1985) Clio and the economics of QWERTY. *The American Economic Review* 75, 332–337.

FAO (2006) *Global Forest Resources Assessment. Progress Towards Sustainable Forest Management.* Food and Agriculture Organization of the United Nations, Rome.

FAO (no date a) Project TCP/YUG/2902(A): Institutional Development and Capacity building for the National Forest Program. Food and Agriculture Organization of the United Nations, Rome. Available at: http://www.fao.org/regional/SEUR/ProjMore_en.htm (accessed 21 October 2010).

FAO (no date b) Project GCP/FRY/003/FIN: Forest Sector Development in Serbia. Food and Agriculture Organization of the United Nations, Rome. Available at: http://www.forestryprojectserbia.org.rs (accessed 21 October 2010).

Glück, P., Avdibegović, A., Čabaravdić, A., Nonić, D., Petrović, N., Posavec, S., Stojanovska, M., Imočanin, S., Krajter, S., Lozanovska, N., Marić, B., Milijić, V., Mrkobrada, A. and Trninić, S. (2009) *Final Report on the Research Results*, Volume 1: *Main Results.* Project: Research into the Organizations of private forest owners associations in the Western Balkan region (PRIFORT). European Forest Institute.

Hobbes, T. (1963) *Leviathan.* Meridian Books, New York.

Hogl, K., Pregernig, M. and Weiss, G. (2005) What is new about new forest owners? A typology of private forest owners in Austria. *Small-scale Forest Economics, Management and Policy* 4, 325–342.

Hume, D. (1978) *A Treatise of Human Nature*, 2nd edn. Oxford University Press, Oxford.

INDUFOR (2007) *Support to the Establishment and Development of Associations for Local Forest Owners (ALFOs).* Final Report. Helsinki.

Kadovic, R., Aleksic, P., Tomovic, Z., Medarevic, M. and Orlovic, S. (2008) Basis for development of National Forest Action Program. FAO/GCP/FRY/003/FIN Project 'Forest Sector development in Serbia', Belgrade. Available at: http://www.forestryprojectserbia.org.rs (accessed 21 October 2010).

Kitchoukov, E. and Stoyanov, N. (2002) Privatisation of forestry enterprises in Bulgaria – possibilities, development and problems. In: *Privatisation in Forestry*, 11–14 April 2002, Belgrade, Serbia, Volume II. Faculty of Forestry Belgrade University, Serbia and Institut fur Forstokonomie Albert-Ludwigs-Universitat Freiburg, Germany.

Knight, J. (1992) *Institutions and Social Conflict.* Cambridge University Press, Cambridge.

Konôpka, J. *et al.* (1999) Analýza vývoja a súčasného stavu lesného hospodárstva Slovenskej republiky (1990–1998). *Lesnícke štúdie* 55, 30–33.

Marwell, G. and Oliver, P.E. (1993) *The Critical Mass in Collective Action: A Micro Social Theory.* Cambridge University Press, Cambridge.

Marx, K. (1971) *A Contribution to the Critique of Political Economy.* Lawrence & Wishart, London.

Mendes, A.M.S.C. (2006a) Forest owners' organizations in Portugal: are the infant going to survive? In: Wall, S. (ed.) *Small-scale Forestry and Rural Development: The Intersection of Ecosystems, Economics and Society.* Proceedings of IUFRO 3.08 Conference hosted by Galway-Mayo Institute of Technology, Galway, Ireland, 18–23 June 2006. Galway-Mayo Institute of Technology, Galway, pp. 289–304.

Mendes, A.M.S.C. (2006b) Private forestry and forest policy reforms in Portugal in the context of increasing risk of forest fires. In: Ayanoglu, S., Günes, Y., Coskun, A.A. and Herbst, P. (eds) *Legal Aspects of European Forest Sustainable Development*, Proceedings of the 8th International IUFRO RG 6.13.00 Forest Law and Environmental Legislation Symposium. Department of Forest Law, Faculty of Forestry, Istanbul University, pp. 205–220.

Mendes, A.M.S.C. (2008) The role of institutions in forest development: the case of forest services and forest owners' associations in Portugal. In: Cesaro, L., Gatto, P. and Pettenella, D. (eds) *The Multifunctional*

Role of Forests – Policies, Methods and Case-Studies, EFI Proceedings 55, 2008. European Forest Institute, Joensuu, pp. 105–116.

Mendes, A.M.S.C., Størdal, S., Adamczyk, W., Bancu, D., Bouriaud, L., Feliciano, D., Gallagher, R., Kajanus, M., Mészáros, K., Schraml, U. and Venzi, L. (2006) Forest owners' organizations across Europe: similarities and differences. In: Niskanen, A. (ed.) *Issues Affecting Enterprise Development in the Forest Sector in Europe*. University of Joensuu, Joensuu, pp. 84–104.

Milijić, V. (2007) Structural and social characteristics of private forest owners' associations. Association of Forest and Wood processing Engineers and Technicians of the Republic of Serbia, Belgrade, *Forestry* 3–4, 119–128.

Milijić, V., Nonić, D., Radosavljević, A. and Grujicić, I. (2007) Associating private forest owners as a contribution to rural development of Serbia. In: Proceedings of the International Conference *Integral Protection of Forests, Scientific Technological Platform*, Forestry Institute of the Republic of Serbia, Belgrade, pp. 69–75.

Ministry of Agriculture, Forestry and Water Management (2003) A strategy for rural development in Croatia. Draft paper.

Ministry of Agriculture, Forestry and Water Management (2005) Analiza stanja privatnih šuma u Republici Hrvatskoj [Analysis of the situation in private forests in Croatia].

Nichiforel, L. (2007) Assessing rent seeking processes in Romanian private forestry from a property rights perspective. Master Thesis submitted in partial fulfilment of the degree of Master in Environmental Governance. Albert-Ludwigs University, Freiburg.

Nonić, D. (2004) Organization of forestry in transition: relations of state forestry institutions and private forest owners. Doctorate dissertation. Faculty of Forestry, University of Belgrade, 232 pp.

Nonić, D. and Herbst, P. (2005) Fragen des Privatwaldsektors in Serbien. *Forstzeitung* 7, 32–33.

Nonić, D. and Milijić, V. (2008) *Status Quo Analysis: Private Forestry in Serbia and its Role in NFP/NFS Process*. CEPF, Lover print, Sopron.

Nonić, D., Tomić, N., Marković, J., Herbst, P. and Krajcić, D. (2006) Organization of private forest owners in Serbia compared to Austria, Slovenia and other Central European countries. *Forstwissenschaftliche Beiträge Forstpolitik und Forstökonomie* 5, 95–106.

Nonić, D., Milijić, V., Marković, J. and Radosavljević, A. (2008) *Organization of Private Forest Sector in Serbia (Private Forest Owners and Small and Medium Enterprises): Legal and Political Aspects*. IUFRO Division 6: Social, Economic, Information and Policy Sciences, Research Group 6.13.00, Forest Law and Environmental Legislation International Symposium, Sarajevo, Bosnia and Herzegovina.

North, D. (1990) *Institutions, Institutional Change and Economic Performance*. Cambridge University Press, Cambridge.

Olson, M. (1971) *The Logic of Collective Action. Public Goods and the Theory of Groups*. Harvard University Press, Cambridge, Massachusetts.

Pregernig, M. and Weiss, G. (1998) Forest policy in Austria: policy making by the sector for the sector. Paper presented at the European Regional Meeting on *The Underlying Causes of Deforestation and Forest Degradation in Europe*, 28–29 October 1998, Bonn, Germany.

Rametsteiner, E., Aldrian, A., Bauer, A., Eberl, W., Sekot, W., Wagner, S. and Weiss, G. (2005) Country report Austria to Cost E30: economic integration of urban consumers' demand and rural forestry production. In: Jáger, L. (ed.) COST E30 Economic Integration of Urban Consumers' Demand and Rural Forestry Production. Forest Sector Entrepreneurship in Europe: Country Studies. *Acta Silvatica & Lignaria Hungarica* Special Edition 2005.

Schmithüsen, F. and Hirsch, F. (2009) *Private Forest Ownership in Europe*. Geneva Timber and Forest Study Paper 25. United Nations Economic Commission for Europe/ Food and Agriculture Organization of the United Nations – Timber Section, Geneva.

Smith, A. (1969) *The Theory of Moral Sentiments*. Liberty Classics, Indianapolis, Indiana.

Smith, A. (1976) *An Inquiry into the Nature and Causes of the Wealth of Nations*. Liberty Classics, Indianapolis, Indiana.

Smith, V.L. (no date) *Experimental Methods in Economics*. Interdisciplinary Center for Economic Science, George Mason University, Fairfax, Virginia. Available at: http://www.neuroeconomics.net/pdf/materials/455.pdf (accessed 21 October 2010).

Spencer, H. (1969) *Principles of Sociology*. Archon Books, Hamden, Connecticut.

Stoyanov, N. and Kitchoukov, E. (2002) Status of Bulgarian forestry in the period of transition to market economy. Paper presented at the International Conference *Privatization in Forestry*, 11–14 April 2002,

Belgrade, Serbia. Faculty of Forestry, University of Belgrade and Institute fur Forstökonomie, Albert-Ludwigs-Universität, Freiburg.

Stoyanov, N. and Stoyanova, M. (2003) Characteristic, organization and management of private forests in Bulgaria. Paper presented at the International Workshop *Forest Operation Improvements in Farm Investment*, 9–14 September 2003, Logarska Valley, Slovenia. Joint FAO/ECE/ILO Committee on Forest Technology Management and Training.

Stoyanov, N. and Stoyanova, M. (2005) Forest and forest products country profile: Republic of Bulgaria. Geneva Timber and Forest Discussion Paper 38. UNECE, Geneva and FAO, Rome.

Weber, M. (1978) *Economy and Society*. University of California Press, Berkeley, California.

Weber, M. (1989) *The Protestant Ethic and the Spirit of Capitalism*. Unwin Hyman, London.

Weiss, G., Hogl, K., Rametsteiner, E. and Sekot, W. (2007) Privatwald in Österreich – neu entdeckt. *Schweizer Zeitschrift für Forstwesen* 158, 293–301.

Williamson, O. (1975) *Markets and Hierarchies*. Free Press, New York.

Williamson, O. (1985) *The Economic Institutions of Capitalism*. Free Press, New York.

6 The Contribution of Leader+ to the Implementation of Innovative Forest-related Projects

Diana Feliciano,* Bill Slee, Gerhard Weiss, Anne Matilainen and Thomas Rimmler

Abstract

A key criterion of the EU's Leader approach is the demand for innovative actions to promote rural development. Given that the main objective of the COST Action E51 was the development of knowledge that enables integration of innovation and development policies for a more effective and sustainable development of the forest sector, this chapter makes an assessment of the extent of Leader's support for innovative forest-related projects. To achieve our aim, the lists of Leader+-funded projects in Austria, Finland, Portugal and Scotland were analysed and the forest-related projects identified. A questionnaire was sent to the Local Action Groups that had approved forest-related projects. Questionnaires were completed by either Leader+ beneficiaries or project managers; 47 questionnaires were returned from a total of 108 sent out. In the total of the four countries, only 3.7% of the projects funded by Leader+ were classified as forest-related. There is evidence from the respondents that Leader+ was implemented in a way that supported bottom-up initiatives, a feature that is seen important for integrated rural development. New networks that were created by Leader+ lasted beyond the duration of funding. The risk-friendly and innovation-oriented character of the Leader approach is, however, not strongly valued by the respondents, so our findings provide no evidence of whether Leader+ supported especially novel and innovative types of projects.

6.1 Introduction

Since the late 1970s, there has been an emerging awareness of the particular problems of remote rural areas with high dependency on primary sector activities such as agriculture and forestry, which include low or negative rates of growth, depopulation and high costs of public service provision. Many of those regions were designated as agriculturally less favoured and subsequently under the disbursement of structural funds as Objective 1 or Objective 5b regions. The new strategies for economic and social revival of rural areas included the diversification of economic activities, especially into tourism and leisure, the promotion of local products and new markets, the development of human resources and the support of national and international cooperation, and at the same time promoted the participation of local stakeholders in the creation of a rural Local Action Group (LAG) plan (Serafim, 1999). In 1988, the document 'The Future of Rural Society' presented the European Commission's new approach to rural development (European Communities

* Corresponding author.

Commission, 1988). The document considered the challenge of economic and social cohesion in an enlarged European Community (EU) of very pronounced regional diversity, the unavoidable adjustment of farming in Europe to market circumstances and the implications of this adjustment for farmers, farm workers and rural economy in general, and the protection of the environment and the conservation of the EU's natural assets. Also in 1988, the reform of the Structural Funds resulted in the creation of a budget for special programmes named Community Initiatives. The aim of one of these Initiatives was to support disadvantaged regions in promoting a favourable environment for business development, directed to the needs of the domestic market and intensive competition. Within this context, the Leader Initiative was launched, which was enabled by the Communication to the Member States no. 91/C 73/14, 19 March 1991 and co-funded by the European Agricultural Guidance and Guarantee Fund (EAGGF). According to Ray (1998), the announcement of the Leader Initiative was a signal that EU rural development policy was to include further local targeting of Structural Funds policy.

The main objective of Leader was to deliver innovative solutions, assuring the best possible link between actions among the many sectors of the economy. Leader is characterized as being a place-based development policy, building on the area's specificities, encouraging participatory decision making at the local level through facilitation of activities and training of local communities. Thus, Leader was based on a strong ethos of local participation at appropriate spatial scale in both design and implementation phases (Ray, 1998). The Leader approach is based on seven specific features in order to achieve sustainable rural development across Europe (Table 6.1).

After two earlier programming periods, Leader 'I': 1991–1993 and Leader 'II': 1994–1999, Leader+ was implemented in 2000–2006 as a pilot programme for innovative strategies for rural development. For the period 2007–2013, Leader is now included in the rural development policy as a cross-cutting theme of the thematic axes of Pillar 2 of the Common Agriculture Policy (CAP). Of all rural development schemes in Europe, Leader is the scheme that explicitly nurtures innovation. Ray (1998) stated that at the heart of Leader was an ethos to explore

Table 6.1. The seven features of the Leader approach.

Feature	Small description
Area-based approach	Actions should suit real needs and local competitive advantage
Bottom-up approach	Local actors participate in decision making about the strategy and in the selection of the priorities to be pursued in their local area
The Local Action Group	The LAG is in charge of identifying and implementing local strategies, making decisions on the allocation of funds and managing them
The innovative approach	Leader should stimulate new and innovative approaches to the development of rural areas. Innovation is encouraged by allowing LAGs to decide on the actions they want to support. Innovation may mean the introduction of a new product, a new process, a new organization, or a new market
The integrated and multi-sectoral approach	The local development strategy must integrate several sectors of activity
Networking	Includes the exchange of achievements, experiences and know-how between Leader groups, rural areas, administrations and organizations involved in rural development within the EU, whether or not they are direct Leader beneficiaries
Cooperation	It involves a LAG undertaking a joint project with another Leader group or a group taking a similar approach, in another region, Member State, or even third country

Source: http://www.elard.eu/en/Leader (accessed 18 February 2010).

innovative, i.e. experimental, forms of local rural development. The rationale behind the focus on innovation is the perceived need for actions that promote local resources in new ways, actions relevant to local development that are not covered by other development policies or actions to create new products, new processes, new forms of organization, or new markets. Such actions should provide new answers to the weaknesses and problems of rural areas.

Rogers (1995) defines innovation as an idea, practice or object that is perceived as being new by an individual or other unit of adoption. He states that 'newness' of an innovation may be expressed in terms of knowledge, persuasion or a decision to adopt. Being completely innovative is not easy. The same author considers that 'the perceived newness of the idea for the individual determines the reaction to it and if the idea seems new to the individual, it is an innovation'. He also states that an innovation does not always have to be associated with new knowledge. In rural areas, innovation may imply the transfer and adaptation of innovations developed elsewhere, the modernization of traditional forms of know-how or finding new solutions to persistent rural problems, which other policy interventions have not been able to solve in a satisfactory and sustainable way (ELARD, 2011). To the Commission, 'an initiative of the local actors who bring a new solution to the specific challenges facing the area' was considered an innovative project (European Commission, 1994). In the Leader context, actors are the LAGs, which then work together with the individuals and organizations that implement the projects. Innovation has evolved over three stages of the Leader approach before the current scheme. In Leader I, innovation lay principally in the overall method; in Leader II innovation was sought actively in individual projects; and in Leader+ innovation resided principally in the strategy. In practice, perhaps because Leader was less risk averse than many other public funding streams, innovation has remained a feature of many Leader projects, as has the entire so-called Leader process. The aim of Leader+, the most recent to be completed, was to encourage and support rural actors to consider the long-term potential of their local region. The funding was allocated for the development of projects and investments in regions based on the characteristic bottom-up approach with strong local involvement. The central actors were LAGs, which defined the local development programme and the local funding priorities regarding the Leader+ programme, in line with national programmes.

In 1988, the European Union (EU) recognized that the development of forests and woodlands, and the various activities associated both with timber processing and with related services, represents a promising niche for rural development. In Europe, most of the less-favoured regions are covered by large areas of forest and the relationship between forest cover and regional development opportunities emphasizes the importance of forest resources for regional employment and income in those regions (Hyttinen *et al.*, 1999). In Agenda 2000, forestry development is considered a very important part of the EU policy for rural development. Policies that can be implemented to increase the contribution of forests towards rural development have been investigated by several authors (Gluck and Weiss, 1996; Koch and Rasmussen, 1998; Slee and Wiersum, 2001; Forest Europe Growing Life, 2009). In 2003, the Vienna Resolution of the MCPFE called for the strengthening of innovation and entrepreneurship in the forest sector. Forests can be seen to fit into most of the four Leader+ themes. Forests are natural resources, so theme 1, 'Making the best use of natural and cultural resources, including enhancing the value of Natura 2000 sites',[1] includes them. Forests contribute to 'improve the quality of life in rural areas' (theme 2), since they offer opportunities of employment and income not only in the traditional forest sector but also in the tourist and recreation sectors (Elands *et al.*, 2004). Forests also contribute to increase the attractiveness of rural areas as places to live. Themes 3 and 4 'Adding value to local products' and 'Using new know-how and new technologies to make products and services in rural areas more competitive' connect to enhancing the potential for rural development linked to the immense variety of wood and non-wood forest products in rural areas.

This study, as part of the COST E51 project and working group, aimed at identifying and analysing innovative cases for the integration of forests into a wider concept in rural development processes using the example of projects supported by the EU Leader Initiative. For the analysis, Leader+ was chosen since the mainstream Leader is not yet completed (2007–2013). The countries included in the study are Austria, Finland, Portugal and Scotland.

According to the Organisation for Economic Co-operation and Development (OECD) regional classification of rural areas,[2] the four countries are in the top 11 OECD countries with the highest share of the territory classified as 'predominantly rural' (Finland is ranked fifth, Austria ninth, Scotland tenth and Portugal 11th). Of the four countries, Finland is the country with highest percentage of forest cover and Portugal is the country with the highest percentage of private forest area, followed by Austria. The high percentage of non-industrial private forest (NIPF) in Portugal is also noticeable (Table 6.2).

To assess the contribution of Leader+ in the implementation of innovative forest-related projects, the following research questions were proposed:

1. Did Leader+ support innovation in forest-related projects?
2. Was Leader+ an important source of funding for innovation in forest-related projects?

6.2 Methodology

To undertake an assessment of the forest projects funded by Leader+ in Austria, Finland, Portugal and Scotland, the next two steps were followed.

6.2.1 Research into the Leader+ databases and the classification of forest-related projects in the four countries

Funding is awarded by LAGs, which take decisions on community-driven projects

Table 6.2. Some characteristics of the four countries.

	Portugal[a]	Scotland[b]	Austria[c]	Finland[d]
Area (thousand km^2)	92	79	84	338
Population (million inhabitants)	10.6 (2008)	5.1 (2006)	8.3 (2008)	5.3 (2008)
Predominantly rural area (OECD, 2008) (%)	74	75	80	89
Forest area (%)	38	17 (2004)	47 (2002)	86.5 (2008)
Private forest area (%)	93.4 (1995)	53 (1999)	80 (2004)	69 (2008)
NIPF area (%)	86.9 (1995)	n.a.[e]	50 (2004)	60 (2008)
NGO area (%)	n.a.	1.1 (1999)	n.a.	n.a
Communal forest area (%)	5.4 (1995)	0	n.a.	6 (2008)
State forest area (%)	1.2 (1995)	50 (1999)	20 (2004)	26 (2008)

[a]Population: World Bank estimates (http://www.worldbank.org/, accessed 24 February 2010); Forest data: Mendes *et al.* (2004) and Mendes (2005), in Mendes and Feliciano (2005).
[b]Population: General Register Office for Scotland (http://www.gro-scotland.gov.uk/, accessed 24 February 2010); Forest data: Forestry Commission (2004) and 1995–99 National Inventory of Woodlands and Trees, in Slee *et al.* (2005).
[c]Population: World Bank estimates (http://www.worldbank.org/, accessed 24 February 2010); Forest data: Statistik Austria (2004) and BFW Waldinventur 2000–2002, in Rametsteiner *et al.* (2005).
[d]Population: World Bank estimates (http://www.worldbank.org/, accessed 24 February 2010); Forest data: Finnish Statistical Yearbook of Forestry (2008).
[e]The distinction between industrial and non-industrial private forests is not made in Scotland.
NIPF, non-industrial private forest; NGO, non-governmental organization; n.a., not applicable.

with a wide community benefit. The national Leader+ websites[3] contain databases with LAG contacts, the approved projects, a small description of projects, the entity or person who undertakes the project, the date of approval and the total funding approved. For Portugal and Finland, the national Leader office was contacted since the online database was not up to date. In the four countries, all the Leader+-funded projects were checked and forest-related projects identified.

For the purpose of the COST Action E51, a broad definition of forestry was agreed by the team to classify forest-related projects. This definition is that used by Hyttinen (1994), who stated that from the business economic viewpoint of a single entrepreneur, a broader definition of forestry can be used, including the entire forest owner's forestry-related activities. The forest-related projects identified from the national databases were classified according to the categories below.

Territorial services

Forest services include two types of services – environmental and socio-cultural. Environmental services can be the conservation of soil, water biological diversity, micro- and macro-climatic effects and nutrient cycling. Socio-cultural services can be recreation, tourism and the protection of cultural, aesthetic and scientific values (FAO, 2000). The notion of territoriality in forest services specifies that they are provided as a result of the nature of forestry as a multifunctional land use and that production and management practices necessarily influence the quality and quantity of those services (see Slee, Chapter 8, this volume). For the purpose of this study, we consider that territorial services with respect to forestry include tourism and recreational products, where woodland views, woodland character or woodland identity are an important attribute of the product or service; environmental ecosystem services such as biodiversity, which are associated with forests and woodlands; and supporting ecosystem services such as flood control or soil protection, which impact on neighbouring and downstream areas. We also considered fighting and prevention of forest fires as a territorial service, since the aim is the protection of forest ecosystems.

Innovation in the wood chain

Some Leader+-funded projects in the forest sector potentially relate to forest products linked to local sources of wood or local styles of processing. The forest sector is defined as forestry and forest industry (sawmill, furniture, other wood, and pulp and paper industries) (Hyttinen *et al.*, 2002). Examples of forest-related projects classified as innovation in the wood chain are furniture design, silviculture or wood craftsmanship.

Renewable energy

Wood is a sustainable and low carbon source of energy that can be burned to generate heat or electricity. Wood used as fuel (wood fuel) may be available as firewood (e.g. logs, bolts, blocks), charcoal, chips, sheets, pellets and sawdust. Forest-related projects classified as renewable energy concern those that promoted the use of wood fuel as a source of energy.

Non-wood forest products (NWFPs)

Non-wood forest products include game animals, fruits and berries, honey, fungi and foliage, and value-added products associated with these. We classified NWFPs projects as those concerning the production, processing and marketing of these products. We also included those where the aim was to disseminate information and improve local knowledge about NWFPs.

6.2.2 The distribution of questionnaires

A questionnaire with 17 closed and open questions (plus sub-questions) was prepared by the team and sent out by e-mail to the LAGs or to the beneficiaries of Leader+ (organization/business/person that implemented the projects). The questionnaires were sent to the LAGs whenever the contacts of the beneficiaries were not available from the databases or directly to the beneficiaries if the contacts were available. The number of questionnaires sent out by each of the four countries depended on the number of forest-related projects identified per country. All the LAGs that approved forest-related projects were contacted. In the case of Finland, questionnaires were first sent to the LAGs, but since none responded a web search was undertaken to find the beneficiary's contacts. Questionnaires were then sent directly to the beneficiaries whose contacts were found. In Austria, some questionnaires were answered during a phone call. In the cases of Portugal and Scotland, questionnaires were first sent to LAGs and then to beneficiaries whenever contacts were provided.

Six themes from the questionnaires were chosen for further analysis, according to the following research questions:

- Did Leader support innovation in forest-related projects?

  1. The source of the original idea for the project.
  2. The type of innovation funded by Leader+ and examples of innovations.
  3. The additional number of people employed during the project and that remained employed after it had ended.
  4. The way LAG looked at forest-related projects.

- Was Leader an important source of funding for forest-related projects?

  5. The exclusivity of Leader+ funding and the importance of Leader+ for the viability of forest-related projects.
  6. The extent that Leader+ was helpful for beneficiaries and the way the respondents saw Leader+.

6.3 Results

6.3.1 Results obtained from the analysis of the national databases

As shown in Table 6.3, a total of 12,376 projects were funded by Leader+ in

Table 6.3. Some characteristics of the Leader+ Initiative in the four countries.

	Total (4 countries)	Portugal	Scotland	Austria	Finland
LAGs					
Number of LAGs under Leader+	145	52	12	56	25
Number of LAGs with forest-related projects	76	37	6	9	24
LAGs with forest-related projects (%)	52	71	50	16	96
Funding					
Total Leader+ funding		€279 m	£17 m[a]	€184 m	€170.4 m
Total Leader+ funding in forest-related projects		€4.9 m	£634,000	≥€5 m	n.a.
Projects					
Territorial services	237	27	8	10	192
Innovation in the wood chain	101	33	13	14	41
Renewable energy	12	0	1	1	10
NWFP (includes hunting)	112	46	1	0	65
Other	4				4
Total of forest-related projects	466	106	23	25	312
Total number projects financed by Leader+	12,376	7,234	723	1,730	2,689
Forest-related projects (%)	3.7	1.5	3.2	1.4	11.6

[a]€/£ in 2005 = between 1.4 and 1.5 (http://www.taxfreegold.co.uk/2005forexrates.html, accessed 05 March 2010).
LAGs, Local Action Groups; NWFPs, non-wood forest products; m, million.
Source: National Leader+ databases (Portugal: http://www.Leader.pt/; Scotland: http://ukLeader.org,uk/; Austria: http://archiv.Leader-austria.at/network/projects.html; Finland: http://www.lande2000.fi/; Jumppanen *et al.* (2008).

Portugal, Scotland, Austria and Finland. From these, about 466 were classified as forest-related projects. This corresponds to about 3.7% of the total projects funded by Leader+. Of the four countries, Portugal was the country where the greatest number of projects was approved, followed by Finland. In terms of the total number of forest-related projects, Finland has the greatest number followed by Portugal. In terms of percentage of LAGs that approved forest-related projects, Finland has the highest percentage followed by Portugal. In Finland, the majority of the projects approved are on territorial services, in Portugal on NWFPs and in Austria and Scotland on innovation in the wood chain.

In Finland, forest-related territorial service projects mainly relate to recreation and nature tourism. In Portugal, the majority of projects on territorial services are related to fighting and prevention of forest fires and the main NWFPs projects related to honey production and hunting. In Scotland, projects on the innovation of the wood value chain are usually related to the management of woods and on the use of wood for construction. Territorial services generally relate to mountain biking and forest paths. In Austria, one of the LAGs (Murau), which was exclusively focusing on wood, covered more than half of all forest-related projects. In this LAG, the majority of the projects were related to the use of wood. In total for Austria, more than half of the projects were wood value chain related, while the others mostly related to tourism and nature/forest education. The very limited attention given to renewable energy projects was noticeable.

6.3.2 Results obtained from the questionnaires

From a total of 108 questionnaires sent out in Austria, Finland, Scotland and Portugal, 47 replies were obtained (Table 6.4). These questionnaires concerned mainly projects classified as territorial services and innovation in the wood chain. Four questionnaires concerned NWFPs projects and only one was a renewable energy project.

Questionnaires were completed either by the LAG managers or by beneficiaries in Scotland and Portugal and only by the beneficiaries in Austria and Finland. The project managers completed the questionnaires whenever they knew about the project because of their involvement in the application process. When this did not happen, questionnaires were forwarded and completed by the beneficiaries. The case study projects were carried out by a diverse range of stakeholders. In Portugal, beneficiaries were mainly forest owners' organizations, municipalities and individuals. In Scotland, almost every questionnaire concerned a different stakeholder, e.g. a woodland conservation organization, a rural property management agency, a forestry contracting association and a government department. In Austria, projects were mainly carried by working groups, associations and societies, often formed by forest owners or woodworking companies, municipalities or tourism boards. In Finland, beneficiaries were mainly local forest management associations, most of the time in partnership with individuals or other entities. When asked about the source of the original idea for the project, 30 respondents replied that it was their own idea to add something new to an existing project/scheme (Fig. 6.1). Specialist

Table 6.4. Number of questionnaires obtained per country and type of forest-related projects.

	Portugal	Scotland	Austria	Finland	Total
No. questionnaires obtained	20	6	12	9	47
Territorial services	7	3	3	4	17
Innovation in the wood chain	11	2	9	3	25
Renewable energy	0	1	0	0	1
Non-wood forest products	2	0	0	2	4

advisers, friends and relations were other sources used widely for getting ideas. Other sources included were companies and forestry actors, and the extensive partner network of a wood cluster in Austria.

Thirty-five applicants responded that they did not seek funding for their projects from any other public organizations but 12 applicants did seek external funding. For those that sought other funds, at least nine reported that they did receive other funding. Also, 25 respondents asserted that if Leader+ had not funded the project, it would not have gone ahead. From the 18 that said it would have gone ahead anyway, 17 considered that it would have been slower and/or at a smaller scale and only one respondent answered that it would have been about the same. The reasons elicited for not undertaking the project without Leader+ funding were: lack of own funding, lack of human resources and difficulties in getting funding from other sources. One respondent commented: 'There wouldn't have been enough resources to execute the project along with the organization's normal tasks'.

In general, Leader+ promoted innovation in new products, new processes and new networks with little emphasis on the creation of new markets (Fig. 6.2). Examples of product innovation include the creation of new furniture lines (Austria), a new hiking trail for a region (Austria), producing honey from pine stands (Portugal), the creation of a wood pasture habitat (Scotland), supporting the use of cork in building and furniture (Portugal), the creation of leisure facilities (Austria) or the cooperation between schools and a forest management association resulting in the implementation of forest tracks

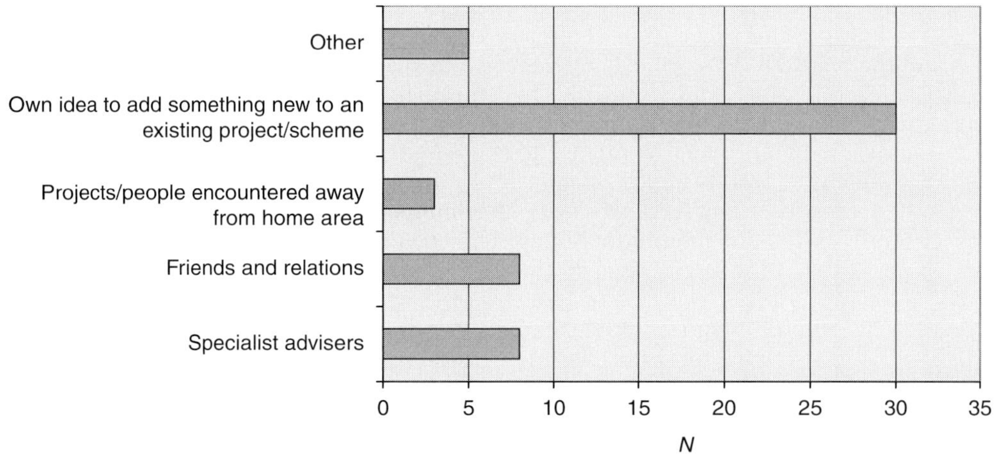

Fig. 6.1. Sources of ideas for the forest-related projects (*N* = number of answers).

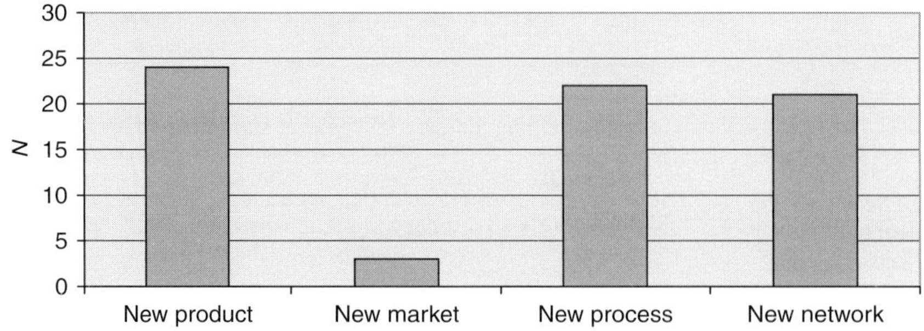

Fig. 6.2. Type of innovation in the forest-related projects (*N* = number of answers).

near to schools (Finland). New processes include the creation of a biomass heating system on a Scottish island, a new way of valuing the 'montado'[4] ecosystem and the creation of local teams to implement projects. Finally, networking was considered several times as an innovation. Examples include networks of leading firms, networks of forest owners' organizations in Austria, cooperation between schools and a *forest management association* in Finland, and cooperation between partners along the whole value chain (in the development of various products and joint marketing). An example of a Portuguese project where networking was emphasized was the implementation of an agro-environmental demonstration site. This project was open to the community, forest owners and technicians, involved partnerships with the World Wildlife Fund and was articulated with other projects related to the ecological recovery of the region. A Finnish local forestry centre wrote: 'The goal was to develop new networks. Now, a few years after the project, cooperation has continued e.g. by lending and renting equipment.'

Leader+ did have an effect on employment of rural areas. As is shown in Fig. 6.3, during the implementation of the projects 33 part-time jobs, 37 full-time jobs and 13 seasonal jobs were created in Scotland, Portugal and Finland. In Austria, no information on jobs creation was supplied by the respondents. This gives an average of 2.18 jobs per project (83 jobs/38 projects excluding Austrian projects). From the total number of jobs created, 25 part-time jobs, 30 full-time jobs and 13 seasonal jobs were maintained after the implementation of the project.

As presented in Table 6.5, 34 respondents considered that Leader+ was definitely helpful in providing funding and

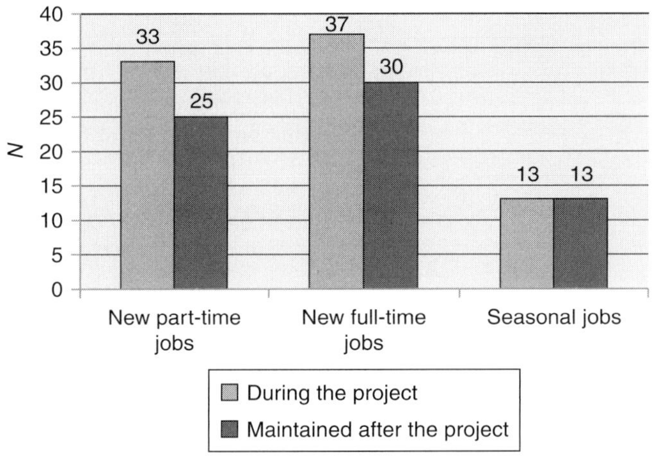

Fig. 6.3. Number of jobs created by the Leader+-funded projects in the four countries (N = number of answers).

Table 6.5. The extent Leader+ was helpful (number of answers).

	Not helpful at all	Not really helpful	Helpful to a limited degree	Helpful to a certain degree	Definitely helpful
Providing funding	2	2	2	7	34
Pre-application advice	1	3	2	16	25
Project aftercare	4	7	5	18	12
Creating new networks	8	8	9	10	10
Tolerance of risk	9	9	12	6	5
Support of non-mainstream innovation	4	6	7	17	11

pre-application advice. The extent to which Leader+ was helpful in creating new networks was less clear. Answers are almost equally spread around the five different categories. In relation to tolerance to risk, Leader+ was considered helpful to a limited degree. In project aftercare and support to non-mainstream innovation, most respondents considered it helpful to a certain degree.

Thirty-three respondents also asserted that Leader+ was distinct from other programmes/funding frameworks in terms of the support given and most of them considered it as more personal/approachable, more helpful and more local/accessible (Table 6.6). Nine respondents did not consider Leader+ distinct.

Because the questionnaires were completed by both LAG staff or beneficiaries, a decision was made to explore further the differences between the answers given only by beneficiaries and probe if it matched with the previous results. We found out that there are not many differences if only the answers from beneficiaries are considered.

Finally, it was asked whether the LAG looked favourably on forestry-related projects. Respondents (beneficiaries and LAG managers) rated Leader+ in the top two classes (Fig. 6.4).

6.4 Discussion and Conclusion

Overall, the Leader+ Initiative funded a relatively small number of forest-related projects as a proportion of the total number of projects in each country. From the four countries, Portugal and Austria are those where the percentage of forest-related projects was lowest. In the case of Portugal, from the applicants' point of view, the fact that it was the Ministry of Agriculture managing the programme may have led them not to consider Leader+ as a possible source of funding for forest-related projects. A personal communication (15 January 2010) from a member of staff of a Portuguese LAG stated that they would like to have had more applications on forest-related projects but that simply did not happen.

Table 6.6. Distinctiveness of Leader in terms of the support given[a] (number of answers).

Grade	1	2	3	4	5	
Less personalized	0	0	2	5	24	More personalized
Less helpful	1	0	3	5	22	More helpful
Less local/accessible	1	1	2	2	25	More local/accessible

[a]Answers given by the Local Action Group managers or beneficiaries.

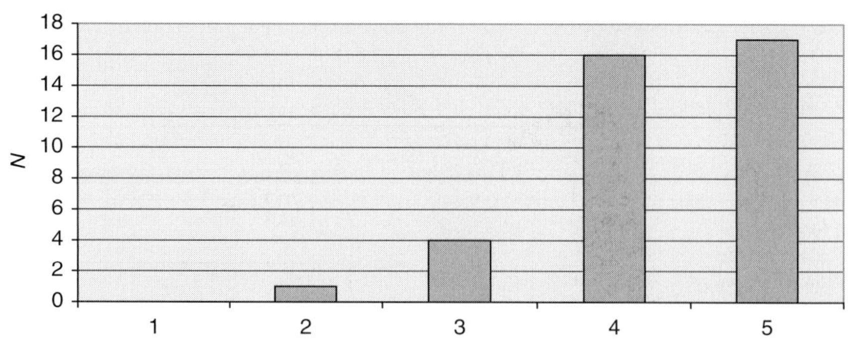

Fig. 6.4. The extent to which the LAG is seen to look favourably on forest-related projects (N = number of answers). 1, It did not favour forest project at all; 5, it was very favourably disposed to forestry projects.

Interestingly in Finland, Leader+ funded mainly recreation activities and projects focused on non-wood forest products. However, it must be noted that given the nature of the Leader+ programme in general, the larger private forest or wood processing companies were not part of the target groups for the programme. Traditionally, other well-developed national policy support measures and programmes (e.g. National Forest Policy, the Finnish Funding Agency for Technology and Innovation (Tekes) funding) have been used to support the traditional 'mainstream' forest sector. Leader and Leader-like funding have been used mainly in the development of the rural communities and in the diversification of rural micro-businesses rather than in larger-scale industrial development.

In Portugal, with the third highest percentage of forest cover after Finland and Austria, the number of forest-related projects funded by Leader+ was considerable, even though the percentage of forest-related projects funded in relation to the total number of projects was low. Examples of forest-related projects concern the dissemination of information about fighting and prevention of forest fires and the minimization of the risk of forest fires. These actions would be needed anyway given the high risk of forest fires in Portugal, but Leader+ may have been the only source of funding available to support them. It is not a coincidence that the forest owners' organizations were the principal bodies implementing these actions. These organizations emerged in the 1990s and Leader+ was probably the most available source of funding to buy equipment and to promote the forest services provided by them. These projects may not be considered innovative but at the time represented a new concept of forest management and advice to forest owners, especially in north and central Portugal. Other common forest-related projects were the establishment of montado fairs and seminars in the south. The reason for these projects was the great importance of the montado ecosystem for the cork sector and for conservation purposes. Portugal is the world leader in the production, processing and exporting of cork and the montado is highly valued for sheltering diverse game animals and protected birds and mammals.

A significant number of the Scottish projects related to conservation rather than the productive management of woodlands for timber. It appears that conservation bodies were better able to exploit the funding opportunities created by Leader+ than forestry organizations.

In Austria, the small amount of forest-related projects may be explained by the fact that the Leader+ programme was implemented in the policy domain of 'regional development' and not 'rural development' (Weiss, 2005; Weiss *et al.*, 2007). This implied that the agricultural administration was not involved in the implementation of the programme. As a consequence, agricultural and forestry actors hardly knew about this funding possibility. In the current programming period, the mainstreamed Leader programme is under the Ministry of Agriculture, Forestry, Environment and Water management and it is expected that agricultural and forestry interests are more strongly considered. The typical projects that can be seen as examples for forest-related projects in Leader+ are the LAG 'Sauwald' with their project 'Baumkronenweg' (Canopy Walkway) as an example for territorial services, and the LAG 'Murau' with the project 'Holzwelt Austria/ Holzzeit' (Wood World Austria/Wooden Times). From all 25 forest-related projects, 16 were implemented by this exclusively wood-related LAG.

From the analysis of the responses of the questionnaires some general conclusions can be drawn:

1. Leader+ supported the creation of new products, the development of new processes and the creation of new networks but it is not clear whether Leader+ supported especially novel and innovative types of projects. Respondents valued the funding and application support, but not particularly the risk-orientation or the support of non-mainstream ideas. Therefore, the risk-friendly and innovation-oriented character of the Leader approach is not strongly supported by our findings.

2. Leader+ did have an effect on rural employment. There is evidence that full, part-time and seasonal jobs were created, and the overwhelming majority of these jobs were sustained beyond the end of the project.

3. Leader+ triggered the implementation of forest-related projects that built on local actors' ideas to add something new to an existing project or scheme. This confirms that innovation in some forest-related projects in rural areas may imply little more than the transfer and adaptation of innovations developed elsewhere or the modernization of traditional forms of know-how.

4. In relation to the distinctiveness of Leader+, most respondents asserted that it was more personalized, more helpful and more local/accessible than other programmes. This confirms that Leader+ was implemented in a way that supported bottom-up initiatives, a feature that is seen important for integrated rural development.

5. From the answers to the questionnaires, it is not clear whether Leader+ was an important and alternative source of funding for forest-related projects. Most of the respondents asserted that the project would not have gone ahead without Leader+ funding, this suggesting that the type of projects supported fell outside the aims of other funding programmes. But most of them did not seek funding for their projects from other public organizations and this may be evidence that they were not expecting their projects to be funded by other programmes. From the applicants that sought funding from other public organizations, most were successful, suggesting that Leader+ was not the only source of funding for the type of forest-related projects considered.

6. Although the percentage of forest-related projects funded by Leader+ in the four countries was low, most of the respondents considered that Leader+ looked very favourably on forest-related projects. This may reflect respondent bias since both beneficiaries and project managers would be expected to rate Leader's impacts highly. A deeper look into the responses revealed that one member of the staff of a Portuguese LAG rated only three out of five for this question and that in Finland responses to this question ranged from two to five out of five, indicating a degree of willingness to criticize the LAG for its non-engagement with forestry. To draw a more robust conclusion, questionnaires should have also been sent to applicants who did not get their forest-related projects approved and an assessment of the number of forest-related projects not approved should have been made.

The various components of the Leader process fit comfortably with projects relating to forestry, especially when building on forestry projects that are outside the mainstream wood supply chains. The general principles of the Leader approach – the partnership building with local actors, the locally grounded plan, the territorial approach and the creation of network initiatives involving a range of actors connect naturally and easily to the development of territorial goods and services relating to forestry. Leader+ thus emerges as a modest but useful policy initiative for supporting new forest institutions and a range of forest-related actions in the development of non-wood forest products, value-added products and environmental services. There is evidence from the respondents that the Leader approach was valued by beneficiaries and, in some cases, that the new networks created by Leader lasted beyond the duration of funding. These results provide some vindication for the Leader approach and its mainstreaming in the 2007–2013 funding period, although it seems likely that Leader's funds will tend to support projects outside the core wood supply chain and large-scale forest industry and support development more in the field of territorial goods and services rather than wood products. Whether this represents a customization of different funding sources to different facets of forest-related innovation, or an undue emphasis on territorial goods and services cannot be ascertained without further research. What is clear is that the Leader process is valued by its recipients and those who animate it, but has limited reach into the forest sector as a whole.

From a policy perspective, it is important that forest owners or those in partnership with forest owners in wider development projects are fully aware of Leader's potential to provide support to rural development projects. We recommend that the trade press disseminate information, ideally by highlighting successful Leader projects, and that project officers and LAGs make efforts to engage with the forestry community. There is no doubt that many network projects of the type favoured by Leader involve forest owners. What remains less certain is that the forest owners' networks of knowledge connect sufficiently close to the Leader initiatives.

Notes

[1] *Natura* 2000: It is an ecological network of protected areas in the territory of the EU.
[2] OECD classification of rural areas: *Rural community*: When the population density is under 150 inhabitants. *Predominantly rural regions* (PR): >50% of the population live in rural communities. *Significantly rural regions* (SR): 15–50% of the population live in rural areas. *Predominantly urban regions* (PU): <15% of the population live in rural communities.
[3] Portugal: http://www.Leader.pt/; Scotland: http://ukLeader.org.uk/; Austria: http://archiv.Leader-austria.at/network/projects.html; Finland: http://www.lande2000.fi/.
[4] 'Montado' – an ecosystem characteristic of southern Portugal, which combines agriculture, grazing and forestry. The main tree species composing this system are Cork oak and Holm oak.

References

Elands, B.H.M., O'Leary, T.N., Boerwinkel, H.W.J. and Wiersum, K.F. (2004) Forests as a mirror of rural conditions; local views on the role of forests across Europe. *Forest Policy and Economics* 6, 469–482.

ELARD (European LEADER Association for Rural Development) (2011) The Innovative Approach. Available at: http://www.elard.eu/en/innovativeapproach (accessed 5 January 2011).

European Commission (1994) Leader II Guidelines, 94/C 180/12 (online). Available at: http://www.abdn.ac.uk/irr/arkleton/documents/sweden5.pdf (accessed 18 February 2010).

European Communities Commission (1988) The Future of Rural Society – The Commission communication transmitted to the Council and to the European Parliament on 29 July 1988 (online). Available at: http://aei.pitt.edu/5214/01/001713_1.pdf (accessed 18 February 2010).

FAO (2000) Development of national-level criteria and indicators for the sustainable management of dry forests of Asia: workshop report. Bhopal, India. 30 November–3 December 1999 (online). Available at: http://www.fao.org/docrep/003/x6896e/x6896e0e.htm (accessed 20 December 2009).

Finnish Statistical Yearbook of Forestry (2008) *Official Statistics of Finland-agriculture, forestry and fishery*. Finnish Forest Research Institute, Vantaa.

Forest Europe Growing Life (2009) Ministerial Conference on the Protection of Forests in Europe. Vienna 2003 (online). Available at: http://www.mcpfe.org/eng/Commitments/Ministerial_Conferences/Vienna_2003/ (accessed 05 January 2010).

Gluck, P. and Weiss, G. (eds) (1996) Forestry in the context of rural development: future research needs. *EFI Proceedings No. 15, European Forest Institute*, Joensuu.

Hyttinen, P. (1994) *Metsatalouden Kannattavuus*. Tapion taskukirja. Gummerus, Jyvaskyla, pp. 428–435.

Hyttinen, P., Ottisch, A., Pelli, P. and Niskanen, A. (1999) Forest related resources, industries services and know-how in the border regions of the European Union. EFI Working Paper 2, European Forest Institute, Joensuu.

Hyttinen, P., Niskanen, A., Ottisch, A., Tykkyläinen, M. and Väyrynen, J. (2002) Forest related perspectives for regional development in Europe. European Forest Institute Research Report No. 13. Brill, Leiden.

Jumppanen, A., Törmä, H., Zawalinska, K., Kolehmainen, J., Pulkkinen, R., Korteniemi, J., Pohjola, T. and Latvala, M. (2008) Maaseudun kehittämisohjelmien teema-arviointi 2000–2006. Maa- ja metsätalousministeriön julkaisuja 7/2008.

Koch, N.E. and Rasmussen, J.N. (1998) *Forestry in the Context of Rural Development: Final Report of COST Action E3*. Danish Forest and Landscape Research Institute, Horsholm.

Mendes, A.M.S.C. and Feliciano, D. (2005) Portugal. In: *Acta Silvatica and Lignaria Hungarica: special edition COST E30 Economic integration of urban consumers' demand and rural forestry production. Forest sector entrepreneurship in Europe: country studies*, Volume II, pp. 555–642.

OECD (2008) OECD rural policy reviews. Scotland, UK Assessment and Recommendations (online). Available at: http://www.scotland.gov.uk/Resource/Doc/212557/0056531.pdf (accessed 21 December 2009).

Rametsteiner, E., Aldrian, A., Bauer, A., Eberl, W., Sekot, W., Wagner, S. and Weiss, G. (2005) Austria. In: *Acta Silvatica and Lignaria Hungarica: special edition COST E30 Economic integration of urban consumers' demand and rural forestry production. Forest sector entrepreneurship in Europe: country studies*, Volume I, pp. 17–70.

Ray, C. (1998) Territory, structures and interpretation – two case studies of the European Union's LEADER I Programme. *Journal of Rural Studies* 14, 79–87.

Rogers, E.M. (1995) *Diffusion of Innovations*, 4th edn. The Free Press, New York.

Serafim, M.R. (1999) *Outras mediações Estado/sociedade, as parcerias no programa Leader.* DGDR, Col. Estudos e Analises, Lisbon.

Slee, B. and Wiersum, K.F. (2001) New opportunities for forest-related rural development in industrialized countries. *Forest Policy and Economics* 3, 1–4.

Slee, B., Ingram, J., Cooper, R., Martin, S. and Wong, J. (2005) The United Kingdom. In: *Acta Silvatica and Lignaria Hungarica: special edition COST E30 Economic integration of urban consumers' demand and rural forestry production. Forest sector entrepreneurship in Europe: country studies*, Volume I, pp. 725–776.

Weiss, G. (2005) Tanno – a brand for regional fir products. In: Rametsteiner, E., Weiss, G. and Kubeczko, K. (eds) Innovation and entrepreneurship in forestry in central Europe. EFI Research Report 19, European Forest Institute, Joensuu, pp. 131–132.

Weiss, G., Martin, S., Matilainen, A., Vennesland, B., Nastase, C., Nybakk, E. and Bouriaud, L. (2007) Innovation processes in forest-related recreation services: the role of public and private resources in different institutional backgrounds. *Small Scale Forestry* 6, 423–442.

7 How to Support Firm Competitiveness in Timber Industries? Clusters as Policy Means in Four European Countries

Thomas Rimmler,* Roger Coppock, Roland Oberwimmer, Andreja Pirc, Stjepan Posavec and Gerhard Weiss

Abstract

The cluster approach has been used as a guiding framework for policy action, which, by definition, nurtures the existing strengths and growth potential of functionally interrelated and geographically co-located groups of specialized firms cross-sectorally linked by technologies, markets or value chains. The cluster approach has been used to define cluster-based policy agendas in innovation policy and traditional policy fields. The approach offers a wider view on the needs and benefits of improving coordination across policy sectors. The cluster concept has been used widely to reveal and tackle the needs to improve coordination between actions, which fall under the remit of different policy sectors. Cluster promotion policies are often delivered by discrete action singled out in particular action programmes, which are targeted on pre-selected functionally and geographically delimited groups of firms and industries. We report on the following discrete cluster promotion activities in European timber industries: the Finnish Living Cluster 'Lahden alueen asumisklusteri', the national Scottish 'Forest Industries Cluster', the Croatian North-West Timber Cluster 'Drvni klaster zapadne Hrvatske', and the Styrian Wood Cluster 'Holzcluster Steiermark'. Key findings are that cluster promotion, instead of covering all aspects of competitiveness, is being focused towards selected fields of strategic priority. Cluster strategies are clearly differentiated in this respect. Cluster promotion activities are targeted narrowly on corporate business network formation and management or more broadly on reshaping industrial districts into innovation-enabling business environments. Strategies may be mixed both to ensure immediate economic benefits to firms and to develop the cluster's innovative capacity and growth potential in the medium and long term. Clusters operate at the national or regional level. Cluster strategies are more or less purposely and formally integrated and aligned with regional development strategies. The recommendations for how to improve policy efficacy and efficiency include catalytic funding, exploiting value chain linkages for leverage effects, improved strategic alignment and replacing 'wishful thinking' by formal analysis of real cluster potentials.

7.1 Introduction

7.1.1 A cluster approach to firm competitiveness

The cluster approach has been used as a guiding framework for policy action, which nurtures the existing strengths and growth potential of geographic concentrations of specialized firms in adjacent industries vertically or cross-sectorally linked by complementarities and interdependencies based on related technologies, markets or value chain processes.

Single firms or industries are seldom isolated but interlinked with and highly

* Corresponding author.

dependent on their business environment. The cluster approach recognizes this dependency and suggests that a policy has to widen scope of action from individual firms and industries towards interlinked groups of firms and industries. From the viewpoint of policy action, an understanding of these linkages is a precondition for synergy potentials to be revealed and captured. Porter's diamond model has been offered among a multitude of other approaches as a framework for observations about the business environment of groups of firms and industries in terms of synergetic value chain linkages across different industries, demand and factor conditions, and firm strategies with a major impact on their competitiveness (Porter 1990/98, 2001). While the diamond model provides a description of a cluster in terms of value chain transactions for the production of similar products and production-related factor conditions, in a cluster approach to innovation, essential linkages are those by which information and knowledge is interchanged between actors.

Cluster policy – or cluster-based policy[1] – is a notation used to separate a cluster-based approach to redefining policy agendas in traditional policy fields. Compared with a sectoral approach the cluster approach offers a wider view on the needs and benefits of improving coordination and leverage of policies in different fields. A cluster-based policy may be defined with different ends and targets for the different policy areas concerned. Clusters and cluster policies may be defined on different levels or by referring to different aspects of cluster development: a cluster as a policy target may be defined on a macro-, meso- or micro-level, or by its thematic, technological, sectoral or regional focus. The respective policy agendas as well as its leading actors accordingly may be assigned on different levels to different policy fields, namely innovation policy, technology policy, small and medium-sized enterprise (SME) policy, industry policy or regional policy. Policy actions in these different policy fields usually are not mutually exclusive but being properly designed and coordinated complement and reinforce one another (Gilsing, 2000; Roelandt *et al.*, 2000; Nauwelaers, 2003; Andersson *et al.*, 2004; Reid and Peter, 2008).

In the context of innovation policy, the cluster concept has been instrumentalized as a guiding framework for a policy, which aims at creating and strengthening environments conducive to innovation. Because of the perceived importance of the business environment for the innovativeness especially of smaller firms, cluster formation has been the focus of a policy characterized as a cluster approach to innovation policy (Raines, 2001; Nauwelaers, 2003).

Cluster policy, because it cuts across various sectoral policies and industries, requires coordination of action between different sectors and on different levels. Whereas policies conducive to a favourable institutional environment are usually the responsibility of national governments, policies concerning supportive physical, communication and knowledge infrastructures and related strategies for action are usually the domain of both national and regional governments. On the regional level, action is led by regional authorities and implemented within the scope of regional development policy (Raines, 2001; Fromhold-Eisebith and Eisebith, 2005). For a cluster policy to be successful, it must support and be coordinated with strategies of promoting firms' innovative capacity on the local level (den Hertog *et al.*, 1995). On this level, among the main tasks of cluster policy are facilitating inter-firm networks and strengthening small-scale industrial districts or strategic micro-clusters. In practice, these tasks are implemented either by governmental or private industrial actors. In both cases, activities are explicitly focused on discrete groups of firms singled out for special policy attention (Bergman and Feser, 1999; Fromhold-Eisebith and Eisebith, 2005; Feser, 2008). Discrete regional or local level cluster promotion activities, which make explicit reference to the cluster concept, are often delivered by discrete action singled out in particular action programmes. These are targeted on pre-selected localized groups of industries and led by cluster initiatives (CIs; Sölvell *et al.*, 2003).

7.1.2 Localized clusters and cluster promotion

A cluster contributes to firms' competitiveness by providing material as well as immaterial benefits (Storper, 1995, 1997; Steiner and Hartmann, 2002; Martin and Sunley, 2003). By being co-located, firms derive material benefits from transportation and transaction cost savings. Being characterized by specialized factor conditions, a cluster offers firms specialized inputs and skilled labour, which both contribute to productivity gains. These advantages are generated by collective infrastructure services or realized by buyer–seller market transactions. A cluster supports especially firms' innovative capacity in terms of special competences available in the fields of research, training and business services (Rosenfeld, 1997).

For the flow of information and knowledge, linkages other than those related to formal market transactions may be important. Immaterial benefits may not be transmitted in formal business transactions but may be the results of learning processes, which are unintended and informal. Fostering these processes helps to exploit experience-based, tacit knowledge for continuous improvements. Learning processes may otherwise be related to labour mobility, corporate spin-offs and entrepreneurship. Knowledge may be exchanged also non-interactively by observations, which may be facilitated by spatial proximity and social contacts (Granovetter, 1983; Asheim and Isaksen, 2002; Harmaakorpi, 2004; Haahti *et al.*, 2005; Rocha and Sternberg, 2005; Malmberg and Maskell, 2006).

It is argued that it is all these learning processes that provide firms located in a cluster with the advantage to exploit knowledge more rapidly than competitors (Eisenhardt and Martin, 2000). Action aimed to strengthen the innovative capacity of firms, therefore, must include collaboration, trust-building and enhanced communication (Sölvell *et al.*, 2003). Action, however, can rarely be confined exclusively to supporting the emergence of unintended and spontaneous learning but must cover also conscious action to initiate and intensify learning processes. Public knowledge producers such as universities, research institutes and bridging agents can provide major contributions to these processes.

The innovative performance of single firms is determined by factors internal to the firm but also by factors located outside firm boundaries. Firms derive advantage from easy access to specialized resources and from collaboration with co-located and functionally related businesses. Localized learning in contrast to learning between different locations not only reduces transaction cost of knowledge exchange but allows knowledge to be shared, which is tacit and which requires direct interaction (Cohen and Levinthal, 1990).

Small firms are characterized as being highly dependent on external resources and therefore as incomplete economic agents. These firms also typically show narrow networking patterns, which make them highly dependent on resources located in their immediate proximity. In fact, the quality of firms' business environment has a most important impact on their innovative capacity (Oinas, 1998; Capaldo *et al.*, 2004).

Innovation support measures, especially those in response to needs of small firms, have to be based on a broad understanding of innovation, and to comprise other than technological knowledge and competencies. Accordingly, a cluster provides further support for actions aimed to reinforce small businesses' managerial capabilities. The respective actions comprise business planning, developing skills and diffusion of good practices in management and work organization (Arnold and Thuriaux, 1997; Bender and Laestadius, 2005; Tanvig *et al.*, 2005; Hirsch-Kreinsen, 2006; Virkkala and Niemi, 2006; STRAT.AT Plus, 2007–2013).

Small firms often lack the capabilities that allow them to access resources and capabilities located outside firm borders. Therefore, a cluster has to provide 'bridging' structures, which help them find and use appropriate knowledge. The respective discrete actions comprise screening and processing of relevant outside information, brokerage and matching services that facilitate access to research and development (R&D) services, as well as cooperative business network building (Lester and Sotarauta, 2007).

Network building in terms of strong inter-firm ties allows the management of the risks arising from sharing strategic knowledge and making considerable financial commitments to innovation. Cooperative network facilitation is among the major activities of CIs. The major challenge is to overcome the hindrances to cooperation. These are created by deficiencies in skills, trust and the awareness of business opportunities (Boekholt and Thuriaux, 1999; Roelandt and den Hertog, 1999; Sölvell *et al.*, 2003).

A strategy that is aimed at enhancing the local business environment of firms includes the upgrading of local factor conditions with respect to human capital or other advanced input factors. Actions may also be targeted at attracting inward investment by industrial place marketing, developing local anchor firms and promoting new business formation (Salmi *et al.*, 2001; Andersson *et al.*, 2004). Investments of firms that locate or expand their activities in the cluster may initiate catalytic effects, which are channelled by production and knowledge flow linkages. These firms may have an excellent capacity to absorb scientific knowledge and therefore can function as technological gatekeepers.

Investment in local knowledge infrastructure constitutes another strategy to eliminate structural weaknesses of a local business environment (Salmi *et al.*, 2001). In the case of 'less attractive' peripheral regions because of size and resource restrictions, strategies to fill in local 'supply' gaps have to comprise tapping of external knowledge by cooperation with extra-regional actors, such as universities, research and educational institutions.

In the system approach, innovation is the outcome of an institutionally distributed process, which involves the activities of many independent actors. Viewed from a system perspective, therefore, innovation performance most importantly constitutes a coordination problem (Schienstock and Hämäläinen, 2001; Koch *et al.*, 2003). Public agents are usually entrusted with the coordination and alignment of activities in research, education and training, and the development of the respective infrastructures by steering processes of priority setting (Georghiou *et al.*, 2003). Most important for localized clusters are processes on the regional and subregional level. CIs may participate in these processes as the representatives of small businesses. The dialogue between industry, knowledge producers and governmental agents ensures that actions for business environment upgrading are aligned by a commonly agreed strategy and that mismatch between supply and business needs can be avoided (Morgan, 1995; Leydesdorff and Etzkowitz, 1998; Smedlund *et al.*, 2005; Kuitunen and Niinikoski, 2008).

7.2 Cluster Policy and Promotion Practices in Four European Countries

7.2.1 Croatia

Cluster policy

Clusters have been a key priority area for a policy of promoting the competitiveness of the national economy (see Clusters – alliance to success, 19 April 2006, http://www.mingorp.hr). The political foundations for institutionalized cluster support had been laid in 2003 with the Croatian Competitiveness Initiative. Clusters were nominated as priority areas, in both terms of specific fields of competences and their fields of application. The timber industry was nominated as sector of outstanding national concern. Since then the national government has adopted a leading role in developing framework conditions conducive for innovation and for getting positive dynamics on the way in key industrial sectors. CIs have been launched as means to enhance strategy-led interaction between the knowledge sector – the education and research institutions – and the respective industries. Concerning the timber industry the policy target has been to reshape industrial districts into innovation-enabling business environments. CIs are being used as bridging agents between the industries and the other institutional sectors.

For conformity with the new cluster-based approach, appropriate changes of the national policy context were codified lately by a number of agreements and programmes.[2] Institutional adjustments were continued by setting up an industry-led 'National Center for Clusters' with the tasks to coordinate cluster formation and to recruit partners especially for designing their marketing functions (Ministry of Economy, Labour and Entrepreneurship, 2007). Croatia has also joined cross-national policy initiatives for defining innovation and cluster policies suitable under conditions of central and eastern Europe (CEE) countries.[3]

As a consequence of adopting the cluster-based policy approach, actors in different institutional spheres predominantly responsible for shaping favourable cluster environments have increasingly started to embark the process.[4] Clusters have started to develop into platforms for closer interaction between actors in governmental, industry and the educational and research sectors on sectoral, regional and local levels (National Competitiveness Council, 2004). New cross-institutional structures – such as CIs, incubators and regional development agencies (RDAs) – have been established and assigned to operate at the interfaces between different institutional sectors and policy levels as mechanisms for the deepening of respective policy integration.

A number of CIs were established in the key industrial sectors of the Croatian economy. Concerning clusters in the timber industry, the first discrete localized cluster promotion activities were launched in 2003. The cluster projects were initiated by the United States Agency for International Develpment (USAID), a common pattern in transition economies. In 2009, there were seven active clusters related to primary and further timber processing and furniture manufacturing.[5] The largest among these clusters – more accurately, strategic corporate networks – comprises 24 member companies. The dominant motives for cooperation besides marketing have been joint R&D, vocational training and joint input procurement. Base funding for establishing the clusters was provided by firm members in two out of seven cases, while the rest received finance out of the state budget.

The North-West Timber Cluster 'Drvni klaster zapadne Hrvatske'

The cluster started operations as a strategically managed and formally organized cooperative network of independent firms in 2006. During the last few years, a regional centre for incubators, free zones and development agencies was established. Being set up as a public/private partnership, the cluster encompasses ten private companies and four public support organizations.[6]

The cluster is vertically differentiated, as it comprises processing of semi-final and final timber products made from locally sourced raw materials (solid oak, cherry, walnut, maple and ash). Its core products comprise furniture and interior design products for different private and public customer groups.[7]

The focal points of cluster promotion activities are developing and marketing of new products for the cluster's pre-selected market segments. Supportive activities are research and product development, market intelligence and sales promotion.[8] The cluster has been committed also to transnational cooperation in order to tap into state-of-the-art external knowledge that supports its competitive strategy and enhances firms' innovative capacity in the field of product design and thereby their ability to develop and bring higher valued products to the market.[9] This indicates, first, a shift from a competitive strategy of merely exploiting low factor costs towards stronger emphasis on product customization and innovation. It indicates further a redefinition of environmental conditions of strategic importance for the cluster, which formerly were limited to basic factor endowments, towards more advanced knowledge-based competences. As a consequence, the cluster is going to strengthen research–industry interaction.[10]

7.2.2 Scotland

Cluster policy

In the UK, cluster policy was proclaimed and legitimated in the late 1990s as an overarching approach for innovation and technology policy, regional development and enterprise policy, especially concerning SMEs.[11] This new approach in UK economic policy was based on the earlier Scottish experience.[12] In 2000, the Scottish forest industries were defined as one among seven key industrial sectors in the Scottish Government's Economic Strategy.

The decision was preceded by preparatory work of mapping competitive strengths and opportunities and developing a cluster strategy and a 4-year action plan. In 1998, Scottish Enterprise,[13] working with the Forestry Commission, set out a strategy for the realization of the major opportunity presented by the doubling in the domestic wood fibre supply over the next few decades.

The cluster strategy was updated[14] and a second, more comprehensive, strategy paper named 'Roots for Growth' presented 2 years later. The opportunities identified now were much wider, comprising also sawmilling and value-added engineered timber products, panel board manufacture and their downstream value-added operations. This comprehensive strategy framework was targeted towards integration to 'what has been traditionally a somewhat conservative and disjointed industrial sector'. The paper set out a strategy and action plan for the national forest industries over a 5-year period. The first activities under the new strategy were launched end of 2000 and the first conference reporting on achievements was held in December 2001.

Scotland's economic development policy is implemented on a regional level by 22 regional development agencies (LECs) publicly installed. These actors form part of the two enterprise networks, Scottish Enterprise and Highlands and Islands Enterprise. Although the actors responsible for regional development can therefore be identified, the question remains if and how far the notion of 'cluster initiative' in the sense of managed or 'organized' regional clusters (Sölvell *et al.*, 2003) is appropriate for describing Scottish cluster policies and how regional clusters are linked with regional development policy. Admittedly, in setting up the cluster strategy, an issue of secondary importance was the rural dimension, despite the opportunities for promoting rural development, given the large concentration of forestry employment in rural locations.

The Scottish forest industries cluster

The cluster was initiated in 2001 and Scottish Enterprise was entrusted with setting up and operating the cluster. The cluster comprises the Scotland (NUTS1) region of the UK. Its functional coverage extends across national borders, because of production sites of large firms being located outside the Scottish borders and because there are large production facilities within national borders, which are strongly linked with externally based owner companies.

Cluster firms are geographically dispersed, in the sense that firms that are more strongly functionally interrelated are scattered across different locations rather than co-located in specialized regions. The cluster is functionally differentiated across and along different value chains: its companies range from forest owners and forest managing agents to primary processors, and secondary processors in the pulp and paper, sawmilling and panel sector.

The cluster comprises a large number of small firms, with very few large national and transnational players. There are around 385 companies in the Scottish forestry sector, of which only seven are classed as large businesses. Most are SMEs or microbusinesses employing fewer than 25 staff. The large businesses and those at the larger end of the SMEs, which have been most actively engaged with the cluster, account for 60–70% of the timber processed. Besides transnational players, this group comprises indigenous sawmillers and forest management companies, all located at the upper end of the value chain.[15]

A most important objective and activity area has been to upgrade the cluster's knowledge base by cross-sectoral cooperation. An early gap in provision of support for business was the need to reinforce the strategically relevant knowledge base of the cluster and to strengthen the linkages between science and the industries in specific fields of knowledge. Consequentially, with help from industry, the RDAs, the Forestry Commission and Napier University, the Centre for Timber Engineering (CTE) was established as a public–private partnership in the strategic field of knowledge about the wood and timber properties of Sitka spruce as the main commercial forest tree species. Scottish Enterprise encourages knowledge transfer and commercialization through a range of mechanisms. Scottish Enterprise, for example, runs a series of R&D programmes that serve as avenues for technology transfer.

Another of the main roles of the cluster is to provide networking opportunities through conferences, seminars, exhibitions and other media, and thereby to create social capital, enable information exchange and promote cooperation between firms. Collaboration enables firms to make use of knowledge that would not be otherwise accessible. Other important objectives and fields of cluster activities are to facilitate dialogue with policy makers, exert influence on policy making, develop governance and funding schemes for improved firms' responsiveness, and initiate joint efforts with authorities for improved infrastructure services especially in public transportation.

7.2.3 Finland

Cluster policy

Clustering is in the strategic focus of Finnish regional development policy in the sense of policy action aiming at strong and specialized regional growth centres. These centres are hubs of clusters of expertise and business activities containing a strong research community, internationally successful companies, a labour market offering a specialized workforce and a supportive public service infrastructure. Specialized growth centres are the outcome of an innovation-based regionalized industrial policy aiming to create innovation-enabling localized framework conditions for economic growth. These specialized growth centres are located in mid-sized and large city regions.

Finnish regions enjoy far-reaching regional autonomy. To comply with their responsibilities for a coordinated and bottom-up approach, the regions set up on a regular basis and as a joint effort of regional stakeholders a regional business strategy, which is a medium-term plan (4–6 years) endorsed by the municipalities as a common guide for developing the regional industries. In this context, industrial clusters are used as concepts to define and – concerning resource allocation decisions – prioritize focus areas of regional development strategies. These strategies define the targets and tasks of strategic partnerships between central, regional and local governmental agencies, venture capital investors, public–private organizations, the industry, and the higher education and research sectors as stakeholders.

Many features of a regionalized cluster approach to innovation policy can be observed in government-led regional development strategies.[16] The two main policy instruments applied to foster cluster-based regional development are the Centre of Expertise Programme (CoEP) and the Regional Centre Programme (RCP). There are 21 regional centres of expertise approved for the period 2007–2013 to participate in the CoEP. These units are assigned by their special fields of expertise to one or several of 13 nationally significant competence clusters (http://www.oske.net/en/competence_clusters/). Among these clusters, there are two with a special focus on timber products manufacturing. The regional centres of expertise, being the operational units of these knowledge-based clusters, are located in mid-sized and large city regions. They cover all the NUTS3 regions on the Finnish mainland (http://en.wikipedia.org/wiki/Regions_of_Finland). The programme functions as a means of multi-level integration of regional policies, central government and sector policies.

The RCP is an instrument on the subregional level (NUTS4) for turning mid-sized cities and small provincial towns into hubs for the development of new industrial districts located outside strong growth centres. Their industrial base consists mainly of small enterprises operating in local markets and only of few anchor firms operating in export markets. Based on the set of already established actors, the RCP aims at upgrading the subregional 'knowledge infrastructure' by external complementary expertise. Whereas sectoral policy coordination is the target of so-called broad regional policy, the RCP is an instrument of narrow regional policy aiming at strategies and measures tailored to the situation of specific regions.

For both programmes, the national government by their sectoral ministries provides regionally co-financed basic funding.[17] Funding covers the cost of running cooperative networks, of upgrading organizational competences and of processes related to the formulation of regional development strategies and action programmes. The implementation of these action programmes requires separate project funding, which is provided by various sources including state budget, venture capital agencies, municipal and regional sources, and EU structural funds.

The Lahti region Living Cluster

The cluster was established in 2007 and the Lahti Region Development Agency was assigned as cluster facilitator. Starting from 2009, responsibilities were consigned to the incubator Lahti Science and Business Park Ltd. The cluster is a traditional industrial cluster located in the NUTS4 region of Lahti City (http://en.wikipedia.org/wiki/Lahti). Twenty-four per cent of the Finnish furniture industry, and construction companies, which are amongst the largest in Scandinavia, are located in the region. The cluster comprises affiliated industries in modern timber-framed building construction, builders' carpentry and joinery, and structural timber products manufacture. The cluster comprises a large number of small firms and a smaller group of large export-oriented companies, which create the major share of value-added and account for most of the engagement with the cluster.

There are no barriers to cluster membership nor any formal registration required to access 'club goods'. The cluster has a catalyst role by steering the funding decisions of third parties. The strategy for developing the cluster is coordinated with the strategies in different policy fields. The respective decisions are targeted at the reinforcement of regional actors' competences in relevant fields and at filling gaps in the regional knowledge infrastructure. Based on commonly agreed and coordinated strategies, higher education institutions are committed to strengthen the supply of education, training and innovation services in the fields of strategic importance for the cluster. The national CoEP supports the strategies for developing the key business clusters in the region.

The innovation services that are organized by the cluster include tools, which are tailored for different groups of firms on the basis of their innovative capacity. On the one hand, they target firms with the role of spearheads for the implementation of strategies, which are geared to the rejuvenation of the cluster. On the other hand, they comprise measures aimed more broadly to improve firms' business performance in the short and medium term. Whereas these measures respond to the needs to rapidly solve firms' practical problems, but also to explore and exploit new ideas, spearhead projects should contribute to path-breaking innovations.

7.2.4 Austria

Cluster policy

Clusters are an important element of Austrian innovation and industrial policy, by which the country intends to follow up the renewed Lisbon strategy on growth and employment (Commission COM, 2005a,b), as reflected in the National Reform Programme (NRP; Austrian Reform Programme, 2008–2010), the National Strategic Reference Framework for Cohesion Policy (NSRF; STRAT.AT Plus, 2007–2013) and the regional cohesion policy programmes (Regional Development Programmes, 2007–2013). The Community's Strategic Guidelines for growth and jobs for the period 2005–2008 (see http://ec.europa.eu/archives/growthandjobs/pdf/integrated_guidelines_en.pdf) refer to business clusters as instruments to establish attractive framework conditions for enhanced competitiveness. The Austrian Strategic Reference Framework consequentially highlights clusters as a means to reinforce research and development (RTDI) activities and to actively develop industrial districts as business locations (Austria Cohesion Policy, 2007–2013). In developing a national cluster policy, Austria has reacted to the initiatives launched by the Community under the Seventh Framework Programme for Research and Technological Development (FP; http://cordis.europa.eu/fp7/home_en.html) and the Entrepreneurship and Innovation Programme (EIP; http://ec.europa.eu/cip/eip/index_en.htm).

In Austria, depending on the context, a cluster can be an industrial or technological cluster, a corporate network, economic complex, centre of excellence or technopole. With respect to the timber sector, the main policy focus is on the upgrading and repositioning of mature clusters. All existing Austrian industrial clusters (CIs) developed out of regional or local policy initiatives. 'Wood clusters', which have organized activities in seven out of nine provinces, are among the largest in terms of numbers of partner firms (Clement and Welbich-Macek, 2007; Clement *et al.*, 2009).

The main actors of CIs are provincial governments. As organized discrete cluster policy actions, they are targeting on a particular set of SME-dominated industries or businesses. At that level, the core issues of cluster promotion are how corporate networks are brokered, how firm-level activities are integrated in support programmes, how cooperation in knowledge transfer is intensified and how visibility in international markets is improved. Concerning regional development strategies and action programmes, the contribution of CIs is to involve the business sector in the process and to facilitate dialogue for improved coordination with research and other actors in relevant fields.

As public–private joint ventures, CIs rely on public as well as private sources of finance. Their regular operational budget originates from the public budget of the region (base funding) and from their private partners: chambers of commerce, industrial federations and trade associations. Regional funding is channelled through RDAs or their subsidiaries. Other sources are participation fees, sponsoring and compensation received for services. CIs receive national aid. Base funding has to be supplemented by project funding from different sources. Acquisition of third-party project funding is a major concern of CIs. Cluster advisory boards – to fulfil their function as an interface between the institutional spheres – represent governmental agencies, industries and the research sector, and are responsible for the strategic planning and coordination of cluster activities.

Aside from regional CIs, there have been other initiatives with the target to upgrade regional innovation systems and to foster knowledge-based cluster formation. The most important during the last decades have been those financed under various regional infrastructure promotion programmes, which provided federal financial support for the foundation of regional Competence Centres and Impulse Centres. These infrastructures offer tangible as well as intangible support for innovative companies. Impulse Centres fulfil technology transfer and incubator functions, offer business services and provide economic impulses for the region by industrial recruitment. Competence centres are platforms for industry-led collaborative R&D partnership projects with tasks ranging from project formation and management to knowledge transfer. The centres act as 'nodes of knowledge-based clusters' defined by a thematic rather than regional or sectoral focus. Being initiated and established in regions' fields of strengths, they support the implementation of the regions' innovation-driven development strategies. The initiatives have been usually triggered and co-financed by the provinces, partly involving the Community Structural Funds.

The Wood Cluster Styria

The CI is a strategically managed and formally organized cooperative network of small and medium-sized businesses in timber processing, timber products manufacture and affiliated industries and supporting organizations. The CI was established top-down as a formal institution liable to its shareholders for organizing cluster promotion activities. The declared goal was to foster linkages between firms in the timber industries and affiliated industries located upstream and downstream in their value chain. The cluster is liable for the implementation of the regional development strategy of the federal state of Styria for forestry and forest-based industries as identified and prioritized key areas of regional strength. Similarly to other regional clusters in Austria, the role of the CI is to act as an interface between industry, research and policy. In correspondence with this role, the CI's major tasks are: to intensify R&D activities and cooperation in the industries of its sectoral focus; to speed up research-based knowledge transfer, especially in high-tech fields, into marketable products (top-down); to link the supply of support with businesses' private efforts to develop and to bring to the market innovative business ideas (bottom-up); and to articulate the needs for policy action to develop innovation-enabling framework conditions for the specific needs of the cluster.

The CI was instigated by the provincial government and started operations in 2001. At that time the Styrian wood cluster contributed by about 30% to the total sales value of forest industries' Austria-based production. The industries continue to be one of the top job providers in the region. Preceded by a feasibility study that confirmed excellent prospects for growth, the cluster was prioritized as a key strategic area in regional economic development policy. The CI receives base funding to recover its fixed costs from cluster partners and also from public budgets. The other source is service fees. The cluster collaborates with research institutes and participates in research projects that are organized as multi-actor industry–science partnerships. It initiates research in fields of relevance for the cluster. Relevant are those with a significant commercial potential for the cluster in the medium and longer term. Other fields of cluster activities, besides research, are industrial recruitment and new business formation. The cluster provides comprehensive technical and business training. The cluster has started to establish cross-border collaboration focused on business partners and markets in the CEE region.

7.3 Analysis

7.3.1 Cluster approach – observations on the state of implementation

For Croatia, clusters constitute a previously inexperienced approach for which the final shape is still to be determined. There is insufficient information collected by using cluster mapping procedures or other documented methods for reliably identifying clusters and their potentials. This may feed 'wishful thinking'. The need for the new approach, none the less, is admitted.

Putting theory to practice is impaired by a number of things: compared with other EU economies, countries of transition tend to have weaker public institutions and a larger gap of trust between them and the private sector. The lack of common purpose, management experience and strategic alignment are among the reasons offered as an explanation that cluster performance often does not rise to expectations. The urgent need for education and training to ensure professionalism in cluster management seems not to be always recognized.

In Austria, clusters have been identified and prioritized as fields of regional strength in the federal states' economic development strategies. Clusters are well established with a legacy of success. It is reported that the clusters could be used better to intensify interaction between the national innovation system and the production sector.

7.3.2 Different levels and aspects of cluster policies

For the Scottish case, the notion of managed or 'organized' regional clusters seems inappropriate for describing the focus of cluster policy, which is characterized better as a mega- than a meso- or micro-approach. In contrast to the Finnish case, where clusters are used as concepts to define the focus areas of regional development strategies, for Scotland clusters are focus areas of national strategies, which are targeted at national key industries and their production chains. Concerning the location of their timber industries Scotland and Finland both exemplify peripheral regions, where a 'thin' local basis usually restricts the pursuit of full-fledged regional cluster strategies.

7.3.3 Institutionalizing cluster promotion

In the countries where regional clusters exist, cluster promotion takes different forms. In Croatia and Austria, cluster promotion is formally organized into legal entities, which are run by professional management teams. Strategic cluster management may also be entrusted to regional development companies – as in the Finnish case – or organized as regional public/private partnerships – as in the Austrian case – with partners sharing responsibilities. This may be advantageous, namely for the coordination and integration of industry and regional policies. The Scottish case is an example of a national development agency, which operates through a network of local companies. Cluster promotion, instead of broadly covering all aspects of competiveness, is being oriented towards fields of strategic priority.

7.3.4 National coordination of regional strategies

In Finland and Austria, where cluster policies are characterized by a strong regional orientation, the challenge is to confine national support to regional focus areas qualified to develop sufficient critical mass by more rigorous national coordination across regions. In Austria, the issue has been taken up recently by the NRP and the Austrian Council for Research, Technology and Innovation (2010). Also, a national cluster platform has been installed as an instrument for correcting coordination failures.

Finnish innovation policy has already addressed the question of national coordination of regional strategies by its CoEP. Although successful within regions, the programme has not yet met the expectations for a more clarified division of labour between regions. In Finland, timber products are prioritized fields of regional strength in 18 out of 34 regions under the RCP. However, there have not been any clear linkages between these subregional cluster strategies and they seem to be weakly coordinated and aligned with each other. Besides, there is a large number of areas, which are not covered by any national cluster programme. To compensate for the deficiency, cooperation platforms bringing together the sector's national and regional stakeholders – such as the Centre for Timber Products and the Timber Finland Programme – were established.

For Scotland and Croatia, it is not obvious how the export-oriented strategies for the industrial sectors of national priority are aligned with regional development targets.

7.3.5 Integrating inward investment policies with regional development targets

A common target of cluster policy is to enhance the attractiveness of clusters as a business location. For a strategy that relies on inward investments to be sustained, however, attention must be paid to anchoring inward investments into the regional economy. Large externally based firms or their local subsidiaries may provide business opportunities to supplier networks of indigenous SMEs, if properly supported by supplier development programmes. Respective measures may consist of brokerage and matching services, on-site audits and the development of supplier upgrading schemes. Whereas the Scottish strategy is to attract large foreign investors, in Austria the focal point is on peer-to-peer producer networks of small indigenous firms, and in Finland on small firm production networks under the strategic and organizational leadership of domestic 'flagship' companies.

7.3.6 Variation in strategic orientation

In Scotland and Croatia, strategies continue to be targeted on maximizing the use of abundant forest timber growth in fields of traditional technologies and timber product markets although strategic focus is shifting towards knowledge-based success factors. In Finland, more recent strategies are targeted to rejuvenating the cluster at the fringes of its traditional core industries through path-breaking innovations. Spearhead projects cover issues that are cutting across different technologies and fields of competences in related industries. For this kind of strategy, innovation networks made up of highly innovative companies are the main driving forces ('spearheads') in the cluster. Production and distribution networks and marketing alliances are the focus of resource- and production-based strategies, as in the cases of Scotland and Croatia.

7.3.7 Upgrading local factor conditions

It is a common feature cutting across all cases that actions aimed at enhancing the cluster's knowledge base consider especially improving framework conditions with respect to human capital. A major policy focus in all cases is to upgrade local factor conditions by providing training facilities and to adjust training schemes to specific firm requirements. Moreover, action is targeted on upgrading and broadening the cluster's skill set, and on supporting firms' recruitment of skills.

7.3.8 Cross-border cluster cooperation

Concerning the international dimension of national cluster policy, it is argued that cluster strategies must put stronger focus on cross-border and transnational cooperation. Here, the initiative of six CEE regions to cooperate in the field of research and innovation under the leadership of the Styrian Wood Cluster Initiative can be mentioned. International cluster networks for the sharing of experience and best practices have been running for some time under the auspices of the European Cluster Alliance. In Finland, the Lahti region participates in one of these networks.

7.3.9 Fostering interaction and cooperation between firms

It is surely a valid assumption, also more largely than for the cases studied, that private firms hesitate to join a CI if they are requested for commitments of which the real benefits are not perceivable. Firms usually are not enthusiastic about cooperation unless they find themselves with their 'back against the wall'. Also the fear of free-riders may impede collective action. This may be expected especially in the case of top-down policy initiatives, because to preclude criticism of discrimination, these are committed to large membership instead of selective inclusion of partners.

7.4 Key Findings and Policy Recommendations

7.4.1 Key findings

Cluster policies are differentiated because clusters are in different stages of their life cycle or differ in their strategy and targets:

- Cluster strategies may be targeted: (i) on input–output linkages (with their focus on procurement, production and distribution network building); (ii) on the availability of basic factors of production, such as raw materials and skilled labour; or (iii) on providing and upgrading advanced input factor conditions, such as referring to capabilities of knowledge creation and transfer (with their focus on R&D and innovation network building).
- Innovation strategies may have their target either: (i) on the mobilization and exploitation of already available resources and competences for innovations, which, regarding their impact on the cluster, are more incremental than radical by nature ('catching-up' strategy); or (ii) on the development and exploration of new resources and competences for innovations, which are both radical in the sense of new-to-the-market and important for repositioning the cluster ('first-mover' strategy). The cases of Austria and Finland represent mixed strategies in these respects.
- Strategic cluster targets can be described as cluster expansion (Scotland, Croatia) and cluster rejuvenation (Finland, Austria).
- Export-oriented expansion strategies based on industrial recruitment and foreign direct investment may be more or less purposely and formally integrated in regional and rural development strategies. Austria and Finland is suggested as an example for the former, Scotland and Croatia for the latter.

7.4.2 Policy recommendations

- Efficacy and efficiency of policy action can be increased: (i) by applying the principle of catalytic funding; (ii) by exerting leverage through value chain linkages; (iii) by coordination and integration of policy action, whereby avoiding overlapping and improving strategic alignment; and (iv) through replacing 'wishful thinking' by cluster mapping procedures.
- Cluster policy formulation and cluster management professionalism can be improved through strengthening trans-regional and transnational cluster cooperation for experience sharing, learning and dissemination of best practices.
- Practices, however, should not be transferred and adopted blindly, but one should be aware of the different institutional framework conditions in which they are applied.
- Alignment of national cluster expansion strategies with regional development policy targets can be enhanced through anchoring inward investments in the regional economy by means of linkages with cooperative networks of indigenous suppliers.
- A policy to tackle firms' low participation in collaborative actions must be conscious of the fact that for its decision to join a collaborative project, firm participation is strongly motivated by immediate economic gains.
- A top-down cluster formation policy, which is committed to large cluster membership, may cause free rider problems.

Acknowledgements

Thanks to Vesa Ijäs, Chief Development Manager, Lahti City Science and Business, and Sandra Hižak, Cluster manager-director, DRVNI KLASTER d.o.o., for their support.

Notes

[1] Concerning the notation 'cluster policy' and 'cluster-based policy', based on the literature examined, we could not observe that the concepts would denote different phenomena. For conciseness, we prefer to use the term cluster policy instead of cluster-based policy.
[2] The Strategic Development Framework Programme for 2006–2013; the Strategic Coherence Framework Programme 2007–2013; the Science and Technology Policy of the Republic of Croatia 2006–2010; the Government Cluster Programme.
[3] The country joined a communication platform for EU cluster initiatives, initiated cross-border cooperation between universities, research institutes and private companies, and participates in developing joint approaches to enter global markets and to upgrade cluster management practices.
[4] In the last few years, regional chambers of commerce, educational facilities, local authorities and regional development agencies are becoming increasingly involved.
[5] Slavonski hrast (the Slavonian Oak Cluster), Drvni klaster sjeverozapadne Hrvatske (the Timber Cluster of Northwestern Croatia), Hotel plus, Tehnointerijeri, In Parket, Gorski kotari, Drvni klaster (the Timber Cluster).
[6] Three regional Chambers of Commerce located in the districts of Varaždin, Čakovec and Krapina, and the College for Technology and Design in Varaždin.
[7] Important competencies include the fabrication of bended and laminated timber and metal materials and their remanufacturing into flat line and upholstered furniture.
[8] In recent years, specific tasks have been: to foster cooperative business networking; to search for business partners; to disseminate information among member firms; to organize joint marketing campaigns; to cooperate with clusters in other regions; to initiate and manage multiple-actor R&D projects; to organize education and training; to initiate and manage EU funded activities; and to communicate with stakeholder groups and policy makers.
[9] The cluster participates in the international ECO mobel wood design programme (a MED programme) and is member of the national Wood First! project.
[10] The cluster will act as a centre of expertise for timber products businesses with the main emphasis on R&D project activities.
[11] In 1998, clusters were identified as a core area of economic development in the UK Government's Competitiveness White Paper entitled 'Our Competitive Future – Building the Knowledge Driven Economy'. An 'overarching approach' in this context denotes especially the coordination of the regional development policies implemented by the Local Enterprise Companies. Source: Innova Cluster Mapping Project, Country Report: United Kingdom (available at: http://www.clusterobservatory.eu/index.html#!view=documents;mode=all;sort=name;uid=;id=). DTI (now BERR) provided a practitioners' guide as a road map for cluster policy implementation prepared by ECOTEC (A Practical Guide to Cluster Development, DTI, available at: http://www.dti.gov.uk/files/file14008.pdf).
[12] According to J. Edgar, Scottish Enterprise, 'the emergence of the Scottish cluster approach was not initially followed in the rest of the UK'. Source: Clusters and priority industries – A Scottish perspective. Presentation at the Conference organized by the eCluster consortium held 18–19 January 2007 in the Veneto Region.
[13] Scottish Enterprise is Scotland's national Economic Development Agency, set up in 1990 by the UK government. Scottish Enterprise is responsible for implementing economic development policies (including skills and learning) for lowland Scotland through a network of 12 Local Enterprise Companies, which are public sector organizations. Scottish Enterprise reports to the Scottish Executive and through it to the Scottish Parliament. Its operating strategy is set out in Smart, Successful Scotland: Ambitions for the Enterprise Networks, published by the Scottish Executive. Available at: http://www.scottish-enterprise.com/publications/shareddocs/sss.pdf
[14] By a partnership of Scottish Enterprise and industry representatives of the Forest industries development council.
[15] Source: draft paper provided by Roger Coppock.
[16] For information about Finnish cluster policy and cluster programmes read: Europe INNOVA Cluster Mapping Project/European Cluster Observatory, country report: Finland. Available at: http://www.clusterobservatory.eu/upload/Policy_Report_Finland_20080116.pdf
[17] National government funding is based on an obligatory agreement between regions and national government represented by the Ministry of the Interior about the use of earmarked funds.

References

Andersson, T., Schwaag Serger, S., Sörvik, J. and Wise Hansson, E. (2004) The Cluster Policies Whitebook. IKED International Organisation for Knowledge Economy and Enterprise Development. Available at: http://www.innovating-regions.org/download/The_Cluster_Policies_Whitebook_-_IKED.pdf

Arnold, E. and Thuriaux, B. (1997) Developing firms' technological capabilities. Report to OECD, Technopolis, Brighton, UK.

Arthur, W.B. (1989) Competing technologies, increasing returns and lock-ins by historical events. *Economic Journal* 99, 116–131.

Asheim, B. and Isaksen, A. (2002) Regional innovation systems: the integration of local 'sticky' and global 'ubiquitous' knowledge. *Journal of Technology Transfer* 27, 77–86.

Austria Cohesion Policy (2007–2013) European Union. Regional Policy. Leaflet, 2 pp. Available at: http://ec.europa.eu/regional_policy/atlas2007/fiche/at_en.pdf

Austrian Council for Research, Technology and Innovation (2010) Position paper: Strategy 2010 – Perspectives for Research, Technology and Innovation in Austria. Available at: http://www.rat-fte.at/strategy-documents.html

Austrian Reform Programme (2008–2010) 2nd Austrian Reform Programme for Growth and Jobs 2008–2010. Available at: http://www.era.gv.at/space/11442/directory/11486/doc/12128.html

Bender, G. and Laestadius, S. (2005) Non-science based innovativeness – on capabilities relevant to generate profitable novelty. Proceeding paper contributing to the conference Low-tech as misnomer: the role of non-research-intensive industries in the knowledge economy. Brussels, 29–30 June.

Bergman, E.M. and Feser, E.J. (1999) Industrial and regional clusters: concepts and comparative applications. The Web Book of Regional Science. Available at: http://www.rri.wvu.edu/WebBook/Bergman-Feser/contents.htm

Boekholt, P. and Thuriaux, B. (1999) Public policies to facilitate clusters: background, rationale and policy practices in international perspective. In: OECD 1999. Proceedings Boosting innovation. The Cluster Approach. OECD, Paris, pp. 381–412. Available at: http://www.clusterbg.net/content/library/EN/Boosting_Inovations_Cluster_Approach.pdf

Capaldo, G., Landoli, L. and Ponsiglione, C. (2004) Value creation in small firms: a competence-based approach. Conference Proceedings Value creation in entrepreneurship and SMEs, 20–23 September, University of St Gallen. Available at: http://www.kmu.unisg.ch/rencontres/RENC2004/Topics/Capaldo_Renc_2004_Topic_A.pdf

Clement, W. and Welbich-Macek, S. (2007) A success story: 15 years of Austrian cluster initiatives (Erfolgsgeschichte: 15 Jahre Clusterinitiativen in Österreich). Final report on the request of the Austrian Federal Ministry of The Economy and Labour (BMWA). Available at: http://www.clusterplattform.at/index.php?id=77&f=1

Clement, W., Pamminger, W., Walter, E. and Bauer, A. (eds) (2009) Austrian clusters: inventory and perspectives. Final Report, July (Cluster in Österreich: Bestandsaufnahme und Perspektiven. Endbericht Juli. 4C foresee – Management Consulting GmbH, Wien. Available at: http://www.foresee.at/studien.php

Cluster Policy Report Austria (2007) Results for the mapping of cluster policies in Europe reported as a part of the Europe INNOVA Cluster Mapping Project. Available at: http://www.clusterobservatory.eu/index.php?id=42&nid=

Cohen, W. and Levinthal, D. (1990) Absorptive capacity: a new perspective on learning and innovation. *Administrative Science Quarterly* 35, 128–152.

Commission COM (2005a) Communication to the spring European Council. Working together for growth and jobs. A new start for the Lisbon Strategy. Brussels, 2 February 2005, 34 pp. Available at: http://ec.europa.eu/archives/growthandjobs/pdf/COM2005_024_en.pdf

Commission COM (2005b) Communication from the Commision to the Council and the European Parliament. Common actions for growth and employment: The Community Lisbon Programme. Brussels, 20 July 2005, 10 pp. Available at: http://ec.europa.eu/archives/growthandjobs/pdf/COM2005_330_en.pdf

den Hertog, P., Roelandt, T.J.A., Boekholt, P. and van der Gaag, H. (1995) Assessing the distribution power of national innovation systems. Pilot study: the Netherlands. OECD National Innovation Systems Project. Phase I. Country reports.

Eisenhardt, K.M. and Martin, J.A. (2000) Dynamic capabilities: what are they? *Strategic Management Journal* 21, 1105–1121.

Feser, J. (2008) On building clusters versus leveraging synergies in the design of innovation policy for developing economies. In: Blien, U. and Maier, G. (eds) *The Economics of Regional Clusters. Networks, Technology And Policy*. Edward Elgar, Cheltenham, UK, pp. 185–206.

Fromhold-Eisebith, M. and Eisebith, G. (2005) How to institutionalize innovative clusters? Comparing explicit top-down and implicit bottom-up approaches. *Research Policy* 34, 1250–1268.

Georghiou, L., Smith, K., Toivanen, O. and Ylä-Anttila, P. (2003) *Evaluation of the Finnish Innovation Support System*. Publications 5/2003. Ministry of Trade and Industry, Helsinki.

Gilsing, V. (2000) Cluster governance: how clusters can adapt and renew over time. Paper prepared for the DRUID PhD Conference, Copenhagen, January. Erasmus University, Rotterdam. Available at: http://www.druid.dk/conferences/winter2000/gilsing.pdf

Granovetter, M. (1983) The strength of weak ties: a network theory revisited. *Sociological Theory* 1, 201–233.

Haahti, A., Madupu, V., Yava, U. and Babakus, E. (2005) Cooperative strategy, knowledge intensity and export performance of small and medium sized enterprises. *Journal of World Business* 40, 124–138.

Harmaakorpi, V. (2004) Building a competitive regional innovation environment – the regional development platform method as a tool for regional innovation policy. Doctoral dissertation. Available at: http://lib.tkk.fi/Diss/2004/isbn9512270110/

Hirsch-Kreinsen, H. (2006) Low-technologies: a forgotten sector in innovation policy. Paper presented at the ProAct Conference Innovation Pressure, 15–17 March, Tampere. Available at: http://www.pilot-project.org/publications/publications.html

Koch, P., Norgren, L. and Oksanen, J. (2003) Goodnip – Good Practices in Nordic Innovation Policies. Part 1: Summary and Policy Recommendations. STEP REPORT 06/2003.

Kuitunen, S. and Niinikoski, M.-L. (eds) (2008) Knowledge clusters as builders of innovation communities: about the value of the Centre of Expertise Programme (Osaamisklusterit innovaatioyhteisöjen rakentajina. Näkökulmia osaamiskeskusohjelman lisäarvoon). Työ- ja elinkeinoministeriö, Osaavien keskusten verkosto OSVE, Helsinki, 120 pp. Printed publication can be ordered from the Ministry of Employment and the Economy at osaamiskeskusohjelma@tem.fi.

Lester, R.K. and Sotarauta, M. (ed.) (2007) Innovation, universities, and the competitiveness of regions. *Tekes. Technology Review* 214.

Leydesdorff, L. and Etzkowitz, H. (1998) The triple helix as a model for innovation studies. *Science & Public Policy* 25, 195–203.

Malmberg, A. and Maskell, P. (2006) Localized learning revisited. *Growth and Change* 37, 1–18.

Martin, R. and Sunley, P. (2003) Deconstructing clusters: chaotic concept or policy panacea. *Journal of Economic Geography* 3, 5–35.

Ministry of Economy, Labour and Entrepreneurship (2007) Croatian Exporting Strike (Hrvatska izvozna ofenziva). Strategy 2007–2010. Available at: http://www.hup.hr

Morgan, K. (1995) The learning region: institutions, innovation and regional renewal. *Regional Studies* 31, 491–503.

National Competitiveness Council (Nacionalno vijeće za konkurentnost) (2004) Preporuka za povećanje konkurentnostihrvatske. Available at: http://www.vlada.hr/aktualne_teme_i_projekti/aktualne_teme/konkurentnost/55_preporuka_za_povecanje_konkurentnosti_hrvatske

Nauwelaers, C. (2003) Background paper on cluster policies. Prepared for the Trend Chart Policy Workshop Innovative Hot Spots in Europe: Policies to promote trans-border clusters of creative activity, 5–6 May, Luxembourg.

Oinas, P. (1998) The Embedded Firm? Prelude for a revived geography of enterprise. *Acta Universitatis Oeconomicae Helsingiensis* A–143.

Porter, M.E. (1990/98) *The Competitive Advantage of Nations*. The Free Press, New York.

Porter, M.E. (2001) *Clusters of Innovation: Regional Foundations of U.S. competitiveness*. Council of Competitiveness, Washington, DC.

Raines, P. (2001) The Cluster Approach and the Dynamics of Regional Policy-Making. Regional and Industrial Policy Research Paper No. 47. European Policies Research Centre, University of Strathclyde. Available at: http://citeseerx.ist.psu.edu/viewdoc/download?doi=10.1.1.131.7017&rep=rep1&type=pdf

Regional Development Programmes (2007–2013) Österreich. Available at: http://ec.europa.eu/regional_policy/atlas2007/index_en.htm

Reid, A. and Peter, V. (2008) Sectoral innovation systems: the policy landscape in the EU 25. Final report of the project Sectoral Innovation Systems in Europe: Monitoring, Analysing Trends and Identifying

Challenges (SYSTEMATIC) as part of the Europe Innova initiative. Available at: http://www.europe-innova.org/servlet/Doc?cid=9894&lg=EN

Rocha, H.O. and Sternberg, R. (2005) Entrepreneurship: the role of clusters. Theoretical perspectives and empirical evidence from Germany. *Small Business Economics* 24, 267–292.

Roelandt, T.A. and den Hertog, P. (1999) Cluster analysis and cluster-based policy making in OECD countries: an introduction to the theme. In: OECD 1999. Proceedings Boosting Innovation. The Cluster Approach. OECD, Paris, pp. 9–18. Available at: http://www.clusterbg.net/content/library/EN/Boosting_Inovations_Cluster_Approach.pdf

Roelandt, T.A., Gilsing, V.A. and van Sinderen, J. (2000) Cluster-based innovation policy: international experiences. Paper presented at the 4th Annual EUNIP Conference, Tilburg, 7–9 December. Available at: http://publishing.eur.nl/ir/repub/asset/824/rm0012.pdf

Rosenfeld, S.A. (1997) Bringing business clusters in the mainstream of economic development. *European Planning Studies* 5, 3–23.

Salmi, P., Blomqvist, K.-M., Ahola, J. and Kyläheiko, K. (2001) Industrial districts and regional development: towards a knowledge-based view. Telecom Business Research Center. Lappeenranta University of Technology. Working Papers 7. Available at: http://www.tbrc.fi/pubfilet/TBRC_500000372.pdf

Schienstock, G. and Hämäläinen, T. (2001) Transformation of the Finnish Innovation System: A Network Approach. Sitra, Report No. 7. Available at: http://www.sitra.fi

Smedlund, A., Ståhle, P. and Köppä, L. (2005) Bridging organisations and shared governance – (Välittäjäorganisaatiot ja jaettu johtajuus). In: Koskenlinna, M., Smedlund, A., Ståhle, P., Köppä, L., Niinikoski, M.-L., Valovirta, V., Halme, K., Saapunki, J. and Leskinen, J. (eds) *Bridging organisations (Välittäjäorganisaatiot – moniottelijat innovaatioita etsimässä).* Technology Review (Teknologiakatsaus), 168/2005. Tekes, Helsinki, pp. 19–52.

Sölvell, Ö., Lindqvist, G. and Ketels, C. (2003) The Cluster Initiative Greenbook. Available at: http://www.cluster-research.org/greenbook.htm

Steiner, M. and Hartmann, C. (2002) Material and immaterial dimensions of clusters. Cooperation and Learning as infrastructure for innovation. Regional Science Association 42nd European Congress, Dortmund, 27–31 August.

Storper, M. (1995) The resurgence of regional economies, ten years later: the region as a nexus of untraded interdependencies. *European Urban & Regional Studies* 2, 191–221.

Storper, M. (1997) *The Regional World.* Guilford, New York.

STRAT.AT Plus (2007–2013) Strategic Report for Austria pursuant to Article 29 Regulation (EC) No 1083/2006. Available at: http://www.oerok.gv.at/fileadmin/Bilder/3.Reiter-Regionalpolitik/2.EU-SF_in_OE_07-13/2.1_Nationale_Strategie/STRAT.AT/STRAT.AT_Report_EN_HP1.pdf

Tanvig, H., Lindegaard, K., Sørensen, J.F.L., Stoye, M., Vatn Jensen, M., Virkkala, S., Niemi, K., Aradóttir, E., Fraas, M., Pedersen, T.E., Mariussen, Å., Persson, L.O., Larsen, K. and Pettersson, Å. (2005) Innovation system and the periphery (ISP). Final report. January. Available at: http://vefir.unak.is/isp/default.htm

Virkkala, S. and Niemi, K. (eds) (2006) Peripheral localities and innovation policies. Learning from good practices between the Nordic countries. Nordic Innovation Centre. PLIP project. Available at: http://www.nordicinnovation.net/prosjekt.cfm?Id=1-4415-47

8 Innovation in Forest-related Territorial Goods and Services: an Introduction

Bill Slee

Abstract
The study of innovation with regard to territorial services is often complicated by the public or quasi-public good character of many territorial services connected to forestry. Although firm-level innovation can and does occur in relation to territorial goods and services, much innovation occurs among networks of social and economic actors in bounded territorial areas. The conceptual background to the study of innovation in territorial goods and services necessarily embraces both sectoral innovation systems and spatial or regional innovation systems. Further issues 'muddy the waters' of innovation systems with respect to territorial goods and services related to forestry: the large public sector involvement with forestry in Europe can create 'crowding out' effects, whereby innovation in the public sector inhibits rather than enables opportunity in the private sector. On the other hand, intervention by public bodies may be needed to bring together diverse partners and provide bridging mechanisms to link multiple actors in development projects. This introductory chapter explores the extent to which different innovation systems interact in creating framework conditions in which innovation can take place with respect to territorial goods and services in the forest sector.

8.1 Introduction

The investigation of innovation in relation to forest-related territorial goods and services requires: (i) an understanding of the context in which we are exploring this issue; (ii) an understanding of what exactly comprise forest-related territorial goods and services; (iii) what comprises innovation; (iv) how, if at all, policy for rural development embodies contemporary ideas about innovation within its framework structures and practices; and (v) what the barriers to innovation in forest-related territorial goods and services might be.

At the centre of our interest is the connection between policy for innovation, policies for territorial goods and services in forestry and positive outcomes for rural development. First, we ask, is there evidence of policy coordination and integration between innovation policy, forestry policy and rural/regional development policy and, second, has such integration been effective in delivering positive outcomes?

The territorial context is also important where public or quasi-public goods from the land use industry are evident and where forestry confers a beneficial 'shadow' on a completely unrelated economic agent (Slee, 2005a). Where, for example, forestry provides an attractive tourism resource, the benefits may be derived by tourist operators neighbouring the forest area rather than forest owners. However, the forest owner provides a valued green infrastructure in which other economic activity relating to the leisure-related consumption can be pursued. In this case, the economic performance of the tourism business is contingent on the existence of the forest, but no transfer of resource takes place from tourist business to householder. This could be described as a spatially bounded spillover effect, which can be likened to a halo around the forest.

At present, in the shadow of a financial crisis of enormous magnitude and with the world facing a profound anthropogenic threat from climate change, there is a sense in which the established values of the unfettered market may be ripe for review. If a more interventionist form of regulation is endorsed, particularly with respect to the externalities associated with climate change, the old, low-technology primary industries could become the scene for innovation associated with sustainability enhancement in ways that were quite simply unthinkable a few years ago. Further, there is clearly a need to justify the considerable payment from the public purse to the farming community for the delivery of public support for multifunctionality, which is a response to the widely used public goods policy rationale (OECD, 2001).

8.2 The Context

Many authors have commented on changes in the nature of the rural economy in recent decades. Indeed, it is recognized by Mahroum *et al.* (2007) that rural areas can be the setting for highly innovative development activity. The idea of a post-productivist rural economy has been actively promoted by some (Ilbery and Bowler, 1998), most especially in relation to agriculture, but also in relation to forestry (Mather, 2001; Mather *et al.*, 2006) as a conceptual explanation for a shift away from a policy-driven, production-dominated rural economy to one where environmental concerns and diversification into service-related business activity were prevalent processes. However, the idea of post-productivism has been attacked by Evans *et al.* (2002), who accuse its users of implying a naïve dualism between productivist agriculture on one hand and post-productivist agriculture on the other, when in practice elements of both types of land use are widely evident.

A second conceptual peg on which to hang ideas about the provision of territorial goods and services in the forest sector is the idea of multifunctionality (van Huylenbroek and Durand, 2003). This concept is less disputed conceptually than the idea of post-productivism and implies the delivery of a range of services from land used primarily for agriculture or forestry. It has become embroiled in political discourses about the support for a European model of agriculture and is seen by some as a back-door way of continuing to support agriculture from the public purse in a world pressured by market-liberalizing tendencies (Garzon, 2005). None the less, the OECD (2001) endorses the concept and frames it very much within a neoclassical economic conception of public goods. As in the case of post-productivism, multifunctionality has recently been framed within an agricultural context but can readily embrace other productive rural land uses such as forestry. Indeed the term may have a much longer 'pedigree in forestry' (Dieterich, 1953) and only lately transferred into an agricultural arena. The term 'multi-purpose' has also often applied to forestry, with a history predating the consideration of multifunctionality in agriculture.

A third conceptual peg is the Ecosystem Services Framework developed by the Millennium Ecosystem Assessment (MEA, 2005). The MEA argues that the environment delivers supporting, regulating, provisioning, cultural services. In a sense, the MEA's notion of ecosystem services can be seen as an ecocentric rather than an economistic assertion of the multiple functions of the environment or indeed any biologically centred land use. The explicit identification of functions such as soil protection, biodiversity maintenance, water quality and quantity regulation acknowledges some of the multiple roles of a productive rural land use such as forestry.

A final contextualizing conception of relevance to territorial goods and services is the recognition of a tendency towards diversification of rural land use management away from conventional production. It is widely recognized that commodity production has often been unprofitable for primary producers and that new economic opportunities may exist for those who have historically been rather single-minded primary producers. These have been explored in an agricultural context by Slee (1989) and from a forestry perspective by Mantau *et al.* (2001). Both Slee and Mantau recognize that many, but by no means all, of the possible diversifications away from conventional primary production have public good characteristics. Mantau stresses the importance of framing appropriate institutional arrangements for marketizing these quasi-public goods. Subsequent literature (e.g. O' Connor *et al.*, 2006) emphasizes the shift in the rural economy from a process of modernization of productive forces of primary land use towards a new paradigm of rural development based on diversified endogenous enterprise.

Rather than fretting over concepts, it may be better to explore the facts. There is widespread evidence of a shift towards what we might term economies of consumption rather than production in rural areas (Slee, 2005b). Associated with this is a discernible shift in the primary land use sector towards a greater interest in the scope for developing new enterprises in farm and forest sectors. These connect to both market and policy drivers and have led to the increased provision of leisure-related services, other commercial opportunities for adding value to the primary product base and the development of new institutional or policy means for the enhanced provision of public goods or other key ecosystem services.

8.3 Territorial Goods and Services in Forestry

The term 'territorial goods' (or territorial goods and services) requires definition. The term is not in common parlance and there are at least two dimensions of territoriality that require acknowledgement. First, there is a notion of territoriality closely connected to product differentiation, where the locality of production is embodied in the character of the product. Second, there is another notion of territoriality in which multifunctional land use is responsible for provision of a range of products or ecosystem services, many of which have public good characteristics.

Territoriality is an attribute of a product or service that is linked inextricably to the place of production. It manifests itself most obviously in the Protected Designation of Origin (PDO) and in Protected Geographical Indication (PGI) designations for food and drink products in the European Union (EU). The origins of these schemes lie in national designations relating to the protection of quality wines and the underlying belief, more widespread in southern Europe than the north, that food and drink product carries place-specific attributes, which are a function of the '*terroir*' – the particular combination of biophysical, technical and cultural elements that manifest themselves in a particular product. Article 2.1.a of Council Regulation 510/2006 (EC 2006) notes that the scheme will operate where 'the quality or characteristics of which are essentially or exclusively due to a particular geographical environment with its inherent natural and human factors'. A number of forest-based products such as chestnuts and forest mushrooms are included in specific designations (Box 8.1). These EU regulations create Property Rights for the producers of the designated product.

Over and above the EU schemes, there are many private, non-governmental organization (NGO) and national schemes that seek to link production with place. The Countryside Agency in England ran an 'Eat the View' campaign, which endeavoured to support regionally specific foodstuff production in high-quality environments. Many public sector organizations at regional level support food initiatives which, with or without European designation, seek to link production and place in the eyes of the consumer.

Box 8.1. The Borgotaro Mushroom trail – Emilia-Romagna.

This trail was developed with support from the EU Leader initiative to give a thematic identity to a regional, development initiative built around the *Boletus* mushrooms, which are found widely in the area. The trail provides those visiting the area with a range of information, and links them to places for gastronomy, for the sale of regional food and to recreational facilities (http://www.stradadelfungo.it/). It recognizes the place-specific cultural significance of food and in a coordinated way and links this to new development opportunities. It is a good example of a new network as an institutional innovation, linking together existing actors for mutual benefit. This particular initiative connects very closely to the idea of place specificity of the product to the extent that the initiative makes use of the EU's PGI indication for four species of *Boletus* found in the area.

There are many food and drink products with strong regional identities. Sometimes new forms of innovation related to drawdown of corporate profits for the enhanced delivery of environmental goods and services may be possible. A good example of this is the UK's Royal Society for the Protection of Birds receiving *c.*€0.50 for every bottle sold of the brand Black Grouse whisky. The funds then support the conservation of this threatened woodland species. This approach has also been used to promote environmentally friendly farming, both through organic production with its purported environmental benefits (and a strict regulatory regime) and more specifically though milk products associated with wildlife friendly farming. The drawdown of funds for corporate social responsibility, as in the use of voluntary carbon sequestration in new woodland planting, is a further illustration of how corporate funds can be directed into environmental goods and services, potentially with a place-based identity.

It is a small step to assert territoriality as an attribute of regional products in general, rather than just food and drink, and package other products and services with a regional label. The most obvious example of this is in the case of tourism products. Place-specificity clearly matters to many tourists. They visit places because of their biophysical and cultural assets. Accommodation with a prime view or a seafront location often commands a premium price. Whole regions are given particular identities linked to natural biophysical features, such as the ancient Caledonian pinewoods of Speyside or Deeside in the Cairngorms National Park (Fig. 8.1). The Cairngorms region clearly deploys the Scots pine in its iconography. Thus, it can be seen that place matters at a range of spatial scales and that forestry can, but does not always, contribute to shaping that space.

Particular places are also associated with particular cultural attributes. Whether with the wooden churches of the Ukrainian Carpathians, the archaeological heritage of stone circles of the Chalk Downlands of England or Dartmoor, or the Gaelic language and music of western Ireland or north-west Scotland, it is evident that these cultural attributes are valued by both local people and visitors. They can be valorized in a variety of ways, many of which are explicitly associated with particular places.

The second type of territorial good/service comprises the multiple ecosystem services that arise because of multifunctional rural land use (Harvey, 2003), the value and volume of which are shaped by the production practices in the land management regime. Whether the ecosystem service is biodiversity, soil quality protection or watershed management, it is self-evident that production and management practices will influence the yield of these environmental services. Here the territoriality is linked inextricably to the socio-ecological specificities of land use (Box 8.2). Most often, these multiple ecosystem services are at best only partly reflected in market prices and their delivery is contingent on well-designed policy incentives.

Fig. 8.1. Early 20th-century tourist map of the Cairngorms, Scotland (now a national park), showing use of Scots pine (*Pinus sylvestris*) in iconography.

Box 8.2. The Alpannonia long-distance hiking trail – Austria.

This project supported both by EU Leader and Objective 1 support comprises a 100-km trail targeted at hikers for pleasure, which descends from the high mountains through the Alpine pasture region and forested mountains to rolling hills in wine regions until it ends at the edge of the Pannonian plane in Hungary (http://www.alpannonia.at/). The project aims to waymark the trail and also coordinates suppliers of services (accommodation, gastronomy, etc.), initiates new attractions (nature trails, museums, forest education sites, etc.) and coordinates marketing (website, flyers, events, etc.). The project illustrates the types of challenge often connected with the development of landscape attractions for tourism: there are many suppliers and many beneficiaries who are difficult to organize. How to direct/redirect profits from tourism to a central organization that provides services such as planning, coordination, marketing, quality assurance, etc., as well as to single land-owners that provide landscapes and their management, maintain trails, run attractions and suffer disturbance, all have to be addressed.

In summary, territorial goods and services with respect to forestry are variable and can be seen to include:

- non-timber forest products such as game animals or birds, fruit and berries, fungi and foliage and value-added products associated with these;
- wood products where those products are linked to local sources of wood or local styles of processing (e.g. of furniture design, wood craftsmanship);
- tourism and recreational products where woodland views, woodland character or woodland identity is an important attribute of the product/service (Fig. 8.1);
- environmental ecosystem services, such as biodiversity, which are associated with forests and woodlands; and
- supporting ecosystem services such as flood control or soil protection impacting on neighbouring and downstream areas.

In the case of territorial goods and services in forestry, the forest/woodland attribute can range in significance from modest to all-embracing. Much, of course, depends on the consumer. For some, the woodland characteristics may be central and intrinsic to the experience; in other cases, it may form a backdrop to other activities such as skiing, visiting historic monuments, playing golf or cultural activities. There is a tendency to regard the forest woodland setting primarily as a place for tourism. However, evidence from eastern England (Slee *et al.*, 2004) indicates that residential choices are frequently made on the basis of the locational attractiveness of tree-rich regions.

The multifunctionality of forests and woodland creates considerable scope for providing a wide range of territorial goods and services to meet the demand of different types of tourist or recreationist. However, as in any multifunctional situation, there is a need to consider the extent to which more of one attribute compromises the availability of another attribute. This may be especially the case where the primary motivation of the forest owner is timber production, in that the single-minded pursuit of timber may compromise biodiversity, landscape and a range of supporting and regulating services, and limit recreational and tourist opportunities.

8.4 Innovation – Competing Theoretical Perspectives on the Nature of Innovation

A considerable literature has developed around regional innovation systems (Cooke *et al.*, 2004). Regional innovation theory has been built on the disparate foundations of different disciplines to understand the nature and drivers of development in geographically bounded contexts. At various times, the idea of an innovative milieu (Camagni, 1995), where firms are bonded by strong social capital; the triple helix (Leydesdorff and Etzkovitz, 1998), where partnerships between the state, the private sector and research organizations are major drivers of innovation; untraded interdependencies (Storper, 1995), where tacit knowledge links actors together; or a cluster (Porter, 2000), where spatial proximity of firms creates a platform for competitiveness and innovation, serve as recognition by a range of authors from a range of disciplines of the importance of regions as entities within which innovation occurs. Further, according to Lawson (1997), 'there has been a movement in focus away from input–output-type studies to a consideration of the less tangible absorptions and exchanges between agents and the qualitative nature of rules, conventions which allow agents to act in capable-innovative ways'.

In this section, we accept the broad definition that embraces product, technological, marketing mix or policy-related innovation. We also recognize that innovation is something that occurs within mutable institutional contexts. Often those institutional contexts are framed by supply chains, clusters or networks within a particular sectoral context but increasingly they are found within a regional context. Innovation can thus take on a predominantly cross-sectoral (horizontal) perspective in relation to a local area or region, or may take on a sectoral context and be driven at regional or national scale. Both sectoral and spatial innovation systems can be observed. In some cases, hybrid forms are also evident, with sectoral clusters operating at regional scale.

The nature of innovation has been discussed elsewhere in this volume. Undoubtedly, a transition has occurred from what might be thought of as natural resource-based regional dynamics, when underlying natural resources such as coal, steel or oil drove economic growth, to a situation where the knowledge economy has become a major driver of economic development and where a 'creative class' (Florida, 2002; McGranahan and Wojan, 2007) develops new ways of using knowledge to create economic advancement (Cooke *et al.*, 2004). The nature and definition of the creative class have been much discussed. Hansen *et al.* (2005) develop Florida's ideas arguing that a constellation of factors are important in shaping a region's capacity to innovate. Talents, tolerance and a technological foundation are required to deliver economic growth. These are human resource factors that are quite separate from business-friendly policies that have often been actively promoted as means of regional development and they call for an entirely different form of supportive intervention.

Much of the discussion of innovation is in the context of dynamic urban areas associated with high-technology, knowledge-based industries. However, there is evidence that some types of footloose economic activity will move to attractive rural locations precisely because of the attractiveness of these regions to the creative class. Thus Persson and Westholm's (1994) observations on the emergence of a new mosaic-like pattern of growth in Sweden can be seen as a reflection of the attractiveness of certain places to creative talents and the networks that underpin their evolution.

However, the knowledge base has a number of components (Hansen *et al.*, 2005), including synthetic knowledge (technology-based), analytical knowledge (science-based), symbolic knowledge (creative-based) and artisanal knowledge (endogenously based). Synthetic knowledge tends to drive innovation in classic regional industrial clusters producing incremental innovation, which comprises more development of products than radical innovation. Analytical knowledge tends to be more important in particular industries like biotechnology and such industries tend to locate close to leading universities. Symbolic knowledge is associated with the creative industries and the arts and the product is often ephemeral. The extent to which these different types of knowledge are formal or tacit varies considerably and this may impact on their role in nurturing innovation. Where tacit knowledge dominates, the scope for outsiders to bring new talents and ideas into an area may be more limited.

8.5 Rural Development and Innovation

Rural areas of Europe with their historic dependency on low-technology industries have not been seen as the most innovative of areas. The prevalent imagery of rural areas has been one structured around backwardness, strong bonding capital of the indigenous community and a tendency towards suspicion of outsiders and innovators (Hansen *et al.*, 2005). The prevailing image of innovation has been in urban-located knowledge-based industries rather than in low-technology industries in rural areas. However, it is outside the high-technology knowledge-based innovation hubs that most new jobs are created and in which more prosaic forms of innovation take place. Furthermore, if the warnings from many experts of a combination of energy and food shortages begin to bite, the traditional rural industries may again become an arena ripe for investment and innovation.

The prevalent image of rural areas as backwards and traditional may also need tempering in the light of evidence of in-migration into rural areas and the emergence of major differences in growth potential between rural areas dominated by traditional industries and those where the creative class has established a platform, which tend to be areas where there are high levels of natural amenity (McGranahan and Wojan, 2007).

A second type of innovative development to have emerged in the last two decades has been labelled 'neo-endogenous' rural development (Ray, 2006), whereby development is initiated through a set of bottom-up processes, albeit bottom-up processes where a degree of bridging capital to an external world seems also to be important. The neo-endogenous model of development can be seen as a socially grounded and place-centred approach to development based on valorizing local assets including culture and land-based products (Box 8.3). This approach to development and the explicit support of innovation are embedded within the EU's Leader approach. Leader had its origins as an experimental initiative to support an alternative model of development in the early 1990s but has subsequently been mainstreamed to constitute one of the four axes of the current rural development plans.

In the last few years, there has been something of a reconsideration of European thinking on rural development. DG Regional Policy has been asserting the importance of exploring the dynamics of city regions as development hubs. Marino (2009) has argued:

> The old hierarchical relationship between rural and urban areas – where rural areas were viewed simply as suppliers of food to more developed urban areas – has gone. Rural areas are multi-functional. They still provide resources, both commodities and the less tangible natural and cultural resources we increasingly value such as biodiversity and traditional landscapes. But they are also the location for economic activities such as knowledge-intensive services. This more complex reality should be reflected in the way we conceptualize the relationship between rural and urban areas.

Box 8.3. Incomer-driven development of new market opportunities, Finland.

Oy Aurinkolehto Ltd is a company producing birch sap products. It was established in Finland in 1996 by a young couple who moved into the area. The business idea was fixed after very thorough research on different kinds of possibilities for successful entrepreneurship in rural areas. The main motivation was to find a source of livelihood that would make it possible for the owners to move from urban area to the countryside. The owners did not have previous experience with birch sap, but they invested a lot of time to collect and analyse information widely, before making the decision to specialize in it. The company developed an innovative collection and production system for industrial-scale production of sap. The system also enables serial production, which makes it possible to produce sap in the large scale necessary, e.g. for export activities. The suitable marketing channels were established through market research and through cooperation with potential clients and by utilizing the entrepreneurs' previous contact networks.

The emergent diversity of rural developed countries is both an opportunity and a challenge as a seedbed for innovation. On the one hand, there has been in-migration of the so-called creative class into some rural areas. Such people's propensity to innovate is well known. On the other hand, the rather protective orientation of new and affluent rural residents has been described as NIMBY-ism – a Not-In-My-Back-Yard approach to new development. Such attitudes are bound up with an amenity-based view of countryside, rather than seeking new ways for developing the natural resource base of rural areas.

8.6 Barriers to Innovation in Relation to Forest-related Territorial Goods and Services

Forestry is often thought of as a mature industry and much of the associated wood supply chain has similar characteristics. As such, it does not seem fertile ground on which innovation might flourish. However, the territorial goods and services associated with forestry are not commodity products. Their niche character, their location-specific attributes and their income elastic demand make them appealing products to the forest owner. It may also be that established institutions such as policy instruments, including subsidies, can also act as barriers to innovation, because they create stability and may impede innovation as a creative solution to a problem. Subsidies for public good delivery could even be regarded as a means of adaptation rather than innovation. However, at a time when sustainability and climate change concerns are rising up the political agenda, forestry's role as a sustainable production system, a renewable fuel source and a sequesterer of carbon adds significantly to the attractiveness of the conventional products of the sector (Box 8.4).

There are several obstacles to innovation in relation to forest territorial goods and services. First, the ownership structure and institutions constitutes a potential barrier; second, the public good character of so many territorial goods and services is an issue that must be addressed; third, the fact that increases in territorial goods and services often come at the expense of traditional forest products may be problematic.

The ownership of forestry is complex and this complexity has increased over time with new types of forest owners, especially environmental NGOs. In many parts of Europe, forestry is characterized by small-scale forest owners. In some countries, there are strong private forest owners' associations that can galvanize action and provide a conduit for innovation. In other countries, forest owners' associations are weak, small private forests are unmanaged and the scope for innovation is deeply compromised by the fragmented ownership structure and divergent interests of forest owners. Where state forests exist, the larger size and public accountability can lead to enhanced delivery of ecosystem services and to leadership and innovation, as is the case with mountain bike developments in the UK (Box 8.5). However, it is possible that public forest sector leadership may crowd out private

Box 8.4. Institutional-driven innovation in Alentejo, Portugal.

The Association Terras de Dentro, located in the Alentejo (Portugal), is a non-profit private organization with a mission to promote integrated development of rural areas. The Leader+ initiative supported this association in the promotion of several projects related to cork oak stands (montados). For example, the projects 'Valorisation of the montado', 'Conservation and valorisation of montado' and 'Environmental Education about the montado' developed actions to increase awareness of cork-stripping, seminars about forest management plans, incentive systems, prevention of forest fires and recovering of burnt forest areas, a workshop about cork products, an exhibition named 'Cork House' and activities in schools about the montado ecosystem. Some of these actions were innovative since it was the first time they were implemented in the region covered by this association.

sector operators in the provision of territorial goods and services.

The public good character of so many ecosystem services associated with forestry creates a formidable obstacle, transferring innovation from the market arena to the policy arena. Although a variety of policy means exist to reward forest owners for public good delivery, these are predominantly rather blunt instruments for dealing with public good delivery. However, Mantau *et al.* (2001) have shown that new institutional arrangements can create room for manoeuvre and thereby scope for innovation. Indeed, it is at an institutional and governance level that the greatest scope for innovation may well occur in relation to the enhanced delivery of forest-related public goods (Box 8.6). The particular case of carbon sequestration has been the subject of significant attention recently, most notably through policies for Reduced Deforestation and Degradation (REDD) but these REDD arrangements relate to developing countries and tropical forests. There is a rather longer history of developed country schemes, which are the subject of a chapter in this volume (Ciccarese *et al.*, Chapter 9).

It is not inconceivable that regional innovation system priorities will clash with sectoral innovation system priorities in relation to forestry. The demands of the forest processing sector tend to be for low cost and consistent supplies of wood raw material for commodity production of timber, particleboard or paper. The demands of the territorial or regional innovation system are more likely to be for multifunctional forestry, which will cross-connect to other activities such as tourism and recreation and possibly creative activities. It is self-evident that a forest monoculture of high

Box 8.5. Mountain biking in the UK and the Forestry Commission: United Kingdom.

The investment by the Forestry Commission and other funding bodies of a number of mountain bike trails has given Scotland global market leadership in the development of such trails. This is almost entirely attributable to the recognition that the infrastructure of the state forestry sector, including its prior commitment to recreational provision and its large areas of forest in upland and mountain areas created a resource that could be exploited for mountain biking. The development of a series of linked sites across the Southern Uplands of Scotland in the Seven Stanes Trail (http://www.7stanes.gov.uk/) illustrates the nature of the infrastructure required. A range of levels of trail are put in with clear indications to users of the challenge that they pose. A website links the trail users to a variety of surrounding businesses. There is unambiguous evidence that such trails are instrumental in attracting in new visitors and enhancing the performance of local economies. Some forest owners are sensitive to the possibility that if the state provides such as service, other forest owners may be unable to commercialize such facility themselves.

Box 8.6. A private sector example of commercial offsetting.

A number of voluntary private sector carbon sequestration schemes have been promoted. **TreeSmart** is one such scheme.

What is TreeSmart?

TreeSmart is a carbon sequestration programme aimed at removing carbon dioxide from the atmosphere that has been created by the transport sector. It does this by helping to establish and maintain eucalypt plantations destined for eventual harvesting and replanting.

Who is TreeSmart?

The TreeSmart programme is run by a private company (TreeSmart Australia Pty Ltd) set up for the express purpose of running the TreeSmart programme on a ‘profit for purpose’ basis. Any profits derived by TreeSmart are re-invested in further tree planting for sequestration purposes. The TreeSmart programme concept was developed by the The Urban Transport Institute and Green & Gold Tree Farms. http://www.treesmart.com.au/Info.html

timber production trees is likely to be less attractive for recreation and biodiversity than a multifunctional forest (Willis *et al.*, 2003). Additional issues such as safety may be a concern, where intensive and dangerous production practices, such as felling, forwarding or winching, take place in close proximity to recreation.

8.7 Conclusions and Discussion

This introduction has made the case for a broad view of what constitutes innovation in territorial goods and services. It explores the concept of a territorial good or service and asserts that the strength of the territorial attribute can be compelling or weak, depending in part on the product/service and in part on the consumer. As well as what might be seen as artisanal products, a case is made for including both leisure-related goods and services, which make significant use of the forest, and the wide array of public goods and wider ecosystem services associated with forestry.

There are many different ways of exploring innovation in a territorial context. The extensive regional development literature acknowledges the importance of innovation in development. There are many different theoretical perspectives. Key findings are that different types of knowledge are important in different sectors, that strong bonding social capital can be a barrier to some types of innovation and that bridging capital may be crucial in connecting forest owners to new opportunities. Although dominated by research into high-technology regions, there is some literature that identifies the key role of innovation in rural areas and also acknowledges a role for the creative class.

The incorporation of thinking and theorizing about innovation into European rural development policy is very limited. The one major area of innovation in European policy for rural areas is the EU Leader initiative. Leader has tended to focus on tacit knowledge and the development of the social economy and what has been termed neo-endogenous rural development. As a policy initiative, it is highly regarded not just for its outcomes but also because of its capacity to nurture innovative cross-sectoral local networks (Secco *et al.*, 2009). Indeed, the outcomes are in part a consequence of innovation in relation to institutional form and governance structures. The scope for forestry to benefit from Leader is explored later in this volume. Although the overarching rural development plans incorporate private forestry, the tendency has been to wrap existing national schemes into the new institutional arrangements rather than innovate with new schemes of support. It is noted that there is a growing tendency to think in terms of city regions and to break down the rural–urban dichotomy that has so long driven European policy. This reorientation to city regions would comprise a substantial innovation if it could be operationalized in new governance structures.

To date, the integration between innovation policy and policy for territorial goods and services has been modest, but not entirely absent. Particularly within the Leader axis of the Rural Development Plan innovation is promoted, but the degree of connection of territorial goods and services in the forest sector to regional innovation systems is very limited.

There are several obstacles and barriers to the development of innovation in territorial goods and services in forestry. These include a fragmented ownership structure, powerful state forestry agencies with the capacity to crowd out innovation in the private sector and the considerable element of public goods in territorial goods and services, which places the obligation on the state to innovate with respect to policy. Any of these constraints can be turned into opportunities for innovation. Strong forest owners' associations can overcome the obstacles of small size and this is clearly an arena for innovation. Finally, the emergence of new policy means can comprise innovation. The massive support from some developed nations such as Norway for REDD provides a case in point. This is an innovation of potentially profound significance in the attempts to address anthropogenic global warming.

Without further empirical investigation, we can only speculate as to whether there are real tensions between regional and sectoral innovation systems. Some of the studies in this volume may yield clues. However, the principal differences between innovation in the wood and territorial goods and services sectors is the increased importance of policy, institutional or governance-related innovation in the territorial goods and services sector. In the wood sector, there is scope for innovation, but it is more likely to revolve around the marketable products, their technologies of production and supply chains to market.

Finally, we should acknowledge the uncertainties of our time and how these might shape platforms for innovation. The enormity of the task in decarbonizing economic activity (or at least substituting renewable carbon for fossil carbon) in attempting to address anthropogenic global warming will almost certainly revitalize the forest sector. But such developments will create new tensions between storing carbon in growing trees and embodying carbon for a limited period of time in sustainable production. We need not only better understand the relative values of sequestered carbon and sustainable wood products, but also develop a more refined picture of the multiple ecosystem services and trade-offs associated with the full suite of territorial goods and services provided by forests. Only then can their true value to society be valorized in new policy means. Herein lies a major challenge for innovation systems: to create effective framework conditions and governance mechanisms to mediate these competing claims, without impeding those parts of the forest sector which work within the normal market constructs.

Acknowledgement

I would like to thank Peter Elsasser for his useful comments on an earlier draft.

References

Camagni, R.P. (1995) The concept of innovative milieu and its relevance for public policies in European lagging regions. *Papers in Regional Science* 74, 317–340.

Cooke, P., Heidenrich, M. and Braczyk, H.-J. (2004) *Regional Innovation Systems* (2nd edn). Routledge, London.

Dieterich, V. (1953) *Forstwirtschaftspolitk.* Parey, Hamburg and Berlin.

European Union (2006) Regulation 510/2006. On the protection of geographical indications and designations of origin for agricultural products and foodstuffs. OJ 31/3/2006.

Evans, N., Morris, C. and Winter, M. (2002) Conceptualising agriculture: a critique of *post-productivism* as the new orthodoxy. *Progress in Human Geography* 26, 313–332.

Florida, R. (2002) *The Rise of the Creative Class – And How it's Transforming Work, Leisure, Community, & Everyday Life.* The Perseus Books Group, New York.

Garzon, I. (2005) Multifunctionality of agriculture in the European Union: is there substance behind the discourse's smoke. Presentation paper no. 20, Center on Institutions and Governance, UC Berkeley, California.

Hansen, H.-K., Vang, J. and Asheim, B. (2005) The creative class and regional growth: towards a knowledge-based approach. CIRCLE Working paper 15/2005.

Harvey, D.R. (2003) Agri-environmental relationships and multi-functionality: further considerations. *World Economy* 2, 705–725.

Ilbery, B. and Bowler, I. (1998) From agricultural productivism to post-productivism. In: Ilbery, B. (ed.) *The Geography of Rural Change.* Longman, London.

Lawson, C. (1997) Territorial clustering and high-technology innovation: from industrial districts to innovative milieux. ESRC Centre for Business Research Working Paper 54, Cambridge University, Cambridge.

Leydesdorff, L. and Etzkowitz, H. (1998) The triple helix as a model for innovation studies. *Science & Public Policy* 25, 195–203.

Mahroum, S., Atterton, J., Ward, N., Williams, A., Naylor, R., Hindle, R. and Rowe, F. (2007) *Rural Innovation.* NESTA, London.

Mantau, U., Merlo, M., Sekot, W. and Welcker, B. (2001) *Recreational and Environmental Markets for Forest Enterprises – a New Approach Towards Marketability of Public Goods.* CAB International, Wallingford, UK.

Marino, L.D. (2009) Urban–rural linkages fostering sustainable development: the rural development perspective. Presentation to Seminar on Urban–rural linkages fostering sustainable development, Brussels, 23 January.

Mather, A. (2001) Forests of consumption: post-productivism, post-materialism and the post-industrial forest. *Environmental and Planning C: Government and Policy* 19, 249–268.

Mather, A., Hill, G. and Nijnik, M. (2006) Post-productivism and rural land use: cul de sac or challenge for theorization? *Journal of Rural Studies* 22, 441–455.

McGranahan, D.A. and Wojan, T.R. (2007) Recasting the creative class to examine growth processes in rural and urban counties. *Regional Studies* 41, 1–20.

Millennium Ecosystem Assessment (2005) *Ecosystems and Human Well-being: Synthesis.* Island Press, Washington, DC.

O'Connor, D., Renting, H., Gorman, G. and Kinsella, J. (eds) (2006) *Driving Rural Development: Policy and Practice in Seven EU Countries.* Van Gorcum, Assen.

OECD (2001) *Multifunctionality; Towards an Analytical Framework.* OECD, Paris.

Persson, L.O. and Westholm, E. (1994) Towards the new mosaic of rural regions. *European Review of Agricultural Economics* 21, 409–442.

Porter, M.E. (2000) Location, competition, and economic development: local clusters in a global economy. *Economic Development Quarterly* 14, 15–34.

Ray, C. (2006) Neo-endogenous rural development in the EU. In: Cloke, P., Marsden, T. and Mooney, P. (eds) *Handbook of Rural Studies.* Sage, London, pp. 278–291.

Secco, L., Pettenella, D. and Maso, D. (2009) 'Net-system' models versus traditional models in NWFP marketing: the case of mushrooms. *Small Scale Forestry* 8, 349–365.

Slee, B. (1989) *Diversified Farm Enterprises* (2nd edn). Farming Press, Ipswich, UK.

Slee, B., Evans, R. and Roberts, D. (2004) Forestry in the rural economy: a new approach to assessing the impact of forestry on rural development. *Forestry* 77, 441–453.

Slee, B. (2005a) The halo effect: a widened perspective on the relationship between forestry and the rural economy. *Economic Studies (Bulgaria)* 14, 10–22.

Slee, R.W. (2005b) From countrysides of production to countrysides of consumption? *Journal of Agricultural Science (Centenary Review)* 143, 1–11.

Storper, M. (1995) The resurgence of regional economies, ten years later: the region as a nexus of untraded interdependencies. *European Urban and Regional Studies* 2, 191–221.

van Huylenbroek, G. and Durand, G. (2003) *Multifunctional Agriculture – a New Paradigm for European Agriculture.* Ashgate, Aldershot, UK.

Willis, K.G., Garrod, G.D., Scarpa, R., Powe, N.A., Lovett, A.A., Bateman, I.J., Hanley, N. and Macmillan, D. (2003) *The Social and Environmental Benefits of Forests in Great Britain. Report to the Forestry Commission.* Social & Environmental Benefits of Forestry Phase 2. Forestry Commission, Edinburgh.

9 Innovative Market Opportunities Related to Carbon Sequestration in European Forests?

Lorenzo Ciccarese, Peter Elsasser,* Antonis Horattas, Davide Pettenella and Gregory Valatin

Abstract

Carbon markets evolving around the United Nations Framework Convention on Climate Change may offer significant innovation potential for the European forest sector. We describe the opportunities that arise (and do not arise) for forest enterprises in the frameworks of various new markets: those directly induced by the Kyoto Protocol market-based mechanisms (such as Emissions Trading, the Clean Development Mechanism and the Joint Implementation), the European regulated carbon market (Emissions Trading Scheme), and the voluntary carbon market. This chapter also takes into account the trade-offs between marketing the carbon sequestration service and traditional timber production, as well as possible influences of changing accounting rules recently being discussed for the post-2012 period.

9.1 Introduction

Air has for long been cited as one of the very few examples of a 'free' good in economic textbooks (e.g. Samuelson, 1970, p. 37), i.e. a good that is so abundant that it need not be utilized in an economic way. However, the accumulation of pollutants and greenhouse gases (GHG) in the atmosphere has turned it into a scarce resource. While air pollution around industrialized cities and acid rain in remote areas have long been considered local and regional phenomena, the discharge of air pollutants such as ozone-depleting substances and GHGs into the atmosphere is indeed a global problem. Thus, the Earth's atmosphere has become a standard example of a global public good, its overexploitation being typical of the allocation problems associated with such goods.

Public goods problems are relevant for forestry in many ways. On the one hand, forests suffer from public 'bads' like air pollution and acid rain. On the other hand, many recreation and protection services provided by forests actually are public goods; their marketability is still restricted to niche markets (Mantau *et al.*, 2001), if they can be marketed at all. As a consequence, such goods may be undersupplied to society; another consequence is that forest enterprises forego income opportunities for some part of their production. From an

* Corresponding author.

institutional economics perspective, one reason for overexploitation and undersupply of public goods is the lack of clearly defined property rights, which leads to market failure. However, the policy measures applied by European countries to remedy such failure have traditionally followed a command-and-control approach, leaving the underlying property rights problem unchanged (e.g. the recent European air quality directive; EU, 2008a).

This differs from climate policy. In the wake of the United Nations Framework Convention on Climate Change (UNFCCC, 1992), institutions and rules have been created that establish new tradeable property rights for specific GHGs. Even though these rules are intricate (and often reserved against the land use sector as a whole; cf. Schlamadinger *et al.*, 2007), they fulfil a necessary precondition for the marketability of climate-related outputs of forestry, and thus offer potentially important innovation opportunities for forest enterprises. To some extent, the impact of these rules is merely indirect, by changing the price relations between wood and competing material and energy sources. Direct influences exist with respect to the carbon sequestration service of forests (and possibly in the future, of wood products). In what follows, we first give a short overview of the additional quantity of carbon that could be sequestered by European forests (with a focus on EU member states). Afterwards we describe the relevant institutional framework, and the opportunities and risks that do (or do not) arise for forest enterprises because of the newly established property rights in various new markets: the intergovernmental market for emission certificates ('Kyoto market'), the Clean Development Mechanism (CDM) and Joint Implementation (JI), the EU Emissions Trading Scheme (ETS), and finally, voluntary carbon markets. We also discuss the trade-offs that exist between marketing carbon sequestration services and traditional timber production, and the impacts that renegotiations of carbon accounting rules might have on the respective property rights.

9.2 Carbon Sequestration Potential in European Forests

The forestry sequestration potential in Europe depends on a multitude of factors: on the growth conditions, age and composition of the forests in the respective countries; on future impacts of climate change itself on growth and decomposition rates as well as on the frequency and intensity of natural disturbances and on other ecological processes of forests (EEA, 2008a); but also on socio-political factors such as the development of carbon markets and changes in other economic sectors that influence the competitiveness of forestry mitigation versus other sector options in achieving climate mitigation goals. Therefore, it is impossible to quantify accurately the additional sequestration potential of Europe's forests; any such estimation has to rely on a multitude of assumptions. However, some estimates may shed light on the order of magnitude that can be expected.

The Fourth Assessment Report of the Intergovernmental Panel on Climate Change (IPCC, 2007) estimated in Europe an 'achievable' sink, by 2040, ranging from 90 to 180 million t of carbon dioxide equivalent ($MtCO_2e$) per year. The assessment considers all options offered by forests and forestry (protecting existing forests, increasing forest area, enhancing carbon density per unit of forest area, developing substitution effect of wood for energy and building material) and assumes a scenario of climate change mitigation policies with a price of US$100 per tCO_2e (IPCC, 2007).

However, under the rules, procedures and restraints on forestry activities set up by the Kyoto Protocol (KP) and the Marrakesh Accords (see below) the actual potential of the forest-based climate change mitigation activities is less substantial, 'relatively small' according to the European Environmental Agency (EEA, 2008b, p. 28). Combining estimates of expected abatement by countries in meeting their respective targets under the KP, the projected use of annual net carbon stock changes under Articles 3.3 and 3.4 during the period 2008–2012 is about 42.4 $MtCO_2$ per year in the EU-15 (EEA, 2009,

p. 106).[1] So far, a total net sequestration of about 8.2 $MtCO_2$ per year of the commitment period is intended from afforestation and reforestation activities (as defined by Art. 3.3 of the KP), the remaining 34.2 $MtCO_2$ by activities according to Art. 3.4 of the KP (mostly by forest management). For the EU-15, this is equivalent to 1.0% of base year emissions, or to 17.0% of the EU-15 reduction commitment of 341.0 $MtCO_2$ per year of the commitment period compared with base-year emissions (EEA, 2009).

9.3 Regulated Carbon Markets in the KP Framework

The background to the international emission market is the KP to the UNFCCC (KP, 1997), which entered into force in February 2005. By January 2010, the KP has been ratified (or accepted or approved, respectively) by 189 countries (plus the European Community as international organization).[2]

Of these, 38 industrialized countries (listed in Annex I to the UNFCCC) have committed themselves to reduce national anthropogenic GHG emissions by specified amounts. Altogether, these countries must reduce emissions by at least 5.2% below 1990 levels within the first commitment period (2008–2012). The individual reduction commitments vary; they are specified in Annex B to the KP. For the EU member states, the common reduction goal is 8.0%[3] (under the condition that the goal for the EU as a whole is not missed, different amounts for EU members are possible and have indeed been negotiated internally; see KP's Art. 4.5).

The KP allows Annex I Parties to add to or subtract from their initially assigned amount, thus augmenting or reducing the level of their allowed GHG emissions over the commitment period, by trading Kyoto units with other (Annex I and non-Annex I) Parties. These additions and subtractions are carried out in conformity with the so-called 'Kyoto mechanisms', which are Emissions trading (ETS), under Art. 17; JI, under Art. 6; and the CDM, under Art. 12. These market-based mechanisms enhance the flexibility of Annex I Parties to meet their GHG emission reduction or limitation commitments, by allowing these Parties to take advantage of lower-cost emission reductions outside their territories.

Practically, Annex I countries have to provide emission allowances for each emitted metric tonne of carbon dioxide equivalent (CO_2e) in the respective commitment periods. Basically, such allowances can stem from four sources (and are termed differently according to source):

- from the initially assigned amount at the start of a commitment period (Assigned Amount Units, AAUs);
- from joint climate protection projects together with other Annex I Parties in the framework of JI (Emission Reduction Units, ERUs);
- from joint climate protection projects together with developing countries in the framework of the CDM (Certified Emission Reduction, CERs); or
- from domestic Land Use, Land-Use Change and Forestry (LULUCF) activities that absorb carbon dioxide (Removal Units, RMUs).

In spite of the different origins and names, all four kinds of allowances mentioned here refer to the same unit (tCO_2e) and are, in principle, interchangeable.[4] In combination with tradability (as warranted by KP Art. 17), the establishment of quantified reduction commitments implicitly creates a system of ownership of emission rights. That is, it creates economically valuable property rights, even though the regulations of the KP alone do not establish concrete rules for operational emissions markets (this has been left to the subsequent negotiation process). Ideally, prices that might evolve in an international allowance market reflect the average marginal costs of emissions reduction of those states that are subject to emission limitations and eligible for emissions trading[5] (the logic is that as long as the technical possibilities for one unit of emission reduction are cheaper than the market price for an allowance, it is more favourable for a state to use these technical possibilities and to sell unneeded allowances accordingly; conversely it is more favourable to buy additional allowances as long as technical emission reduction is more expensive). Therefore, the prices realized in a future international emissions market can be directly interpreted as the monetary value that 1 t of reduced CO_2 emissions has for an economy that tries to comply with its reduction commitment. As long as the reduction is accepted in the accounting procedure, its origin does not change this value, be it from reduced industrial missions or from sequestration in forests.

9.3.1 Forestry in the first commitment period (2008–2012): main features of current agreements

In the negotiation process subsequent to the KP (i.e. in the conferences from 'Kyoto' to 'Marrakesh', 1997–2001), several rules specifically related to LULUCF have been agreed upon for the first commitment period 2008–2012, some of which are quite restrictive for forestry projects:

- *Mandatory accounting of activities related to KP Art. 3.3:* Carbon stock changes and non-CO_2 emissions (CH_4, CO, N_2O, NO_x) between 2008 and 2012 caused by afforestation, reforestation or deforestation since 1990 must be included in the commitments of Annex I countries. A Party included in Annex I that incurs a net source of emissions under the provisions of Art. 3.3 may account for anthropogenic GHG emissions by sources and removals by sinks in areas under forest management under Art. 3.4, up to a level that is equal to the net source of emissions under the provisions of Art. 3.3, but not greater than 9.0 Mt C/year (=33 Mt CO_2e/year), if the total anthropogenic GHG emissions by sources and removals by sinks in the managed forest since 1990 is equal to, or larger than, the net source of emissions incurred under Art. 3.3 (decision 16/CMP.1, Annex §10).
- *Optional accounting of activities related to KP Art. 3.4:* Countries may also elect to include carbon stock changes and non-CO_2 emissions between 2008 and 2012 on areas subject to forest management (or three other land use activities)[6] to meet their commitments. If a country has elected to account for any of these activities, it must account for carbon stock changes on all lands subject to these activities; moreover, if any of these activities have been elected, this choice is binding also for future commitment periods. The amount accountable for forest management is restricted by country specific caps (Table 9.1), which are, in most cases, only a fraction of the anticipated uptake.[7]
- *CDM and JI:* The only forestry activities accepted under the CDM are afforestation and reforestation. A further restriction is the general rule that accountability is limited to 1% p.a. of the involved industrialized country's total emissions in 1990. In order to address non-permanence (which refers to the reversibility of carbon sequestered in forest pools in the biosphere) of afforestation and reforestation project activities under the CDM, different types of CER allowances are issued for forestry activities, all of which expire after a specified time (decision 5/CMP.1). In the JI framework, afforestation and reforestation projects as well as forest management projects are basically accepted.[8] However, the latter underlie the same country-specific caps, which have already been mentioned above. Both JI and CDM projects must lead to measurable as well as verifiable emission reductions (or to a carbon sequestration) that are additional to what would have occurred without the projects.

9.3.2 Monetary value of emission allowances

Bearing in mind that allowances are basically interchangeable (whether originating from GHG emission reductions or from LULUCF sinks), and that forestry counterbalances only

Table 9.1. Projected net carbon stock changes under Arts 3.3 and 3.4 for the first commitment period of the Kyoto Protocol and election of Art. 3.4 activities.

	Art. 3.3	Art. 3.4			Total
	Net carbon stock change 2008–2012 (Mt CO_2/year)	Additional activities elected	Net carbon stock change 2008–2012 (Mt CO_2/year)	Max. allowance for FM[a] 'cap' (Mt CO_2/year)	Accountable sum[b] (Arts 3.3+3.4) (Mt CO_2/year)
Austria	–0.7	None	Not applicable	(–2.31)	–0.7
Belgium	No estimates available	None	Not applicable	(–0.11)	Not estimated
Bulgaria	Not reported	None	Not applicable	(–1.36)	Not estimated
Cyprus	Not reported	Not applicable	Not applicable	Not applicable	Not estimated
Czech Republic	Likely small sink	FM	Likely above max. allowance	–1.17	–1.17
Denmark	–0.288	FM, CM, GM	FM: –3.60; CM: –1.72	–0.18	–2.185
Estonia	No separate estimates available	None	No separate estimates available	(–0.37)	Not estimated
Finland	+1.90 to +2.40	FM	–2.5 to –3.0	–0.59	–0.59
France	–0.84	FM	–67.62	–3.23	–4.07
Germany	No estimates available	FM	–7.3	–4.55	–4.547
Greece	–0.80	FM	–2 to –4	–0.33	–1.13
Hungary	Not reported	FM	Not reported	–1.06	Not estimated
Ireland	–2.24	None	Not applicable	(–0.18)	–2.236
Italy	No separate estimate available	FM	–10.20	–10.19	–10.2
Latvia	Net sink	FM	No estimates available	–1.25	Not estimated
Lithuania	No separate estimates available	FM	No separate estimates available	–1.03	Not estimated
Luxembourg	Net sink	None	Not applicable	(–0.04)	Not estimated
Malta	Not reported	Not applicable	Not applicable	Not applicable	Not estimated
Netherlands	–0.11	None	Not applicable	(–0.04)	–0.11
Poland	Net sink	FM	Likely above max. allowance	–3.01	–3.01
Portugal	–3.36	FM, CM, GM	FM: –0.8; CM and GM: –0.5	–0.81	–4.66
Romania	Not reported	FM, RV	Not reported	–4.03	Not estimated
Slovakia	Net sink	None	Not applicable	(–1.83)	Not estimated
Slovenia	No estimates available	FM	–1.3	–1.32	–1.3
Spain[c]	Not estimated separately	FM, CM	Not estimated separately	–2.46	–5.8
Sweden	+0.60	FM	–15.00	–2.13	–2.13
UK	–2.68	FM	–1.69	–1.36	–4.04
EU-27	**–8.23[d]**		**–31.07[d]**	**–37.44 (–44.92)**	**–47.87[d]**

NB: Consistent with the reporting of emission inventories, a negative sign (–) is used for removals and a positive sign (+) for emissions.

[a]Numbers in brackets added from Decisions 16/CMP.1 and 8/CMP.2 of the Conference of the Parties serving as the Meeting of the Parties to the Kyoto Protocol (not reported in EEA, 2009).

[b]In addition to accounting for FM up to the maximum allowance, Parties may account for removals from FM to compensate net emissions under Art. 3.3. In Finland and Sweden, removals from FM are projected to exceed the sum of emissions under Art. 3.3 and the maximum allowance for FM.

[c]Spain only estimated the aggregated reductions of Arts 3.3 and 3.4 together.

[d]Sums according to EEA (2009).

FM, forest management; CM, cropland management; GM, grazing land management; RV, revegetation.

Source: EEA (2009, Annex, pp. 101–102; modified).

a small part of the Annex I states' emissions, markets for emission allowances can provide useful information in estimating the monetary value of forestry's sequestration service. However, information about the market value of emission allowances in the intergovernmental trade is still not very reliable for several reasons. One of these is the very limited number of completed pilot transactions – this number will probably increase significantly only at the end of the commitment period, when governments know better by how much they will have fulfilled (or missed) their respective emission limits. Furthermore, prices for such transactions are usually the result of bilateral bargaining; they are not necessarily published, and may be influenced not only by the economic scarcity of the allowances, but also by political considerations. Prices in bilateral trading reported by the World Bank have been about €10/AAU in 2008/2009 (Capoor and Ambrosi, 2009, p. 56). It can be assumed that these prices are comparatively low; because of the little market volume they might still be restricted by CER-Prices today (CER due December 2010–2012 are reported to be about €11.50–13.50/CER according to broker information).[9] An AAU auction has been announced several times by the Czech Republic according to press reports,[10] but has still not taken place.

Another source of price information is the ETS of the EU, which started before the first Kyoto commitment period (i.e. in January 2005). Information about the prices realized by the ETS is available on a daily basis;[11] however, it must be noted that the market rules for the products traded there (European Emission Allowances = EUA) are not directly comparable to those for the intergovernmental trade in allowances, although both types of allowances uniformly refer to 1 t CO_2. The main differences are: first, that trading partners in the ETS are individual enterprises rather than governments; and second, that the ETS covers only a part of the national economies' emissions, namely those caused by bigger enterprises in specific energy intensive sectors. Other emitters (and potential suppliers of credits) are excluded from this market. Hence, ETS prices do not reflect the scarcity of emission rights in the whole economy, but only in a subset of it. Keeping these differences in mind, ETS prices can help to approximate monetary values of intergovernmental emission allowances. In the second period of the ETS, which started in 2008, spot market prices have fluctuated between €7.95 and €28.75/EUA, recently stabilizing around €14/EUA (Fig. 9.1). For EUA futures due December 2012, prices are higher by €1–2/EUA.

Estimates of the marginal costs of emission reduction are a supplementary source of market-related value information. Again, the

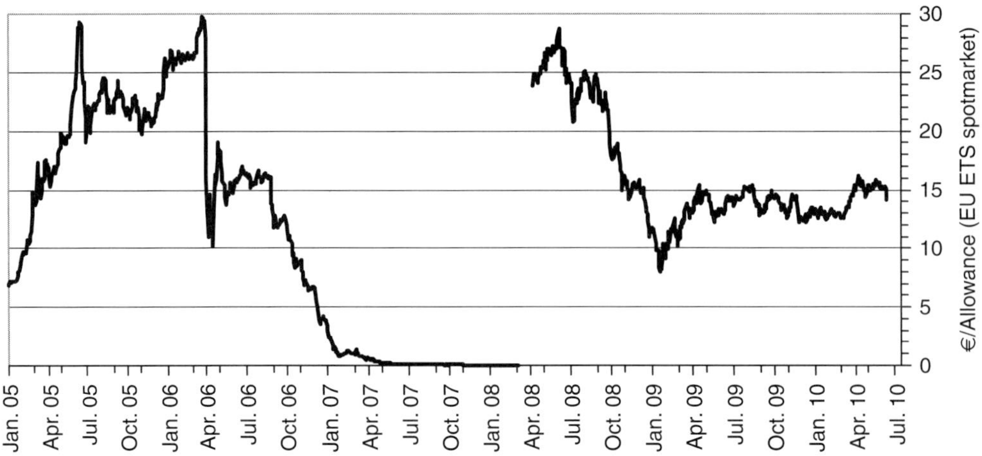

Fig. 9.1. Development of EUA prices since start of the ETS (data sources: Nordpool; EEX).

reason here is that rational decision makers will invest in abatement technology rather than buying additional emission allowances as long as additional investments are cheaper than additional purchases of allowances, and vice versa. Abatement costs depend on the desired amount of emission reduction at a specific time, and the technical possibilities of achieving this goal; the more that trade between agents is restricted, the higher are the overall costs of abatement (because such trade restrictions will exclude the use of some cheaper technical possibilities existing elsewhere). Estimates of avoidance costs based on model forecasts therefore depend on several assumptions about such cost influences. Even before the KP entered into force, several model results had been published (assembled by IEA, 2001), the respective estimates for the first commitment period ranging from US\$6 to US\$36/t CO_2 when trade between all Annex B states was assumed. Newer studies confirm this order of magnitude (Heggedal and Kverndokk, 2007). Holtsmark and Mæstad (2002) have calculated marginal avoidance costs between US\$12 and US\$45/t CO_2 for different EU member states (Sweden being an outlier with US\$97/t CO_2), which would fall to US\$16/t CO_2 under free trade, or US\$26/t CO_2 under restricted trade with 'hot air' (Holtsmark and Mæstad 2002, p. 215). Criqui and Kitous (2003), too, estimated a certificate price of US\$26/t CO_2 for the existing trading possibilities, which would be reduced if additional JI and CDM certificates were included.

Summing up, the market value of 1 t of reduced CO_2 emissions – or of 1 additional t of CO_2 permanently sequestered by forests – can be estimated within the range €10 to €30 for the first commitment period. The reduction in economic activity and thus emissions (as well as the opportunity costs of their avoidance) because of the recent world economic crisis may result in this value being closer to the lower than to the upper bound.

The (maximum) volume that the existing forests in the EU could contribute to a respective Kyoto-based market is, however, restricted by the caps negotiated for KP Art. 3.4 (listed in Table 9.1). Since only two-thirds of the EU countries have elected to account for 'forest management', the total that could be accounted for annually amounts to 37.44 Mt CO_2 out of a total potential contribution of 44.92 Mt CO_2. Multiplied by the value estimates above, the total potential annual value can be put at around €0.37–1.12 million (or €1.9–5.6 billion for the full 5-year commitment period). Afforestation of previously unforested areas (as allowed for under KP Art. 3.3) could further increase this sum. However, reliable volume estimates for this are not available for most EU countries (EEA, 2009, Annex, p. 101).

9.3.3 Implications for the forest sector

The value estimates of the newly established property rights outlined above amount to about 5% of the value-added of European forestry as a whole (cf. UNECE, 2005, p. 121). In this regard, the KP provides significant potential for marketable sequestration services as a possible innovative product of European forestry.

However, several drawbacks exist. First, carbon sinks in existing forests constitute property rights only in countries that have opted for 'forest management' according to KP Art. 3.4, but not in countries that have not. In the latter case, only credits derived from afforestation and reforestation GHG balances would be accountable. Second, it has to be kept in mind that the property rights discussed here belong to states, not to individual enterprises, since the KP regulates legal relationships between states only. For individual enterprises to take a share, political agreement on a revenue-sharing mechanism would be necessary. Although systems are conceivable which would devolve carbon property rights completely to the enterprise level (e.g. by establishing a project-based incentive framework; cf. Schmidtke and Kägi, 2006), several barriers complicate such a solution (Elsasser, 2008). One of these is that the transaction costs for implementation, administration and control of such a project-based system at state level as well as at enterprise level might easily consume most of the net revenues – especially since these revenues are severely restricted by the country-specific caps. Beyond that, the caps also induce (or tighten) distributive problems of a project-based system: for example, the question of whom to include and whom to exclude from revenue sharing (should all forest enterprises receive a proportional share of the distributable benefits even if this share might be negligible because of the cap, or should the procedure be more selective in order to maintain the incentives for the sequestration service?), or the question of how to distribute the liability for permanence of the sinks, if only some part of the benefits produced can be distributed. Because of such problems it seems that most European countries will not establish project-based revenue sharing systems, either because their governments prefer an indirect transfer to the forest sector of the financial benefits generated, e.g. through some form of subsidy, or because they decide not to share carbon revenues with forest enterprises at all.

Even if a state decides in favour of some form or other of carbon revenue sharing, using a forest stand as a carbon sink implies that the trees cannot be harvested and used for timber production, i.e. the production of sequestration services may cause opportunity costs at the state as well as at the enterprise level. Very probably the shape of the marginal opportunity cost curve is convex and increasing with quantity (because with increasing number of stands to be abandoned from wood utilization, the probability increases that stands with high timber values will have to be abandoned). In contrast, marginal carbon sequestration revenues depend only linearly upon quantity, and do not increase further once the national cap (or the maximum amount a single enterprise is entitled to, respectively) is reached. This not only implies that the possible sequestration revenues are restricted by the cap, but also that the balance of revenues and opportunity costs might become negative for higher quantities of sequestration (this is presented schematically in Fig. 9.2).

Parameterizing this figure and identifying an 'optimal' balance of timber harvesting and carbon sequestration for an

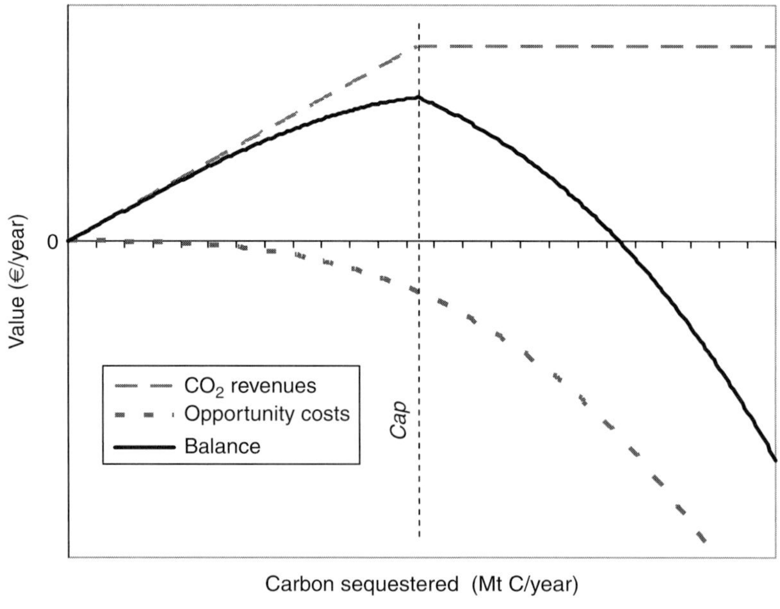

Fig. 9.2. Cumulative curves of marginal CO_2 revenues, opportunity costs and their balance depending on quantity of carbon sequestered (schematic).

enterprise (or for a country or for the EU as a whole) is not possible in a general manner while data about quality, quantity and monetary value of all the forest stands potentially involved are lacking. However, a back-of-the-envelope calculation may help in judging the relevance of such opportunity costs: assuming a mean carbon price of €20/t CO_2, gross revenues of a cubic metre of raw wood used as a carbon sink amount to about €27/m^3 (standing volume). In contrast, average timber revenues can be much higher – in the case of Germany for example, they have been estimated at about €40/m^3 (standing volume, year 2006/2007) (Elsasser, 2008, p. 26 ff.).[12] Hence harvesting and selling the timber may currently be more economical for most harvestable stands than selling the carbon sequestration certificates that might be generated by the same stands.

9.3.4 Future prospects

The intrinsic complexities of LULUCF activities have led to contentious and prolonged debates about the opportunities and merits of their inclusion in the KP's first commitment period (2008–2012). The Marrakesh Accords specify which LULUCF activities are to be included under the KP for the first commitment period and provide rules on how they are to be accounted for (Schlamadinger *et al.*, 2007). Current provisions for inclusion of LULUCF contain tangible limitations and drawbacks that now, with a better understanding of the problems associated with LULUCF accounting, are under renegotiation for the future climate change commitments after 2012.

At the 13th session of the United Nations Climate Change Conference in Bali, Indonesia (December 2007), nations agreed on a 2-year negotiating process, the Bali Roadmap, which included 'tracks' under the UNFCCC and the Protocol, and set a deadline for concluding the negotiations at the 15th session of the UNFCCC (COP-15), in Copenhagen (December 2009). However, a compromise was not reached in COP-15, so that binding agreements for the post-2012 period have been delayed. With respect to LULUCF, a decision was at least drafted (FCCC/KP/AWG/2009/L.15), which requests the Subsidiary Body for Scientific and Technological Advice (SBSTA) to establish a work programme that explores moving from an activity-based approach towards a land-based approach that: comprehensively accounts for all anthropogenic removals and emissions from the land base (living and dead biomass, and soil carbon), as are already required to be reported under the UNFCCC; considers procedures for additional LULUCF activities under the CDM; considers the role of non-permanence; revises reporting guidelines; and revises supplementary methodologies. The annex to the draft decision contains options on accounting for forest management using reference levels or caps, definitions of natural disturbance and harvested wood products, as well as an option for a land-based approach. However, this draft decision has not yet been formally approved by the parties.

Given the current state of negotiations, it remains questionable whether the accounting procedures for forest sinks valid in the first commitment period will be maintained at all in the future. Several alternative accounting rules have recently been in the negotiations for a post-Kyoto period. Krug *et al.* (2009) have demonstrated for the case of Germany that some of the accounting options discussed may even reverse carbon sinks to sources, or sources to sinks, depending on whether the additional carbon sequestration within a given period is compared with the carbon stocks of a base year (gross–net accounting, as is current practice for forest management), or to previous sequestration rates (net–net accounting), and whether the reference is a fixed base year or a moving reference period. Such a change of accounting rules could make the property rights associated with the carbon sequestration service elusive: the associated liability to achieve a specific minimum sequestration rate could even turn such property rights into a burden for a state's forest sector and its enterprises, rather than offering novel income opportunities for forestry.

9.3.5 'Kyoto mechanisms' CDM and JI

Annex I parties to the KP can also meet their obligations (at least partly) by using two kinds of cooperative mechanisms: the CDM and JI. These mechanisms aim at providing cost-effective opportunities, based on the implementation of projects that both reduce or offset emissions abroad. The kind of mechanism under which a project is developed depends on the geographical areas where the planned activities take place. If these take place in a developing (non-Annex I) country, the project undergoes the CDM, and when these take place in another Annex I country, the project undergoes the JI mechanism.

The CDM is today the largest project-based compliance market in the world; both the trading volume and market value of the CDM have grown substantially in recent years (Capoor and Ambrosi, 2009, pp. 31–32). Because CDM projects have to be implemented in a non-Annex I country, the CDM is in general not an instrument suited to forest projects within most European countries; however, it might be relevant for European forest enterprises with possessions abroad. CDM forestry projects are restricted to afforestation (planting trees where none was previously growing) and reforestation (replanting trees on recently harvested forest sites) under the condition that these are additional, i.e. projects that would not otherwise have occurred. If, for example, reforestation is required by management plans, laws or regulations, a reforestation project is not considered 'additional' and therefore cannot earn credits under the CDM. Further restrictions also limit the attractiveness of forestry CDM projects. As already mentioned, the KP restricts the use of CDM by limiting the maximum acceptable amount of credits for forestry CDM projects to 1% of the total reduction commitment of the Annex I country. Thus industrial countries cannot make wide use of (forestry) CDM projects for meeting their carbon targets. Beyond that, the additionality requirement as well as the huge transaction costs potentially associated with a CDM project can severely restrict the economical viability of CDM forestry projects. The CDM views additionality in forestry very narrowly. A given site is viewed as having a given productivity potential. If this site is productive and economically attractive, it most probably does not qualify for the CDM. If on the other hand the site offers only low productivity, a project may not be profitable. In this case, it qualifies more easily for the CDM, but can be uninteresting for investors. As the productivity of sites can be improved (to sequester more carbon) by additional inputs, if a broader approach were taken, even economically viable sites could fit into the CDM framework if credits were provided for the additional carbon induced by the carbon credit over and above that generated in the absence of carbon credits (for a further discussion of additionality problems, cf. Trexler *et al.*, 2006).[13]

JI is in many respects comparable to the CDM, but procedures and project requirements vary in detail. Under both mechanisms, projects in foreign countries are financed by an Annex I country, which can use the generated credits for complying with its own emission reduction commitments. However, JI projects are conducted jointly between two Annex I countries. Therefore they might theoretically be more directly of interest to European forest enterprises – specifically since they are not restricted to afforestation and reforestation only but are, as already mentioned, basically open to sequestration projects in already existing forests (i.e. 'forest management' projects), too. Nevertheless, the JI market in general is still a much smaller market than the CDM market (in the years 2007 and 2008, volumes as well as values of JI market transactions were less than 8% of the CDM transactions; Capoor and Ambrosi, 2009, p. 31).

For forestry projects conducted both under the CDM and under JI, the approval rules developed in the international negotiations cause considerable transaction costs, which may significantly reduce the economic viability of such projects. Basically the standard procedures require two steps for approval (DIW and KPMG, 2003): in advance of a project (step 1), a rather comprehensive concept has to be developed, called the PDD ('project design document'), which must be officially approved, audited, validated (in the CDM) or verified (in JI) and then registered (JI: recognized). During the project (step 2) adequate monitoring and periodical reporting are required; based on these reports the emission reductions of the project can be verified and certified (CDM), or registered in the national inventories, respectively (JI). Only after this procedure can CER (or ERU, respectively) be issued and the project investor reimbursed. The costs associated with these procedures are substantial. According to Kapp (2004), the PDD alone for a forestry project requires a budget of around US$50,000–70,000 (including travel costs, laboratory costs for the analysis of soil and wood samples and costs for the field work). The finalized PDD has then to be checked by an Accredited Independent Entity (AIE); fees for the AIE vary between US$20,000 and US$25,000. Together with other costs and fees, this results in costs of around US$100,000 for step 1. In step 2, the periodic monitoring (once per commitment period) can be assumed to cost around US$30,000 (according to the experience of Kapp, 2004, with large afforestation projects in eastern Europe; Cacho *et al.*, 2004, report similar numbers for Indonesia). Monitoring results must also be verified by the AIE, with the same fees as before. Altogether, the transaction costs for forestry projects can be assumed to amount to US$100,000–250,000 including monitoring. This compares quite well to the transaction costs reported for other (non-forestry) CDM projects (cf. Fichtner *et al.*, 2003; Michaelowa *et al.*, 2003), which sometimes even exceed the production costs of the projects. Although simplified project approval procedures may be used in special cases that reduce transaction costs,[14] the costs of meeting the approval requirements can price many possible forestry projects out of the market, or at least favour large-scale projects over smaller ones because of scale economies (for an analysis under German conditions, see Dieter and Elsasser, 2004).

Because of such problems, as well as the temporary nature of forestry credits under the CDM, forestry projects have not yet gained much importance either under the CDM or under JI. According to UNFCCC, to date only a single afforestation and 12 reforestation projects have been registered under the CDM; another reforestation project is requesting registration.[15] Under JI, a Romanian afforestation project is listed by UNFCCC as the only currently registered LULUCF project.[16] Altogether, forestry makes up less than 1% of all CDM and JI projects.[17] Thus, although forestry 'can make a very significant contribution to a low-cost global mitigation portfolio that provides synergies with adaptation and sustainable development' according to the IPCC's Fourth Assessment Report (IPCC, 2007), 'this opportunity is being lost in the current institutional context and lack of political will to implement, which has resulted in only a small portion of this potential being realised at present'.

9.4 Forestry and the European ETS

Although closely linked to the UNFCCC climate policy, the rules of the ETS differ from Kyoto rules in many respects. Aimed at contributing to meeting EU and member states' emissions reduction commitments under the KP by creating an efficient European market in GHG emission allowances as part of an enterprise-level cap-and-trade scheme (EU, 2003a), the ETS came into operation on 1 January 2005. Currently the world's largest carbon market, accounting for over 70% of the estimated total volume and value of transactions worldwide with transactions valued at over US$90,000 m in 2008 (Capoor and Ambrosi, 2009; Hamilton *et al.*, 2009), the scheme does not cover forestry carbon sequestration at present. This is partly related to the scheme being designed principally as a technological driver for emission abatement by energy and industrial sources (EU, 2003b, section 3.2). However, inclusion of forestry is still being discussed. Moreover, the ETS regulations affect wood markets indirectly, by changing the price relations between fossil and renewable materials and energy sources. Thus, the ETS might still turn out to be important for the innovation potential of European forestry.

Phase I of the ETS (2005–2007 inclusive) covered carbon dioxide emissions from four types of industrial activities. These were energy (combustion installations over 20 MWth, mineral oil refineries, coke ovens), ferrous metals production and processing, building material (cement, glass and ceramics) production, and pulp, paper and board manufacture.[18] Coverage under phase II (which runs from 2008 until the end of 2012) is to be broadened from 1 January 2012 to include civil aviation (EU, 2008b). Phase III (2013–2020) is to be extended further to cover some other industries (e.g. ammonia production, production of soda ash and sodium bicarbonate) and other GHG emissions in some cases – such as nitrous oxide from nitric acid production and perfluorocarbons from primary aluminium production (EU, 2009b, Annex 1) – as well as 'technical' carbon capture and storage activities (as opposed to forest sinks). Currently covering 11,000 installations, the ETS accounts for around 50% of EU CO_2 emissions and about 40% of total EU GHG emissions, and is expected to increase to 43% of total GHG emissions in 2013 (EU-COM, 2009).[19]

Apart from the focus upon emission abatement in power generation and major industrial sources, other reasons given for today's exclusion of forestry carbon credits include still unresolved permanence and leakage issues, high transaction and administrative costs, and remaining uncertainties in quantification, monitoring and verification of emission removals by sinks (EU-COM, 2008a,d). For example, the European Commission argued that emissions from forestry cannot be monitored, reported and verified with the same level of accuracy as emissions currently covered by the ETS, which it viewed as a prerequisite for the scheme's extension to other activities (EU-COM, 2008c, p. 4). Characterized as providing only a temporary store of carbon (which would subsequently be released into the atmosphere), inclusion of LULUCF activities was also not viewed as compatible with the scheme's focus on permanent reductions by emission sources (EU, 2003b, section 3.2). Problems associated with the potential reversibility of carbon sequestration are reported to have been the primary concern of most member states opposing inclusion of LULUCF credits.[20] In addition, there were also concerns that inclusion of LULUCF credits, especially where issued in countries without binding national emissions targets, could undermine power and industrial sector emission reduction activities within the EU (CF, 2008; Tuerk *et al.*, 2008). The lack of common approach by member states to accounting for carbon from forestry activities under the KP is considered a further impediment (EU-COM, 2008a; Wemaëre, 2008),[21] as is perceived non-equivalence of GHG removals by LULUCF sinks with emissions reductions from sources (e.g. CAN, 2006; Spash, 2010, forthcoming).

Under the 2004 'linking' Directive (EU, 2004), from 2008 onwards companies have been allowed to use emission reduction credits from CDM and JI to count towards their ETS targets with the aim of reducing EU emissions abatement costs and contributing to global sustainable development (EU, 2003b, section 1.3).[22] However, with some member states concerned that LULUCF credits would be relatively cheap and inclusion could undermine technology transfer activities (Tuerk *et al.*, 2008), forestry credits were excluded under 3(b) of this directive[23] pending further consideration of issues including the temporary nature of forestry CDM credits (EU, 2004, paragraph 9). The European Commission subsequently argued that including temporary credits from forestry[24] under phase II of the scheme would be incompatible with the aim of enhancing its simplicity,[25] instead serving to increase its complexity (EU-COM, 2006, section 3.1), a view also held in considering potential inclusion under phase III (EU-COM, 2008d).

Inclusion of new types of credit under phase III of the scheme after 2013 is conditional upon negotiation of a satisfactory post-2012 international agreement in the UNFCCC framework (EU-COM, 2009). Article 28 of EU (2009b), for example, commits the European Commission to submitting a report to the

European Council on modalities for including emissions from LULUCF activities within 3 months of signature by the EU of an international agreement to reduce GHG emissions by at least 20% by 2020 compared with 1990 levels. However, the Commission currently considers avoided deforestation and forest degradation credits should only be considered for inclusion after a thorough review of the pilot phase and after 2020. The Commission argues that inclusion of associated credits within the ETS would create serious imbalances between supply and demand, and potentially even lead to the collapse of the scheme (EU-COM, 2008a).[26] Indeed, emissions from worldwide deforestation are estimated to be around three times the total volume of emissions currently regulated under the ETS, which is the major operational carbon trading system in the world at present. Other potential obstacles to inclusion are unclear land tenure and the need to improve national monitoring and verification systems in some cases, as well as permanence and liability issues.[27] In view of these perceived impediments, the Commission has recommended that such forestry credits be recognized for ETS compliance only as a complementary tool and in the longer term (post-2020), providing certain conditions (especially supply/demand balance and liability) are met (EU-COM, 2008b, section 4.2).[28]

The arguments against the inclusion of forestry carbon in the ETS remain contested. It has been pointed out that inclusion could make a significant contribution to climate change mitigation by allowing adoption of tougher overall emission reduction targets (Trines, 2006), and that any impact on EUA prices also depends upon appropriate supplementarity rules limiting the maximum proportion of externally-sourced credits used within the ETS (Eliasch, 2008).[29] Modelling for the Eliasch Review, for example, indicates that if the supplementarity limit is raised simultaneously with adopting more stringent targets, prices would not necessarily fall, with inclusion of international forestry credits having no price impact if most credits continue to be sourced from within the EU (Eliasch, 2008, Tables 11.1 and 11.2).[30] A range of methods is available that could potentially be used to manage non-permanence risks associated with forest carbon,[31] and a variety of approaches that could allow inclusion of temporary credits within the ETS.[32]

Despite strong support for inclusion of at least some elements by certain member states,[33] and in contrast to many other ETSs worldwide (as well as plans to extend the scheme to geological sequestration activities) it may appear unlikely that forestry carbon sequestration will be covered in the near future. This position could rapidly change in the aftermath of a post-2012 agreement on climate change being reached, once more comprehensive international rules of carbon accounting for LULUCF are agreed, or as the science base underpinning quantification is strengthened, and tighter monitoring and verification standards are established and receive wider acceptance.[34] Inclusion of forestry carbon sequestration or establishment of a parallel system of incentives to support it may also be expected to be subject to continuing discussions between member states and the Commission,[35] not least given growing appreciation of its essential contribution to climate change mitigation and that most opportunities are currently being lost in the current institutional context (IPCC, 2007, p. 543; Eliasch, 2008). The extent to which the potential opening up of the ETS to forestry credits could be expected to provide opportunities for EU forest enterprises would also depend on exactly how it was accomplished.[36]

Although not covered directly, the ETS could potentially provide a significant source of funding for EU forestry carbon sequestration activities and measures to avoid deforestation and increase afforestation and reforestation in developing countries. These are among nine activities listed for potential funding using revenues of at least 50% of the proceeds of auctioning EU emission allowances (EUAs) by member states (EU, 2009b, Art. 10).[37] Relatively little used to date,[38] auctioning of EUAs is set to become the primary means of allocating allowances from 2013,[39] with over 50% of total EUAs expected to be auctioned by 2013 and 70%–80% by 2020 (Capoor and Ambrosi, 2009).

Furthermore, exclusion of power plants fuelled exclusively by biomass from the ETS could provide a competitive advantage for biomass-fired plants,[40] increasing demand for woodfuel. This could effectively create a premium for associated carbon substitution benefits of forestry (potentially further increasing prices of wood products for other uses too).[41] However, because of the paucity of woodfuel price data, it is difficult to assess to what extent such a premium already exists, or how it compares with prevailing prices for EUAs.

9.5 Voluntary Carbon Markets

Even before the KP had entered into force, companies, agencies and public administrations as well as individuals had started to contribute voluntarily to a reduction of emissions, in order to diminish their 'carbon footprints'. There are many motivations for deciding to take on voluntary reduction targets. Besides idealistic and ethical motives, pragmatic considerations can also play a role. For example, voluntary emission reductions can increase a firm's competiveness by adapting early to future emission limitation regimes (and by reducing the level of energy consumption at the same time), help to address environmentally oriented customers and open up opportunities for green marketing, and generally demonstrate the corporate social responsibility of a firm (Ciccarese and Pettenella, 2008; Neeff *et al.*, 2009).

A direct way of coping with a voluntary emission reduction strategy is to invest in internal abatement measures such as energy efficiency measures, mobility plans for employees, or to support others in doing so. Beyond that, companies (or individuals) also take advantage of carbon offsets generated by others in external GHG emissions reduction projects if these projects' emission reductions are traded to others. Carbon offset projects can offer investors immediate benefits and more flexibility, and they can be based at a wide range of activities, including wind and solar energy, bio-energy and agriculture and forestry sequestration. Such activities need not (yet) be accepted by the respective governments or by UNFCCC regulations (for example, forest management projects in those states that have not opted for activities according to Art. 3.4 of the KP). With respect to market organization, exchange occurs directly between the buyer and seller without requiring state sanction. This so-called voluntary over-the-counter (OTC) market does not necessarily operate via a formal exchange and needs not be driven by a cap. Additionally, voluntary carbon projects are already being brokered by market facilitators, and there are even stock exchanges that have specialized in trading carbon, most prominently, the Chicago Climate Exchange (CCX).[42] Officially sanctioned voluntary emission reduction and carbon offset purchasing programmes have also been organized in some countries, such as Japan (Keidanren Voluntary Action Plan on the Environment) and Australia (Greenhouse Challenge Plus programme).

The carbon unit traded in the voluntary market is usually termed VER (Voluntary Emission Reduction, or Verified Emission Reduction), equivalent to 1 t of CO_2. The reliability and level of standardization of the projects that generate VERs vary – some are only vaguely defined so that their actual contribution to emission reduction remains rather obscure, while others rely on sound certification criteria and independent verification. Ideally a VER is verified by an independent auditor, but has not (or not yet) undergone the procedures required a regulatory authority for registration, verification, certification and issuance under the KP. Indeed, one of the main problems for further integration of forestry activities into voluntary carbon markets is that project standards vary significantly, in terms of eligibility, additionality, quantification methods, monitoring, permanence risks, quality of certification, costs and fees of certification, validation and verification, and finally, also compatibility with socio-economic and environmental requirements. However, in the last few years, more than a dozen third-party standards and methodologies have been defined for verification of emissions reductions. A number of third-party certification standards specifically designed for voluntary forestry carbon offsets have appeared in response to the growing concerns over the credibility of the investments. Hamilton *et al.* (2010) list ten examples of such forest carbon offset standards, which differ with respect to the methodologies used; some of these accept only some of the possible forestry activities, or are restricted to specific geographic regions. To ensure the robustness of forestry-based VERs, along with associated standards and certification procedures, carbon registries have been designed to track carbon unit transactions and ownership as well as reduce the risk of double or multiple sales. As reported by Ecosystem Marketplace:[43]

> dozens of GHG registries exist around the world, but not all of them serve the same purpose. In general, existing registries can be divided into two different categories: emissions-tracking registries and carbon-credit accounting registries. When dealing with a commodity as intangible as a carbon credit, such registries are crucial, but they have not been prevalent in the OTC market until recently. Some standard providers maintain registries in-house, while others are teaming with registry providers from other organizations. Additionally, some registries remain distinct infrastructure providers unaffiliated with a particular standard.

According to the different quality of VER projects, the respective prices vary significantly, too (in 2008, OTC prices ranged between US\$1.20 and US\$46.90/CO_2e according to Hamilton *et al.*, 2010, p. 15). Two main factors restrict demand (and thus, prices) on these markets: one is that demand is, of course, voluntary; the other is a multitude of risks, which has to be assumed by buyers (e.g. reliability doubts in the case of unverified carbon projects, possible problems related to double counting with the regulated market or related to a lack of additionality, as well as the fact that even if a VER follows sound standards and is independently verified, it may later not be usable in a regulated market if it is not ultimately registered, e.g. as CER or ERU). Buyers therefore tend to pay a discounted price for VERs, which takes the inherent risks into account. On the other hand, prices can be quite high in cases of transactions that are politically influenced or have a donation character, rather than being motivated by profit maximization only (this may be assumed specifically in some bilaterally arranged transactions).

In general, the prices paid for carbon credits on voluntary markets are relatively low compared with regulated markets (for example, the volume-weighted average price of a voluntary carbon credit transacted on the OTC market has been estimated at US\$7.34/t CO_2e for 2008 (Hamilton *et al.*, 2009, p. 15), which is about 30% of the average EUA price in the same year); however, they have shown a steady upwards tendency over the last few years.[44] VERs generated from forestry projects sell at similar or even slightly higher prices. (Hamilton *et al.*, 2010, report volume weighted project developer prices for forestry projects of US\$8.44/t CO_2e.) However, transaction costs (e.g. fees demanded by market facilitators along the value chain) partly consume such benefits.

Total traded volume in voluntary carbon markets was still relatively small in 2008, amounting to just 2.9% of the total global carbon markets (4213.5 Mt CO_2e; Hamilton *et al.*, 2009). None the less, the voluntary market has recently seen rapid growth and is projected to expand even faster in the near future also in terms of volume. The estimated transaction volume in the voluntary market of about 4.0 Mt CO_2e in 2004 rose to 23.7 Mt CO_2e in 2006, 65.0 Mt CO_2e in 2007 and 123.4 Mt CO_2e in 2008 (Hamilton *et al.*, 2009). The aggregate value of these transactions amounted to US\$54.9 million in 2006, US\$335.3 million in 2007 and US\$704.8 million in 2008. Hamilton *et al.* (2009) found that the voluntary market has grown 165% while the Regulated Market experienced a 71% growth over the same period.

Turning specifically to forestry carbon sequestration projects in the voluntary market, several pioneer projects come to mind, which have helped in clarifying concepts and developing forest activities to mitigate GHG emissions. The Scolel Té project[45] in Chiapas and Oaxaca, Mexico, is one very popular example: since 1997, the Fédération Internationale de l'Automobile has supported this project, which helps communities in southern Mexico to develop sustainable land management and better livelihoods, in order to offset the GHG emissions caused by Formula 1 teams during a Grand Prix season (estimated in 1997 to be around 20,000 t of CO_2; Trabesinger, 2007). Likewise, the GHG emissions that directly resulted from the production of the movies *The Day After Tomorrow* and *An Inconvenient Truth* were mainly offset by forestry activities. In 2007, the Vatican allegedly became 'the world's first carbon-neutral state' thanks to an unusual donation from a Hungarian start-up called Klimafa: the company started a project to restore 15 ha of an ancient forest on a denuded Hungarian island to offset the Vatican's GHG emissions (Rosenthal, 2007).

Any type of forest activity can be suitable for the voluntary carbon market – from afforestation and reforestation to conservation or agro-forestry, and from forest management activities intended to maximize carbon sequestration in the living biomass to projects aiming at energy and material substitution, including biomass plants or projects that utilize durable wooden products in place of more energy-intensive products such as concrete and steel. In fact, the list of voluntary forestry projects is quite extensive (while forestry projects approved under the CDM or under JI are still rare, as seen above). From a global perspective, voluntary forestry projects *sensu strictu* originate from three major project types: afforestation/reforestation, improved forest management and REDD (Reducing Emissions due to Deforestation and Degradation [in Developing Countries]). Among these, afforestation/reforestation projects are by far the most common: according to a recent survey (Hamilton *et al.*, 2010), they accounted for 59% of the total volume hitherto transacted in the OTC, followed by REDD projects with 24% of the volume, whereas forest management projects contributed only 8% of the volume (the remaining 9% were mixed projects). With respect to location, the bulk of these projects have been conducted outside Europe – mostly in developing countries in Latin America and Africa (38%), but also in North America (38%). European projects have been of relatively minor importance; Hamilton *et al.* (2010) report that less than 5% of the market volume tracked in their survey originated from European countries.

The small share of European projects is related to several specific problems. One of these is the small size of many European forest enterprises, which boosts transaction costs and inhibits market access. Another problem is the interference with the regulated Kyoto market: since most European countries are listed in Annex I, sequestration by their forests is already covered in the national GHG inventories, and could not free up additional AAUs for these countries. If the same sequestration volume were sold by forest enterprises in the voluntary market, this would lead to double counting. The country where a VER project takes place would therefore have to cancel an equivalent number of AAUs in order to avoid such double counting, which would create opportunity costs for this country (Hamilton *et al.*, 2010, p. 21). These factors add to general concerns and criticisms still levelled at forestry contracts for producing carbon sequestration units (Davies, 2007), which have focused on the lack of rigorous standards in the OTC voluntary market,[46] but also on other possible problems, like non-permanence, leakage, lack of additionality and finally, possible conflicts with social and biological sustainability issues. Quite a lot of work may still be needed to provide the voluntary market with transparent and generally accepted rules for carbon offset investments in forestry.

In spite of such problems, several initiatives currently exist, which are developing forest sequestration projects for the voluntary carbon market in Europe. The German *Prima Klima weltweit* currently administers small-scale afforestation projects in the tropics, but also in Germany, the Netherlands, Latvia and Poland; previous projects have also been conducted in Slovakia, Ukraine and Hungary.[47] Similar approaches exist at regional level, e.g. the foundation 'Forests for Saxony'[48] and the 'Forest Shareholder' initiative,[49] which organize afforestation projects in the federal states of Saxony and Mecklenburg-Western Pomerania, supported by the respective state governments and the European Regional Development Fund; or the 'CarboMark' pilot action,[50] which plans to trade credits from a wide set of agro-forestry activities (including forest management, long lived wood products and urban forestry) in two Italian regions.

9.6 Conclusions

Valuing the environmental services of forests has been a particular concern of European forestry for some years. After new tradable property rights for specific GHG were established in the wake of the UNFCCC, hopes have been raised that the carbon sequestration service of forests could offer innovative market opportunities for forest enterprises. However, European (forest) policy does not seem to be very coherent with regard to the inclusion of forest sinks. Already the diverging attitudes towards the election of forest management according to Art. 3.4 of the KP (Table 9.1) reveal some inconsistency between member states in this regard, which may also be correlated with diverging interests between different EU directorates (specifically those for agriculture and environment), as well as between the respective ministries at national level. Overall, current policies reveal that the prospects for European forest owners to valorize the carbon sequestration service under current regulations and schemes are currently limited.

While an EU decision to include LULUCF carbon credits in its ETS is still lacking, selling LULUCF credits on the regular European certificate market will not be possible in the near future, although a later inclusion of forestry is still under discussion. Chances emerging from the ETS for forest enterprises are thus, for the moment, restricted to improved terms of trade for wood *vis-à-vis* competing fossil fuel intensive materials and energy sources. Direct marketing of sequestration credits by using the flexible mechanisms CDM and JI is possible in some cases,[51] but subject to several restrictions with regard to approval, and it is hampered by significant transaction costs. As yet, the practical use of these mechanisms has been limited.

Under the current institutional framework, LULUCF credits appear in the national GHG accounts and serve to help member states comply with GHG abatement commitments at national level (EEA, 2008b). In this respect, aggregate property rights for carbon sequestration by forests rest implicitly with governments.[52] As long as these property rights are not passed on to the enterprise level, the liability for continued sequestration remains at government level, too, which has to cover natural risks, possibly enhanced wood removals, and deforestation pressure, e.g. because of urban and industrial development. In several European states, whether and how forestry should be given a share in the benefits (and, possibly, also the risks) accruing to treasury because of the sequestration service of forests is still under discussion – with potential options ranging from the establishment of a project based framework (e.g. Schmidtke and Kägi, 2006) to simply using existing subsidy schemes (or not distributing funds to forest enterprises at all).[53] However, these prospects are still uncertain, and most probably the European countries will develop different approaches.

In this situation, voluntary carbon markets offer an attractive alternative for marketing the sequestration service of forest enterprises. Because of the limited prices for carbon units, this alternative may be regarded as an opportunity for additional income rather than an incentive to prioritize carbon sequestration as the main product. It may be particularly appropriate for forest enterprises owning many immature stands, or for enterprises not aiming at wood production only, but at multi-purpose forestry. In any case, two problems remain to be solved: one is the high transaction cost associated with the certification and marketing of carbon units; the other is the continuing lack of market transparency with regard to quality and reliability of voluntary carbon certificates, which in the long run could sap the market participants' confidence in this kind of product. Public institutions could play a crucial role in addressing both these problems, e.g. by informing enterprises and prospective customers about the possibilities of the voluntary markets, by coordinating small-scale projects in order to reduce transaction costs, and by supporting or developing credible and transparent certification schemes, or codes of conduct for carbon forestry projects.[54]

Creating an institutional framework that values forestry carbon sequestration will be important if the forestry sector is to be encouraged to play a greater role in helping cut atmospheric GHG concentrations. It will be important if significant opportunities for climate change mitigation are not to be missed.

Notes

[1] Article 3.3 contains regulations for afforestation, reforestation and deforestation, Art. 3.4 for the management of already existing forests (and three other land use categories) (see chapter 3.1). EU-15 comprises Austria, Belgium, Denmark, Finland, France, Germany, Greece, Ireland, Italy, Luxembourg, the Netherlands, Portugal, Spain, Sweden and the UK.

[2] Source: http://unfccc.int/files/kyoto_protocol/status_of_ratification/application/pdf/kp_ratification_20091203.pdf.

[3] Annex B lists 25 EU member states (i.e. all current EU members except of Malta and Cyprus).

[4] This does not, however, fully apply to LULUCF projects in the CDM framework (see below). In this case, credits with limited validity periods are issued (tCER, lCER) in order to account for the reversibility of biological carbon storage.

[5] The current eligibility list is http://unfccc.int/files/kyoto_protocol/compliance/enforcement_branch/application/pdf/eligibility_list_090711.pdf.

[6] These are cropland management, grazing land management and revegetation. Accounting is relative to carbon stock changes and associated greenhouse gas emissions from these activities in 1990.

[7] The basic idea of introducing these caps was pragmatically to exclude non-human induced carbon stock changes, such as atmospheric fertilization and the inherited age class structures of the forests (decision 16/CMP.1, appendix and footnote 5).

[8] Moreover, special conditions limiting the duration and requiring subsequent replacement of credits (like in the CDM case) are not necessary in the case of JI, because any later loss of carbon previously sequestered in a JI project (e.g. because of storm damage, forest fires or to harvests) will show up in the national carbon balances of the participating Annex I-states.

[9] http://www.nordpool.com/en/asa/Markets/Emissions/EUACER2/. Note that these prices apply to CER from emission reduction projects. Certificates from CDM forestry projects with limited durability (lCER, tCER) sell with a considerable discount because of their temporary nature.

[10] http://planetark.org/ark/51852.

[11] For example Nordpool ASA (http://www.nordpool.com/en/asa/Markets/Emissions/EUACER2/), EEX (http://www.eex.com/en/Market%20Data/Trading%20Data/Emission%20Rights/EU%20Emission%20Allowances%20|%20Spot).

[12] The numbers quoted above refer to gross revenues. Seen from an enterprise's view, gross revenues would have to be reduced by the respective transaction costs for approval (and possibly, control) of the sequestration service at the one hand, and for timber harvesting, transportation and marketing on the other. In the comparison above, it is assumed that both kinds of transaction costs belong to the same order of magnitude.

[13] For a review of different approaches to additionality and baseline determination, see Valatin (2009).

[14] For the CDM framework, simplifications have been agreed upon in COP 9 (Marrakesh accords) for small forestry projects that sequester less than 8000 tCO_2 per year and are realized by communities or persons with low income. In the JI framework, a simplified 'track 1' procedure can be used if both partner states meet certain requirements and have formally chosen to apply the 'track 1' procedure. Still Michaelowa *et al.* (2003) have estimated minimum transaction costs of about €80,000 for JI projects following the track 1 procedure (i.e. about 60% of the costs estimated there for the regular track 2 procedure; see Michaelowa *et al.*, 2003, Table 11).

[15] see http://cdm.unfccc.int/Projects/projsearch.html -> Afforestation and reforestation (as of 25 February 2010).

[16] http://ji.unfccc.int/JI_Projects/ProjectInfo.html (as of 25 February 2010).
[17] This is true in terms of the number of registered projects (total CDM: 2029 [forestry: 13]; total JI: 113 [forestry: 1]) as well as the respective amount of credits (CDM: 342 million CER [forestry 417,000 CER]; JI 23 million ERU [forestry 82,000 ERU]). This relationship does not change substantially even if the projects still awaiting validation/determination are taken into account (source: http://cdmpipeline.org/).
[18] Note that some emissions from the forestry sector (broadly defined) downstream are therefore covered because of inclusion of the pulp, paper and board industries. Of 11,500 installations covered by ETS in 2005, 7% (830) were reported to be pulp and paper mills, with the pulp and paper industry accounting for 2% of allowances in 2005 (Hyvärinen, 2005).
[19] To reduce administrative costs, member states are empowered to exclude small installations emitting up to 25,000t CO2e per year with thermal output below 35MW (and also hospitals) providing equivalent emissions reduction measures (e.g. taxation) are in place. See: EU (2009b, Art. 27).
[20] In particular, exclusion of JI credits relating to LULUCF from the ETS stems partly from the lack of modalities developed relating to non-permanence. It was also argued that non-permanence risks would have to be managed by monitoring LULUCF projects indefinitely (EU-COM, 2008a, pp. 58–59).
[21] Noting inconsistencies in methodologies between member states, the European Commission has argued that 'a coherent LULUCF inventory consistent across the EU and at the level of holdings, member states and the Community does not seem to be attainable in the foreseeable future' (EU-COM, 2008a, p. 60).
[22] CERs and ERUs reportedly accounted for 3.9% and 0.002% respectively of credits surrendered under the ETS in 2008. See: http://europa.eu/rapid/pressReleasesAction.do?reference=IP/09/794&format=HTML&aged=0&language=EN&guiLanguage=en.
[23] See new Art. 11a.
[24] Forestry was apparently viewed as delaying rather than reducing emissions (Meadows, 2007).
[25] Greater simplicity is an aim stated in EU-COM (2005, section 1, paragraph 5).
[26] See also: http://europa.eu/rapid/pressReleasesAction.do?reference=MEMO/08/632&format=HTML&aged=0&language=EN&guiLanguage=en.
[27] This refers to the need for a body to take on the liability of ensuring temporary credits are replaced on expiry if the original entity with this responsibility goes out of business. In particular, the European Commission argues that companies considering closing down might be tempted to sell their permanent credits and replace them with cheaper temporary ones (EU-COM, 2008a, p. 58).
[28] Other potential objections include expectations that forestry carbon would be relatively low cost, leading both to large windfall profits for forest owners and continued investment in high carbon intensive energy infrastructures reducing long-term emissions reduction activities (cf. Neuhoff, 2009).
[29] Note that carbon prices can also vary between different types and qualities of credit related, for instance, to differences in the extent to which they offer ancillary (e.g. habitat/biodiversity) benefits. For discussion of carbon valuation issues, see Valatin, 2010).
[30] According to Eliasch (2008, p. 178), 'The reason for this is that the EUA price is determined by the marginal unit of abatement. It is only at high levels of supplementarity, with the EU buying in 50 per cent of its abatement from outside its borders, that forest credits have an impact on the EU price.' The principal challenge is for countries to adopt sufficiently stringent emissions reduction targets with appropriate supplementarity rules (Eliasch, 2008, p. 174). A tighter target might increase costs to participating EU enterprises, however.
[31] For example, the Eliasch Review recommends adopting a buffer approach, maintaining a reserve of credits (Eliasch, 2008, p. 187). See also Valatin (2010).
[32] For discussion of three potential methods to allow use of temporary credits in the scheme (each involving payment of a 'permanence premium'), see Schlamadinger *et al.* (2005).
[33] For example, France, Spain, the Netherlands and Belgium are reported to have supported inclusion of reduced emissions from deforestation and forest degradation (REDD), with Finland, Denmark, Hungary and Austria opposed (CF, 2008).
[34] Practical issues such as high transactions costs and comparatively small average size may also militate against potential extension of the ETS to many new woodlands. Radov *et al.* (2007, pp. 69, 73) report that third-party verification costs for ETS emissions reductions are typically of the order of £1200 (i.e. about €1400) in simple cases, while new woodlands planted under the Farm Woodland Premium scheme in Britain, for example, have averaged about 4.5ha, so associated monitoring costs could easily exceed revenues from abatement. Harmonizing monitoring and verification protocols for forestry carbon sequestration across the EU may also pose more challenges than for those for emissions reductions in many other sectors, as may deciding which forestry activities to cover, and establishing liability for carbon losses.

[35] Article 24 of EU (2009b) potentially allows for member states to extend emissions trading to other activities and other GHG (subject to approval by the Commission). However, Aasrud *et al.* (2009) report that only types of credits allowed during phase II will be accepted during phase III, continuing the exclusion of forestry Kyoto credits.
[36] For example, Murray *et al.* (2007) suggest that reducing emissions reduction credits issued as a consequence of applying additionality, leakage and permanence discounting could reduce associated revenues by 30–95% in the case of other LULUCF (soil carbon sequestration) projects, depending partly upon length of sequestration.
[37] They could also potentially fit well with emerging climate change adaptation agendas.
[38] Four to five per cent were reportedly auctioned on average during phase II (Capoor and Ambrosi, 2009), with the first auction by a member state (the UK) in November 2008 (DECC, 2009).
[39] On exceptions (e.g. modernization of electricity generation in some cases, district heating and high efficiency co-generation), see EU-COM (2009, Art. 10c and 10a).
[40] Conversely, initial free allocation of EUAs may have had perverse effects in effectively preferentially benefitting fossil fuel powered power plants by transferring them valuable assets for which they could then subsequently charge their customers (e.g. Spash, 2010, forthcoming). The move to auctioning EUAs is expected to reduce the likelihood of companies making windfall profits (Capoor and Ambrosi, 2009; EU, 2009b).
[41] Were this so, inclusion of forestry within the ETS might result in at least partially double-counting these carbon benefits (cf. Radov *et al.*, 2007).
[42] http://www.chicagoclimateexchange.com/. The CCX is a North American legally binding GHG reduction and trading system for emissions and bills itself the first market for voluntary C offset investment in the world. It is a membership-based cap-and-trade system, which members join voluntarily and sign up for its binding reductions policy. Emission reduction requirements vary by membership category. Forest investments are among the nine eligible offset project categories; afforestation and reforestation, but also projects in long-lived wood and in managed forest are accepted by CCX as forestry-based carbon offsets investments. However, projects are only eligible in the CCX if they are located in the USA, or in non-Annex I countries.
[43] http://www.ecosystemmarketplace.com/.
[44] According to Hamilton *et al.* (2009), the average price of a voluntary carbon unit transacted on the OTC market was US$4.10/t CO_2e in 2006, US$6.10/t CO_2e in 2007 and US$7.34/t CO_2e in 2008.
[45] See http://unfccc.int/kyoto_mechanisms/aij/activities_implemented_jointly/items/1785.php.
[46] According to Hamilton *et al.* (2008), only 50% of the voluntary forest-based offset credits are based on independent standards, usually generic and not very rigorous.
[47] http://www.prima-klima-weltweit.de/.
[48] http://www.wald-fuer-sachsen.de/.
[49] http://www.waldaktie.de/.
[50] http://carbomark.org/.
[51] Some EU countries do not allow forestry JI projects within their borders at present (e.g. because of accounting for forestry carbon sequestration being covered under KP Art. 3.3 and 3.4).
[52] This could potentially change, for example, once member states have considered modalities for possible inclusion of LULUCF under existing EU GHG reduction commitments and associated effort-sharing arrangements (e.g. under EU, 2009a, Art. 9).
[53] This discussion is supported by the requirement that EU member states use at least 50% of the revenues generated from auctioning ETS emission allowances for specific emission reduction/sequestration measures (EU, 2009b, Art. 10, No. 3). 'Forestry sequestration in the Community' is explicitly listed as one possible measure.
[54] This approach is currently being developed by the Forestry Commission in the UK, for example.

References

Aasrud, A., Baron, R., Buchner, B. and McCall, K. (2009) *Sectoral Market Mechanisms – Issues for Negotiation and Domestic Implementation.* COM/ENV/EPOC/IEA/SLT(2009)5. OECD/IEA, Paris.

Cacho, O.J., Wise, R.M. and MacDicken, K.G. (2004) Carbon monitoring costs and their effect on incentives to sequester carbon through forestry. *Mitigation and Adaptation Strategies for Global Change* 9, 273–293.

CAN (2006) No Sinks in the EU ETS. Briefing Paper, March 2006. Available at: http://www.climnet.org (accessed 20 January 2010).

Capoor, K. and Ambrosi, P. (2009) *State and Trends of the Carbon Market 2009.* World Bank, Washington, DC.

CF (2008) Europe and Forest Carbon. Available at: http://www.climatefocus.com/newspubs/downloads/REDD_reportEU.pdf (accessed 20 January 2010).

Ciccarese, L. and Pettenella, D. (2008) Compensazione delle emissioni di gas-serra. Gli investimenti forestali di carattere volontario. *Sherwood* 14, 5–9.

Criqui, P. and Kitous, A. (2003) *Kyoto Protocol Implementation. KPI Technical Report: Impacts of Linking JI and CDM Credits to the European Emission Allowance Trading Scheme (KPI-ETS) (Service Contract N° B4-3040/2001/330760/MAR/E1 for DG Environment).* CNRS-IEPE/ENERDATA, Paris.

Davies, N. (2007) The inconvenient truth about the carbon offset industry. *The Guardian*, Saturday 16 June 2007, p. 14. Available at: http://www.guardian.co.uk/environment/2007/jun/16/climatechange.climatechange (accessed 30 May 2009).

DECC (2009) The EU Emissions Trading System: preparing for phase III. (PDF) Available at: http://bis.ecgroup.net/Publications/EnergyClimateChangeDECC/EmissionsTrading.aspx (accessed 20 January 2010).

Dieter, M. and Elsasser, P. (2004) *Wirtschaftlichkeit und Wettbewerbschancen von Wald-Senkenprojekten in Deutschland [Economic Efficiency and Competitiveness of Forest GHG Sequestration Projects in Germany].* Arbeitsbericht 2004/11. Bundesforschungsanstalt für Forst- und Holzwirtschaft, Hamburg.

DIW and KPMG (2003) *Leitfaden für die klimaschutzpolitische Bewertung von emissionsbezogenen JI- und CDM-Projekten (3 Bände und Einführung).* BMU, Berlin.

EEA (2008a) *European Forests – ecosystem conditions and sustainable use.* EEA Report 3/2008. European Environment Agency, Copenhagen.

EEA (2008b) *Greenhouse gas emission trends and projections in Europe 2008: Tracking progress towards Kyoto targets.* EEA Report 5/2008. European Environment Agency, Copenhagen.

EEA (2009) *Greenhouse gas emission trends and projections in Europe 2009: Tracking progress towards Kyoto targets.* EEA Report 9/2009. Office for Official Publications of the European Communities, Luxembourg.

Eliasch, J. (2008) *Climate Change: Financing Global Forests (Eliasch Review).* The Stationery Office, Norwich, UK.

Elsasser, P. (2008) *Wirtschaftlicher Wert der Senkenleistung des Waldes unter KP-Artikel 3.4 und Ansätze zu dessen Abgeltung in der ersten Verpflichtungsperiode [Economic value of the carbon sequestration service of forests under KP-article 3.4 and compensation options in the first commitment period].* Arbeitsbericht OEF 2008/6. von-Thünen-Institut, Hamburg.

EU-COM (2005) *Further guidance on allocation plans for the 2008 to 2012 trading period of the EU Emission Trading Scheme.* COM(2005)703. Brussels.

EU-COM (2006) *Building a global carbon market – Report pursuant to Article 30 of Directive 2003/87/EC.* Communication COM(2006)676 of the European Commission of 13th November 2006. Brussels.

EU-COM (2008a) *Accompanying document to the proposal for a Directive of the European Parliament and European Council amending Directive 2003/87/EC so as to improve and extend the EU greenhouse gas emission trading system: Impact Assessment.* Communication COM(2008) 16 of the European Commission. Brussels.

EU-COM (2008b) *Addressing the challenges of deforestation and forest degradation to tackle climate change and biodiversity loss.* Communication COM(2008)645/3 of the European Commission. Brussels.

EU-COM (2008c) *Proposal for a Directive of the European Parliament and Council amending Directive 2003/87/EC so as to improve and extend the greenhouse gas emission allowance trading scheme of the community.* Communication COM(2008)16 of the European Commission. Brussels.

EU-COM (2008d) *Questions and Answers on the revised EU Emissions Trading System.* MEMO/08/796 of 17th December 2008. European Commission, Brussels.

EU-COM (2009) *EU Action against Climate Change: The EU Emissions Trading Scheme.* Brussels.

EU (2003a) Directive 2003/87/EC of the European Parliament and Council of 13th October 2003 on establishing a scheme for greenhouse gas emission allowance trading within the community, consolidated version (incorporating subsequent amendments). *Official Journal of the European Union* L 275 (25 October 2003), 32–46.

EU (2003b) *Proposal COM(2003)403 of 23rd July 2003 for a Directive of the European Parliament and European Council amending the Directive establishing a scheme for greenhouse gas emission allowance trading within the Community, in respect of the Kyoto Protocol's project mechanisms.* European Commission, SEC (2003) 785, 32–46. Brussels.

EU (2004) Directive 2004/1001/EC of the European Parliament and Council of 27th October 2004 amending Directive 2003/87/EC on establishing a scheme for greenhouse gas emission allowance trading within the community, in respect of the Kyoto Protocol's project mechanisms. *Official Journal of the European Union* L 338 (13 November 2004), 18–23.

EU (2008a) Directive 2008/50/EC of the European Parliament and of the Council of 21 May 2008 on ambient air quality and cleaner air for Europe. *Official Journal of the European Union* L 152 (11 June 2008), 1–44.

EU (2008b) Directive 2008/101/EC of the European Parliament and Council of 19th November 2008 amending Directive 2003/87/EC so as to include aviation activities in the scheme for greenhouse gas emission allowance trading within the community. *Official Journal of the European Union* L8 (13 January 2009), 3–21.

EU (2009a) Decision 406/2009/EC of the European Parliament and Council of 23rd April 2009 on the effort of member states to reduce their greenhouse gas emissions to meet the Community's greenhouse gas emission reduction commitments up to 2020. *Official Journal of the European Union* L140 (5 June 2009), 136–141.

EU (2009b) Directive 2009/29/EC of the European Parliament and Council of 23rd April 2009 amending Directive 2003/87/EC so as to improve and extend the greenhouse gas emission allowance trading scheme of the community. *Official Journal of the European Union* L140 (5 June 2009), 63–87.

Fichtner, W., Graehl, S. and Rentz, O. (2003) The impact of private investor's transaction costs on the cost effectiveness of project-based Kyoto mechanisms. *Climate Policy* 3, 249–259.

Hamilton, K., Sjardin, M., Marcello, T. and Xu, G. (2008) *Forging A Frontier: State of the Voluntary Carbon Markets 2008*. New Carbon Finance/Ecosystem Marketplace, New York/Washington, DC.

Hamilton, K., Sjardin, M., Shapiro, A. and Marcello, T. (2009) *Fortifying the Foundation: State of the Voluntary Carbon Markets 2009*. New Carbon Finance/Ecosystem Marketplace, New York/Washington, DC.

Hamilton, K., Chokkalingam, U. and Bendana, M. (2010) *State of the Forest Carbon Markets 2009: Taking Root & Branching Out*. Ecosystem Marketplace, Washington, DC.

Heggedal, T.-R. and Kverndokk, S. (2007) The cost of greenhouse gas mitigation in Europe – Kyoto and beyond. *IAEE Newsletter* 16, 9–12.

Holtsmark, B. and Mæstad, O. (2002) Emission trading under the Kyoto Protocol – effects on fossil fuel markets under alternative regimes. *Energy Policy* 30, 207–218.

Hyvärinen, E. (2005) The downside of European Union emission trading – a view from the pulp and paper industry. *Unasylva* 56, 39–41.

IEA (2001) *International Emission Trading: From Concept to Reality*. OECD, Paris.

IPCC (2007) *Climate Change 2007 – Mitigation of Climate Change. Contribution of Working Group III to the Fourth Assessment Report of the Intergovernmental Panel on Climate Change*. Cambridge University Press, Cambridge.

Kapp, G. (2004) *Abschätzung des Entwicklungs- und Durchführungsaufwands für forstliche CO_2-Senkenprojekte in Deutschland*. GFA Terra Systems GmbH. Unveröffentlichtes Gutachten im Auftrag der Bundesforschungsanstalt für Forst- und Holzwirtschaft, Hamburg.

KP (1997) Kyoto Protocol to the United Nations Framework Convention on Climate Change. Available at: http://unfccc.int/resource/docs/convkp/kpeng.pdf (accessed 24 August 2005).

Krug, J., Köhl, M., Bormann, K., Riedel, T., Rüter, S. and Elsasser, P. (2009) Options for accounting carbon sequestration in German forests. *Carbon Balance and Management* 4, 1–15.

Mantau, U., Merlo, M., Sekot, W. and Welcker, B. (2001) *Recreational and environmental markets for forest enterprises. A new approach towards marketability of public goods [Forstbetriebe und Vermarktung von Umwelt- und Erholungsleistungen. Ein neuer Ansatz zur Untersuchung der Marktfähigkeit öffentlicher Güter]*. CAB International, Wallingford, UK.

Meadows, D. (2007) Taking a common approach to JI and CDM in the EU-ETS. Presentation at 4th ECCP working group meeting on emissions trading, 15 June. Available at: http://www.ec.europa.eu/environment/climat/emission/pdf/4thmeeting/8b_meadows.pdf (accessed 20 January 2010).

Michaelowa, A., Stronzik, M., Eckermann, F. and Hunt, A. (2003) Transaction costs of the Kyoto mechanisms. *Climate Policy* 3, 261–278.

Murray, B.C., Sohngen, B. and Ross, M.T. (2007) Economic consequences of consideration of permanence, leakage and additionality for soil carbon sequestration projects. *Climatic Change* 80, 127–143.

Neeff, T., Ashford, L., Calvert, J., Davey, C., Durbin, J., Ebeling, J., Herrera, T., Janson-Smith, T., Lazo, B., Mountain, R., O'Keeffe, S., Panfil, S., Thorburn, N., Tuite, C., Wheeland, M. and Young, S. (2009) *The Forest Carbon Offsetting Survey 2009*. Ecosecurities, Conservation International, The Climate, Community & Biodiversity Alliance.

Neuhoff, K. (2009) *Implementing the EU Renewables Directive*. Cambridge Working Papers in Economics CWPE 0913. Cambridge University Press, Cambridge.

Radov, D., Klevnas, P. and Skurray, J. (2007) Market mechanisms for reducing GHG emissions from Agriculture, Forestry and Land Management. Report to the Department of Environment, Food and Rural Affairs by NERA Economic Consulting. Available at: https://statistics.defra.gov.uk/esg/reports/ghgemissions/default.asp (accessed 20 January 2010).

Rosenthal, E. (2007) Vatican seeks to be carbon neutral. *New York Times*, 3 September 2007. Available at: http://www.nytimes.com/2007/09/03/business/worldbusiness/03iht-carbon.4.7366547.html?pagewanted=1&_r=1 (accessed 30 May 2009).

Samuelson, P.A. (1970) *Volkswirtschaftslehre [Economics. An Introductory Analysis]* (4th edn). Bund, Köln.

Schlamadinger, B., Bosquet, B., Streck, C., Noble, I., Dutschke, M. and Bird, N. (2005) Can the EU emission trading scheme support CDM forestry? *Climate Policy* 5, 199–208.

Schlamadinger, B., Bird, N., Johns, T., Brown, S., Canadell, P., Ciccarese, L., Dutschke, M., Fiedler, J., Fischlin, A., Fearnside, P., Forner, C., Freibauer, A., Frumhoff, P., Hoehne, N., Kirschbaum, M.U.F., Labat, A., Marland, G., Michaelowa, A., Montanarella, L., Moutinho, P., Murdiyarso, D., Pena, N., Pingoud, K., Rakonczay, Z., Rametsteiner, E., Rock, J., Sanz, M.J., Schneider, U.A., Shivdenko, A., Skutsch, M., Smith, P., Somogyi, Z., Trines, E., Ward, M. and Yamagata, Y. (2007) A synopsis of land-use, land-use change and forestry (LULUCF) under the Kyoto Protocol and Marrakesh Accords. *Environmental Science & Policy* 10, 271–282.

Schmidtke, H. and Kägi, W. (2006) *Möglichkeiten zur Anrechung von 'Forest Management' in der Schweiz als Senken gemäss Kyoto Protokoll Art. 3.4.* Silvaconsult AG, Winterthur/Basel.

Spash, C.L. (2010) The brave new world of carbon trading. *New Political Economy* 15, 169–195.

Trabesinger, A. (2007) Formula 1 racing: power games. *Nature* 447, 900–903.

Trexler, M.C., Broekhoff, D.J. and Kosloff, L.H. (2006) A statistically-driven approach to offset-based GHG additionality determinations: what can we learn? *Sustainable Development Law and Policy* 6, 30–40.

Trines, E. (2006) *The Integration of LULUCF in the EU's Emissions Trading Scheme*. Report 5625. Swedish Environmental Protection Agency, Stockholm.

Tuerk, A., Streck, C., Johns, T. and Pena, N. (2008) *The Role of Land-based Offsets in Emissions Trading Systems: Key Design Aspects and Considerations for Linking*. Working Paper Climate Strategies, Graz.

UNECE (2005) *European Forest Sector Outlook Study 1960–2000–2020. Main report*. United Nations Publications ECE/TIM/SP/20. UNECE, Geneva.

UNFCCC (1992) United Nations Framework Convention on Climate Change. Available at: http://unfccc.int/resource/docs/convkp/conveng.pdf (accessed 23 August 2005).

Valatin, G. (2009) *Carbon Additionality: A Review*. Discussion Paper. Forest Research, Edinburgh.

Valatin, G. (2010) *Forestry Carbon: Valuation, Discounting, Risk Management*. Discussion Paper. Forest Research, Edinburgh.

Wemaëre, M. (2008) Forestry and land uses: should sequestration be included in the EU ETS? Presentation to European State Forest Association (EUSTAFOR) workshop, Brussels, 26 June. Available at: http://www.eustafor.eu/userfiles/File/6%20Matthieu%20Wemeare.ppt (accessed 20 January 2010).

10 The Role of Networks in Non-wood Forest Products and Services Market Development

Daria Maso,* Anne Matilainen and Davide Pettenella

Abstract

This chapter analyses the role of networks in the development of the markets for non-wood forest products and services (NWFP&S) with reference to experiences gained in northern and Mediterranean Europe (using cases from Finland and Italy). The aim of the work has been to analyse a number of selected case studies within the theoretical framework developed by Varamäki and Vesalainen (2003) in order to demonstrate the relevance of networks in NWFP&S development. In the Mediterranean region, high population density, poverty and low wood productivity of forests have traditionally stimulated an intensive use of all potentially available sources of food, fodder and raw materials. In northern Europe, there is also a strong culture of utilizing forest products and, especially with the recent crises in the wood industry, interest in other uses of forests has increased. The enterprises dealing with NWFP&S are typically small or micro-sized and usually, except in a few cases, the turnover generated by the harvest, production, processing and/or commercialization of NWFP&S is quite limited. In this context, it is fairly evident that the creation of networks among NWFP&S producers, traders or sellers, or with other larger enterprises, can be a successful business strategy leading to positive economic results. The four case studies analysed in the chapter provide evidence supporting this idea. They concern both single-product/-service and multi-product/-service networks and, furthermore, they examine different evolutionary stages (from 'development circle' to 'project group'). These different case studies show which can be the role played by network structures in supporting innovation in NWFP&S marketing initiatives. In fact, it can be concluded that analysed networks: have been able to compensate for the traditional problems that micro-enterprises and small and medium enterprises may face in introducing innovations; can facilitate access to public funding and also to a key development factor such as forest land availability; or they can help in increasing the visibility of the sector.

10.1 Introduction

According to the FAO definition, non-wood forest products (NWFP) 'consist of goods of biological origin other than wood, derived from forests, other wooded land and trees outside forests' (FAO, 1999). In this chapter, the definition is extended to also cover services related to the multiple use of forests like nature tourism, recreation, nature and landscape services (i.e. 'non-wood forest services', NWFS). In Mediterranean Europe, around the southern-central Alps, high population density, poverty and low wood productivity of forests have traditionally stimulated an intensive use of all

* Corresponding author.

potentially available sources of food, fodder and raw materials. However, the non-wood forest products and services (NWFP&S) sector only began to be developed as a properly constituted economic sector in quite recent times. Indeed, its growth became apparent about two decades ago, at the same time as the decrease in conventional wood production (Pettenella *et al.*, 2006). In northern Europe, the NWFP&S as an economic sector has an even shorter history than in central Europe because of the strong traditional importance of the industrial wood sector. However, there is also a strong culture of utilizing forest products in northern Europe and, especially with the recent crises in the wood industry, an increased interest towards other uses of forests. The need to develop this sector is also gaining increasing attention in national and regional forest policies.

The enterprises dealing in NWFP&S are typically small or micro-sized and usually, except in a few cases, the turnover generated by the harvest, production, processing and/or marketing of NWFP&S is quite limited. In addition, the income originating from NWFP&S often constitutes only a part of the total income of the companies and production volumes are typically relatively small. The main reasons for this are scarcity of raw material (in terms of both quantity and seasonal availability), low market transparency, lack of business and marketing skills, small market size and high production and transport costs. Moreover, in many cases the small enterprises have limited knowledge and staff resources to undertake properly all the business areas required for enterprise sustainability. Being small companies, they do not have the possibilities to hire experts for every business area like manufacturing, marketing, etc. (Matilainen *et al.*, 2005). This highlights the need for external information as well as taking part in useful strategic networks in order to integrate activities and knowledge and to reach a critical mass of production. In this context, it is quite evident that the creation of networks among NWFP&S producers, traders or sellers (or, in a broader sense, nature-based small and medium-sized enterprises, SMEs) or even with larger enterprises including those from other sectors, can be a successful strategic behaviour leading to good economic results.

The aim of our work has been to present and discuss some theoretical concepts used in network analysis and to test them in selected case studies in order to demonstrate the relevance of networks in NWFP&S development. In this chapter, after discussing the role of networking and cooperation in NWFP&S marketing, we present four case studies of the NWFP&S networks. They are analysed according to the theoretical framework discussed in the chapter by Maso and Pettenella in this volume (Chapter 3).

In order to consider different geographical contexts and therefore different forest-based economies, two case studies were chosen from Italy and two from Finland. The diversity of the case studies also aims to bring out different network strategies in the sector illustrating the different types of benefits for the companies. All cases have been selected based on their innovative nature of networking in the NWFP sector. In the first case study, the NWFP&S network is based on the exploitation of the *Boletus* mushrooms (both a commodity and a recreational service); in the second the network is organized around an adventure park; in the third the network is connected to Lappish nature-based products; and, lastly, in the fourth case study the network is a nationwide association involved in supplying nature-based products and services.

10.2 Importance of Networking and Cooperation in NWFP&S Marketing

NWFP&S are playing an increasing economic role in the forestry sector in Europe. This is influenced by two factors: first, the recent trends in forest ownership; and secondly, the opportunities for innovation process in the forestry sector. With respect to the situation of forest property, data from 23 European countries indicate that 49.6% of forest land and other wooded lands is privately owned and 50.1% publicly owned (less than 0.4%

is classified as neither public nor private – Hirsch *et al.*, 2007; MCPFE, 2007, suggests similar results). There is a wide variability in the ownership structure between countries. In Austria, France, Norway and Slovenia private forest ownership exceeds 75% of the total forest area, whereas in Bulgaria, Czech Republic and Romania it is less than 25% (Hirsch *et al.*, 2007). Important phenomena currently affecting the ownership structure are land restitution and privatization. As a result, in central and eastern European countries, the number of small forest holdings has recently increased. For example, holdings smaller than 6 ha represent 70% of the total area of private holdings in Poland and 40% in Slovenia. Also, at the European level in general, the number of small holdings is vast and data show that almost 75% of the private holdings are less than 3 ha (Hirsch *et al.*, 2007). Moreover, a process of privatization is under way in many public (mainly State) forests. In all the profit-oriented enterprises, both private and public, the declining prices of industrial wood (UNECE/FAO, 2009) have seriously reduced the potential of income generation from traditional forest land use. To overcome these problems, especially within the constraint of a fragmented ownership structure, an increasing number of forest owners, especially in southern and central Europe, are shifting their economic activities towards NWFP&S that may also provide good economic results from small forest areas. While starting up in this sector, entrepreneurs often come into contact with the network concept (perceived as a need, an opportunity, etc. for their activity). The Community strategic guidelines for rural development (see Box 10.1 for a short excerpt) and the Leader approach[1] are fully taking into consideration these needs and expectations.

Enterprises also show a strong need to find new solutions for differentiating their products on the market and for creating new ideas rather than copying old ones. A useful classification of innovations is based on three categories (Dewar and Dutton, 1986):

- absolute (or radical) new products/services;
- incremental innovation (consisting of improved products and services); and
- subjective differentiation (based on communication strategy, branding, etc.).

The first innovation type is quite rare in the initial stages of the forest products value

Box 10.1. Strategic guidelines for rural development (Source: Council Decision 2006/144/CE Annex).

- Subparagraph 3.1 (Improving the competitiveness of the agricultural and forestry sector), point iv): 'developing new outlets for agricultural and forestry products. New outlets can offer higher value added, in particular for quality products…';
- Subparagraph 3.2 (Improving the environment and the countryside), point v): 'encouraging environmental/economic win–win initiatives. The provision of environmental goods, particularly through agri-environmental measures, can contribute to the identity of rural areas and their food products. They can form a basis for growth and jobs provided through tourism and the provision of rural amenities, particularly when linked to diversification into tourism, crafts, training or the non-food sector';
- Subparagraph 3.3 (Improving the quality of life in rural areas and encouraging diversification of the rural economy), point i): '… Tourism, crafts and the provision of rural amenities are growth sectors in many regions and offer opportunities both for on-farm diversification outside agriculture and the development of micro-businesses in the broader rural economy'; point iii): 'putting the heart back into villages. Integrated initiatives combining diversification, business creation, investment in cultural heritage, infrastructure for local services and renovation can contribute to improving both economic prospects and quality of life';
- Subparagraph 3.4 (Building local capacity for employment and diversification), point i): 'building local partnership capacity, animation and promoting skills acquisition, which can help mobilize local potential'; point ii): 'promoting private-public partnership'; point iii) 'promoting cooperation and innovation'.

chain: different constraints (including the limited resources available for research in small enterprises) limit the introduction of totally new products, while for forest services, there are interesting examples of absolute innovation (e.g. adventure parks, forest museums, forest concerts, forest schools, etc.). When some new forest product or service is discovered and launched on the market, it is generally quite difficult for the small and medium-sized companies to protect it with some kind of patent; consequently, it is quickly copied by other enterprises, and the first enterprise rapidly loses its novelty advantage. Branding, such as the 'Go Ape' adventure brand in the UK (http://www.goape.co.uk), may help in protecting a particular type of adventure activity.

The other two kinds of innovations are prevalent in the forestry sector, and especially in NWFP&S segments. However, in a context where most of the enterprises are small or medium-sized, the research and development activities and the knowledge and experience needed to attain these types of innovation are rarely available within a single enterprise. This highlights the importance of networks, which can provide a way to realize incremental innovations and/or subjective differentiations.

The utilization of non-wood forest products and services often has a solid background in tradition and history. However, when considered as a business sector, NWFP&S is still relatively new. At the moment, the economic role of the NWFP&S is still of relatively minor importance compared to that of wood and timber in many countries. However, the interest in the multiple use of forests and NWFP&S production and commercialization is growing continuously and there are regions where the economic potential of these products and services exceeds the value of timber. Especially the tourism and service sector and related products are growing rapidly, and in Finland, for example, nature tourism is currently growing faster than the tourism sector in general (Harju-Autti and Ryymin, 2007). Other new opportunities like cosmetics from nature utilizing the substances derived from natural products, functional food products and well-being services also provide new product development possibilities to the NWFP&S sector. Innovative activity is constantly increasing in the sector; however, at the moment, the added value (especially for NWFP) is still rather low (Kangas, 2001; Ministry of Agriculture and Forestry, 2002; Rametsteiner *et al.*, 2005).

In general, the smaller the business and the smaller the customer group, the more important the organizational aspects become for production, distribution and also market research and promotion (Pettenella *et al.*, 2006). Networking can improve the business activities of NWFP&S enterprises in the production phase, e.g. in forming joint production chains or subcontracting in order to produce larger volumes or provide more complete service packages. Moreover, the networking with other companies and cooperation during the product development phase are often extremely important in order to generate the required information related to both the production and to the markets. In addition, cooperation in marketing and supplying the products is often needed in order to achieve the necessary publicity. Typically, in NWFP&S sector networks a great deal of cross-sectoral cooperation is needed because of the wide range of NWFP&S products (Luonnontuotealan teemaryhmä, 2001; Aarne *et al.*, 2005).

In the eyes of the public, NWFP&S have a very positive image as environmentally friendly, sustainable and local products and services (Matilainen and Aro, 2002). This creates possibilities for using this image to market different products or services under the same general image and attributes related to it such as environmentally friendly, local, safe, etc. The highest forms of this type of cooperative development are product and service packages sold as complementary products. This requires high organizational efforts and indicates that there is need for organizational innovations (Pettenella *et al.*, 2006). Integration and networks in the NWFP&S sector can be created through producer/grower associations, external institutions, partnerships and coordinated initiatives between public authorities and private operators.

10.3 Four NWFP&S Network Case Studies

The case studies are analysed with the aim of highlighting some main aspects of network organization such as the entrepreneurial input that leads to the network establishment, the network type and strategy, its role in the development of the SMEs and the role of public support. Following Yin (2009), the four cases can be considered explicatory-exploratory case studies. An explanation-building strategy (Trochim, 1989) was applied to analyse them and results are reported below.

Table 10.1 presents the four case studies in a comparative matrix based on the Varamäki and Vesalainen (2003) classification and on the number of products and services involved.

10.3.1 Case study 1: the Borgotaro mushroom trail

The first Italian case study is that of the Borgotaro mushroom trail. It is a network developed among enterprises of the Borgotaro area (located between Tuscany and Emilia Romagna) dealing with *Boletus* mushrooms. It can be considered endogenous in the sense that local enterprises were the main promoters in the innovation process. The network is not just limited to those enterprises working on mushroom processing and commercialization; on the contrary, it also includes others that provide services recognized as complementary for the customers that the mushrooms attract into the area. Table 10.2 provides the main information on this network.

On the basis of the classification proposed by Varamäki and Vesalainen (2003), it seems correct to consider this network a Project Group. The enterprises involved correspond to the basic characteristics of this network type: resources and products of the partners are combined into a joint business; there are written agreements among the partners (they are members of the association, and they signed its rules); partners are closely involved with one another especially because of the need to maintain a common image; resources and skills of partners are different but essentially complementary (mushroom delicatessens, restaurants offering *Boletus*-based menus, hotels providing rooms for mushroom-pickers or for people following the Borgotaro mushroom trail, etc.); the common aim is to achieve new market channels through each participating enterprise.

The network certainly contributes to the development of the SME partners: innovations are introduced by some partners and, where possible, copied by others in a process of continuous improvement (e.g. new system of packaging of fresh mushrooms, new tourist services, new e-marketing initiatives, etc.). Moreover, the whole network is probably constantly searching for ways to innovate and expand itself to attract more and more customers and this surely also reflects on individual partners.

With regard to the role of public administration, Table 10.2 shows that some municipalities are members of the Association and that they pay for their membership, so in this way they contribute financially to the network. Another very important aspect regarding the role of the public administration in this network is that concerning the Protected Geographical Indication (PGI) certification.[2] The economic importance of Borgotaro Boletus relies on the fact that it is PGI-certified and

Table 10.1. Case study comparison matrix.

	Evolutionary stage		
Network dimension	Development Circle	Loose Cooperative Circle	Project Group
Single product or service		Case study no. 3	
Multiple products or services	Case study no. 2	Case study no. 4	Case study no. 1

Table 10.2. The Borgotaro mushroom trail.

Network type	Project Group
Entrepreneurial input	Endogenous
Network structure	Members organized in: Charter Members, Ordinary Members and Dues-paying Members or Honorary Members. Organization: Assembly, President, Vice-president, Board of Management, Board's Secretary and College of the Auditors of Accounts. These administrative-political posts are held by the owners of the enterprises participating in the trail project who do not receive any payment for their work
Non-wood forest products and services involved	*Boletus* mushroom
Starting year	2005
Number of participating enterprises	Enterprises: 62. 15 agri-tourisms/farm businesses; 12 hotels/guest-houses; 8 bed & breakfasts/inns/hostels; 9 cheese, sausage and wine growing and producing factories; 2 educational/teaching farms; 3 museums/private collections; 30 restaurants/pubs; 26 typical products sellers
Coordination	Non-profit association hosted in the offices of the Comunità Montana[a] of the Taro and Ceno Valleys
Role of the network	Promotion of the member enterprises
Fundamental links	The Borgotaro mushroom trail
Public administration role	Some Municipalities are members of the Association. Very important for the PGI certification of Borgotaro Boletus, the basis of the economic importance of the mushroom and the area
Funding	€250/enterprise during the first year; €103/enterprise during the following years; €0.52/inhabitant from the member Municipalities; €2000 from the *Comunità Montana*; Other funding from Leader+
Importance of the public funding on the development of the network	Regional authorities have supported the request of the PGI mark and the Borgotaro mushroom trail organization
Network incomes	Basically advertising and visibility for the member enterprises
Sharing of network incomes (if any)	Each enterprise probably reaches more customers, but the monetary result coming from the links among enterprises is not well defined or quantifiable
Importance of the territory quality and uniqueness	Very high: the Borgotaro area is the basis of *Boletus* production, processing and commercialization. The trail develops inside this specific territory
Future network development	The network is well established and economically successful. No major changes are foreseen

[a]Comunità Montana: territorial government body subject to regional planning. It acts to safeguard the hydrogeological, forest and environmental order of the territory in pursuit of harmonious development of the various economic activities present.
PGI, Protected Geographical Indication.
Information and data concerning this case study come mainly from the Master degree thesis by Matteo Sommacampagna and from the website www.stradadelfungo.it.

the public administration has been the main actor in achieving the certification. This allows it to be distinguished from other *Boletus* and have higher market recognition, be more attractive and, finally, a valuable non-wood forest product to be commercialized and around which a network of local SMEs can develop.

The Borgotaro mushroom trail also received some EU funding from a Leader Project in the start-up phase. However, it is basically working autonomously, financing itself by membership fees. Another important aspect is that this network has also established links with external activities and events such as the Festival of Borgotaro mushroom and the National fair of Albareto mushroom and this should lead to an additional reciprocal development impulse.

10.3.2 Case study 2: the Asiago Adventure Park and complementary services[3]

The second Italian case study is a network still at the development stage. Adventure parks are a relatively new NWFS, consisting basically of high-level roped walkways suspended from trees in the forest. Since the first one opened in 2001, they have been experiencing a great success in Italy, with a big expansion registered from 2004 onwards. In 2008, around 60 adventure parks were to be found in Italy. In most cases, they are in high forests and use the suitable trees (those providing the sufficient requisites of dimensions and stability) as supports for the walkways. Because of their originality, adventure parks rapidly became a very attractive recreational service offered in forests. They are also becoming a Leader service stimulating an increase in the supply of other complementary services. The case study described here is a classic example of such a situation. It should be specified that it is not unique, but has been chosen because of the specificity of the surrounding territory.

This network currently involves an Adventure Park, two hotels and a travel agency and is also developing an Orienteering School, a Nordic Walking Association, a society offering 'river sports' to the general public and a didactical farm. The key links are basically between the Adventure Park and the other partners: the two hotels and travel agency offer their customers a free entrance ticket to the Adventure Park. The entrepreneurial input that is leading to network formation is endogenous, since it comes from the Adventure Park. A general structure of links involving all the partners is still missing. Moreover, a new intervention by the local public administration has begun to materialize in the form of a local Tourist Office for promoting the establishment of cooperation among the tourist-related enterprises on the Asiago plateau by organizing meetings, etc. Table 10.3 provides the main information on this network.

On the basis of Varamäki and Vesalainen's classification, this network could probably be set between the Development Circle and the Loose Cooperative Circle. Characteristics typical of the Development Circle can be recognized, such as the absence of concrete investments, the very informal type of cooperation (mainly based on personal bonds deriving from acquaintance or friendship between partners), etc. But there are also traits associated with the Loose Cooperative Circle, especially the presence of a strategic aim (in this case the reciprocal advertising finalized at enlargement of the market share of the various partners).

However, this is a very young network and its future development is not yet defined. The crucial aspect is probably that of the network coordinator. This role is currently played by the Adventure Park but it is equally possible that other partners (even new ones) could assume it. There is the need for an actor willing to organize at a broader scale and involve more partners, to connect more formally to the tourist proposals of the Asiago plateau area, e.g. through the definition of pre-determined packages of sports activities, accommodation, etc. for weekends and holiday weeks. An enlarged and more diversified suite of services and products will give the network more stability. The Adventure Park managers could assume this coordination role in the future.

The Forest Adventure Park was constructed and is operating with the managers' private funds and is financing itself by entrance tickets. Public support for the initiative consists of the concession free of charge by the local municipality of the forest where the Adventure Park was established.

10.3.3 Case study 3: the gift packages of nature-based Lappish products[4]

The third case study is from Finland and presents a network of nature-based entrepreneurs from Lapland combining their products

Table 10.3. The Asiago Adventure Park and complementary services network.

Network type	Development Circle
Entrepreneurial input	Endogenous
Network structure	Personal agreements among the Agility Forest Adventure Park and the other structures involved
Non-wood forest products and services involved	Adventure park: rope walkways suspended from forest trees
Starting year	The Adventure Park opened in 2007, it started establishing links after that
Number of participating enterprises	Enterprises: 4 (in 2008) + 3 in progress. 1 adventure park; 2 hotels; 1 travel agency; 1 Orienteering School; 1 Nordic Walking Association; 1 society organizing 'river sports'; 1 didactical farm
Coordination	Adventure Park
Role of the network	Economic development of the member enterprises; reciprocal advertising
Fundamental links	The location of all activities on the Asiago plateau
Public administration role	Not in the establishment of the project. Mainly promotional through the consortium (but additional to the members own advertising through their websites, etc.)
Funding	Agreements among enterprises
Importance of public funding on the development of the network	No direct role in financing the initiative; the local municipality offered the forest where the Adventure Park was established free of charge
Network incomes	Basically reciprocal advertising for the member enterprises
Sharing of network incomes (if any)	Each enterprise should reach a larger number of customers, but the monetary result from the links among enterprises is not easily defined or quantified
Importance of the territory quality and uniqueness	Very high: the Asiago plateau is a well-defined area naturally and geographically. It is also a mountainous rural area traditionally relying on timber and cheese production and currently searching for new ideas for economic development based on nature tourism. The territory is used as attraction element
Future network development	The involvement of the three new partners, the organization of joint weekend or week activities, the establishment of links among all the involved partners, and not only between the Adventure park and the others. More in detail, the Orienteering School seems to be very interested in developing activity packages with all the other partners. The Adventure Park is also planning to establish agreements with a didactical farm. The Municipality would also like the Park to offer joint packages together with the ice-stadium

in order to create a high-quality gift package of non-wood and nature-based products. The entrepreneurial input can be considered exogenous, since the idea for this network was proposed by an EU-funded project aiming to enhance rural development, the LUPPO project. The project lasted from 2004 to 2006. Its basic idea was to enhance the business activities of nature-based Lappish SMEs by encouraging the contacts among enterprises and expert organizations operating in the same territory, and by improving the entrepreneurs' business and economic skills. The aim was to form long-term networks and start joint activities. A network of nature-based entrepreneurs was organized within the project in order to create gift packages, containing different nature-based products like jams, candy, tea, etc., which could be sold as a souvenir or as a business gift. Different gift packages were developed in groups of at least two enterprises and they contained at least one nature-based product from each company. All the gift packages had the same format and design layout and were tested by the customers in the LUPPO project. In addition, participating enterprises were provided with a kit of business administrative tools (e.g. a step-by-step pricing tool and a tool helping to specify contracts and contractual usage) to help the micro entrepreneurs improve their business efficiency. For each gift package, the 'leading SME' was chosen, which had the responsibility to organize the logistics and the joint selling to the retailers. Table 10.4 provides information on this network.

On the basis of Varamäki and Vesalainen's classification, this network could be considered a Loose Cooperative Circle. The enterprises involved have the basic characteristics of this network type: they share some kind of resource (in most cases at least the gift package boxes and the organization of the related logistical aspects and that of the legal agreements with the final buyer). Costs related to the common resource are shared. Enterprises can attain a critical mass of sales more easily than when working alone. In most cases, i.e. gift packages, the number of participants is limited and the cooperation concerns two or maximum three enterprises. This number enables benefit to be derived from the common resource and usually allows democratic decision making supported by quite loose contractual bonds (even if there are generally written agreements). The cooperation is generally limited in time and must be re-formalized from time to time. A certain level of trust is needed, especially with respect to product quality, which must be high. In most networks formed around the gift packages, the main advantages SMEs get from this kind of cooperation are a cost saving and an enlargement of the market segment reached with the consequent increase in their turnover. The results seem to be concrete, since cooperation is continuing even 1 year after the end of the project.

The role of public administration has been significant at the beginning of the project in forming the networks, supporting the innovation process and providing the needed resources. The public actors have had a major role in providing funds, as well as knowledge and expertise for product development. However, the role of the public actors slowly decreased towards the end of the development project (LUPPO) and the SMEs gradually took over the administration of the networks themselves.

EU funding was fundamental for the initial establishment of the network. In fact, the LUPPO project was financed by the EU Objective 1 funding. The need for this kind of project had also been recognized in the national (Horizontal rural development programme: Maaseutupoliittinenkokonaiso hjelmavuosille 2001–2004) and regional (Lapin Tavoite 1 – ohjelma, Objective 1 programme of Lapland – and Lapin maakuntaohjelma 2003–2006 – The regional strategy of Lapland 2003–2006) policies and strategies, which enabled the positive financing decision for the project.

10.3.4 Case study 4: the Finnish nature-based entrepreneurship association[5]

The second Finnish case study presents a nationwide network aimed at collecting together the actors of nature-based entrepreneurship for cross-sectoral cooperation (e.g. nature tourism, handicrafts and food products) in order to integrate entrepreneurship, education, development activities and research in the sector. The Finnish Nature-based Entrepreneurship Association, a non-governmental national organization formed by entrepreneurs and development organizations, was launched in 2001. In mid-1990, it became evident, on the basis of several development projects and studies, that the different subsectors of nature-based entrepreneurship (like utilizing herbs, decorative material from nature or highly specialized nature tourism products) are, as typical niche sectors, too small to gain enough publicity or obtain focused expertise to be visible on the markets or in the field of rural development. The need was therefore identified to collect these subsectors together under a wider common structure and theme. The Association was established by using as combining aspects the common resource (nature), the image of natural products and the values of the entrepreneurs (sustainable use of nature). The main tasks of the Association are to increase and improve cooperation between entrepreneurs and

Table 10.4. The gift packages of nature-based Lappish products network.

Network type	Loose Cooperative Circle
Entrepreneurial input	Exogenous
Network structure	Small and medium enterprises (SMEs) producing typical nature-based Lappish products that collaborate in groups of at least two
Non-wood forest products and services involved	Typical nature-based Lappish products, e.g. berry jam, mushroom products, candies, handicrafts, herb tea, etc.
Starting year	2005
Number of participating enterprises	Around 20 enterprises (in 2006) involved in several different gift packages
Coordination	At the start of the project, SMEs were trained and supported by external consultants. On a practical level, one of the SMEs participating in a group performs all the logistics for assembling the gift package and delivering it to the final retailer. The others have to supply the necessary quantity of the established product (at least one product each) to be put in the gift package
Role of the network/project	Production of gift packages, testing of packages' success on the market, participation in fairs and 'study trips' to strengthen and improve the network and its product. Since the project ended, the main role in network organization has been taken by the leading entrepreneur of each gift package
Fundamental links	A gift package of typical nature-based Lappish products intended to be sold to tourists
Public administration role	The network was created in the ambit of an EU-funded project administrated by a public education organization from Rovaniemi. Since the project ended (the end of 2006) there has been no public administration role
Funding	The LUPPO project was funded during 2004–2006 by the European Agricultural Rural Development Funds. Since then the project the network has continued with private funding of SMEs
Importance of public funding on the development of the network	It allowed the implementation of the project (SMEs contacts, training, consultancy on marketing and on legal aspects, e-toolkit for the SMEs, etc.). In the year since the project ended, the network continues using SMEs' funding
Network incomes	The advantages that the SMEs received (and still receive) from the network are: • for the enterprise preparing the gift packages: the surplus of money deriving from selling them; • for other enterprises supplying their nature-based products: the consistently increased quantity of products they sell; • for all the involved enterprises: a reduction in the expenses for contracts; they divide the cost of only one contract instead of each one having to pay the expenses for many small individual contracts
Sharing of network incomes (if any)	• Enterprise preparing the gift packages: the profit deriving from selling them. It is determined as follows: $S = P_{box} - P_1 - P_2 - P_3 - C_{box\ log}$ where: S = surplus; P_{box} = price of the gift package; $P_{1...n}$ = prices of the single goods put in the gift package (these correspond to the prices at which the producing enterprises usually sell them individually) • Other enterprises supplying their nature-based products: the price(s) ($P_{1...n}$) of the product(s) they put into the gift package
Importance of the territory quality and uniqueness	A concept of 'region' exists, in the sense that all products come from Lapland and represent Lappish nature-based products. However, this seems not to be very important in the marketing strategy, probably because of the absence of competitors in the sector. In fact, the gift packages are generally intended as a souvenir/business gift from Finland in general, not especially from Lapland. Potentially high: Lapland could be used as a good 'brand' for product differentiation

organizations, to supervise the interests of nature-based companies, to reinforce and increase a positive public image of sustainable nature-based entrepreneurship and to improve and increase quality and sustainability in the sector. The Association also administrates a databank for nature-based entrepreneurship (www.luontoyrittaja.fi) and provides information services, e.g. by producing a regular newsletter service on topical issues. The focus of the activities is not directly on providing direct marketing or selling services to individual SMEs or SME networks, but on reinforcing the sector by working together. The main role of the Association in many practical level product development activities is primarily to support network creation. At the moment, the network includes approximately 300 entrepreneurs and 100 expert organizations from the nature-based sector who pay the annual membership fee.

The network was established based on the long-term development activities of several expert organizations in the sector. Its origin can thus be considered exogenous. The Association was launched as a result of a development project: Luontoyrittäjyyden imago ja verkostoituminen (the Image and Networking of nature-based entrepreneurship), which was carried out by the University of Helsinki Ruralia Institute in 2000–2002. The development project was funded by four ministries. The first activities of the association were also closely connected to the development projects. Later on the Association started to operate with its own resources. The Association currently operates as a significant national 'umbrella-organization' in the sector connecting not only SMEs, but also the small sectoral SME organizations and other associations. One of the main tasks it has accomplished has been the supervision of the interests of nature-based enterprises. Table 10.5 provides the main information on this network.

The Finnish Nature-based Entrepreneurship Association can be classified as a Loose Cooperative Circle with some characteristics of the Development Circle with respect to the type of network. The members share a joint databank, even if the databank is mainly coordinated by the named members and the office staff of the Association, and therefore cannot be seen as at the disposal of all members. The decision making in the network is democratic and a special structure has been established. This structure is also applied to the principles of using and sharing the common resources, the databank and the Association's other possessions (e.g. exhibition stand). A certain level of trust is also required. The Association is continuously thinking of establishing quality criteria for the network in order to guarantee the quality of the products of the members mentioned in the web pages. The Association has also launched ethical guidelines for its members, in which sustainable development is highlighted in all its dimensions. However, as well as the Association aiming to share publicity costs and supervise the interests of nature-based companies, a further main aim is knowledge transfer and benchmarking.

The role of public administration was significant at the start of the process in establishing the kind of network needed for starting up the Association, in supporting the process and providing the needed resources. The public actors have had a major role in providing funding, as well as knowledge and expertise for organizing the Association and in creating and starting up the activities. However, after the establishment of the Association, the role of the public actors decreased and the SMEs increased their role in the network administration. Nevertheless, there are still public organizations as members in the Association.

The project supporting the establishment of the Finnish Nature-Based Entrepreneurship Association was financed by national funding from four different Finnish ministries. However, previous EU and national funded project work also strongly influenced the establishment of the Association. EU funding has also been important in implementing the first activities of the Association and in creating the services for the SMEs. These project activities have been administrated partly by the Association and partly by other public actors. In this second case, the Association has had the role of cooperator in these activities.

Table 10.5. The Finnish Nature-based Entrepreneurship Association network.

Network type	Loose Cooperative Circle
Entrepreneurial input	Exogenous
Network structure	Small and medium enterprises (SMEs) operating in the nature-based sector including, e.g. nature tourism, food products and handicrafts based on nature
Non-wood forest products and services involved	Nature-based products and services and also support services (e.g. knowledge services) to the entrepreneurs in the sector
Starting year	2001
Number of participating enterprises	Approximately 300 SMEs and 100 development and expert organizations
Coordination	The establishment of the Association was initially coordinated and supported by external consultants. However, since becoming established, there has been a selected board (nine members and personal substitute members) with chair and vice chair. A board member's term lasts 2 years. The Association also has a secretary and an office secretary to organize the practical work
Role of the network	To increase and improve cooperation between entrepreneurs and expert organizations, to supervise the interests of nature-based companies, to reinforce and increase a positive public image of sustainable nature-based entrepreneurship and to improve and increase quality and sustainability and visibility of the sector
Fundamental links	Website, databank, newsletter
Public administration role	Public administration had a major role in establishing the Association, but since then its role has been much smaller. There are still one or two representatives of public actors on the board, but otherwise the Association is administrated mainly by SMEs. However, the Association still works closely with several public actors, e.g. in implementing the development projects, so therefore also has strong connections with the public sector
Funding	Especially at the beginning, the role of public national and EU funding was significant. Currently the Association finances its activities by membership fees as well as by selling information services to other SME associations and by taking part as a partner in different kind of development projects
Importance of public funding on the development of the network	It allowed the implementation of the project, which established the Association and enabled the necessary preparatory work. Initially the activities were also closely connected to different kinds of public funded development projects. Although the Association now receives membership fees and some other income, external funding (in many cases public funding) still plays a role in implementing the activities. The Association therefore also actively seeks out and implements development projects that fit in with its objectives
Network incomes	The Association first of all offers its members access to a wide network of actors in the sector. It also provides an integrated information source on nature-based entrepreneurship to be used by its members and some expert services and tools to improve the business activities. The Association acts as the voice of its members in the national discussions related, e.g. to the national strategies and initiatives. In addition the Association organizes study tours, seminars, short-term education courses and joint exhibition stands
Sharing of network incomes (if any)	The network does not share the income among its members

Continued

Table 10.5. Continued.

Importance of the territory quality and uniqueness	The territory has no specific role, although nationality (the Finnish brand) plays a big role in image building. The aim of the Association is to unite the sector in a general national level, a regional network has also been formed but for organizational purposes
Future network development	The Association has begun developing a joint Internet-based selling system for non-wood forest products. The national visibility of the Association as the main 'spokesman' for the sector, e.g. in strategy and policy creation, has also increased significantly

National and regional policies have also played a significant role in the establishment and growth of the Association. The need to bring the actors together at national level had been recognized (Horizontal rural development programme – Maaseutupoliittinen kokonaisohjelma vuosille 2001–2004 – and in the action plans for nature-based entrepreneurship – Luontoyrittämisen toimintohjelma 1998, Luonnontuotealan nykytilan kuvaus ja kehittämisohjelma vuosille 2000–2006) and in several studies and smaller regional development projects relating to the sector (e.g. Rutanen and Luostarinen, 2000; Luostarinen and Vanhamäki, 2001; Matilainen and Aro, 2002). A working group was also established by the Ministry of Agriculture and Forestry to estimate the sector potential during 1996–1997. Without the strong political interest in the sector at different administrative levels, the role of the Association would probably have remained less significant. At the moment, the Association is recognized as an actor at national level and actively takes part in policy development (e.g. national forest policy).

10.4 Conclusions

Based on the case studies, it can be affirmed that networks have an important role in the innovation process of the SMEs operating in forest areas. SMEs are much more frequent in NWFP&S management than in the traditional wood-based activities. Perhaps more than in wood production and harvesting, the NWFP&S sector is strongly involved in innovative processes and this is one of the reasons why the NWFP&S sector is experiencing growing economic importance, at least in Europe (MCPFE, 2007). The four case studies analysed illustrate the role of network structures in supporting innovation in NWFP&S marketing initiatives. Networks in the four case studies have been able to compensate for the traditional problems that micro-enterprises and SMEs may face in introducing innovations (as in the case of the gift packages of nature-based Lappish products), they can facilitate access to public funding (as in the case of the Borgotaro mushroom trail or the Finnish nature-based entrepreneurship association), and to a key development factor such as forest land availability (as in the case of the Asiago Adventure Park) or increasing the visibility of the sector (as in the case of the Finnish nature-based entrepreneurship association).

As can be seen from the case studies analysed, different strategies of network creation are possible and equally successful. Some of the described networks were developed on the basis of various other public funding programmes. Others developed and/or continue their activities relying only on the network's own funding. In any case, a sort of so-called 'Leader approach' or 'grass roots approach' can be seen also at the basis of the networks developed independently from actual Leader or Leader-like funding programmes. The 'Leader approach' to rural development means that projects must be very specific and designed and implemented by local partnerships in order to be able to respond to peculiar local problems. Practice demonstrates that this approach is generally a winning one for projects focused on business development in the NWFP&S sector, as in some of the examples presented. Many NWFP&S are specifically local, deriving from a local tradition and from the use of local raw materials, or are something new but strictly linked to a specific place and to local people. On the other hand, when the aim is a more general development activity, like gaining visibility and recognition for the sector, a wider approach is necessary in order to gain enough resources to reach the aims of networking. In marketing both area-based (regional, local) and thematic-based (non-wood forest products, nature tourism) networks have been proved to be successful. If the local image is strong enough to distinguish itself on the markets, it can serve as a competitive advantage, e.g. in the case of Borgotaro mushrooms. When the regional image is not strong enough on the markets the products are aimed for, and the marketing arguments are based on other than regional aspects, the thematic networks might be a better solution (e.g. in the case of the Finnish nature-based entrepreneurship association). Thematic networks can also require a wider geographical area than just one region to be effective enough in the niche sector.

NWFP&S remain in a sort of limbo: they are not really part of the traditional forestry sector, nor of the mainstream food or handicraft industry or tourism sector, so they rarely draw on the specific policies or policy measures for these sectors. This is probably because of their small size and the consequently moderate market impact. In addition to independent products, NWFP&S have a great potential as attraction or image product or service for rural areas. They can also be good promoters for local development or complementary products to other sectors (see the examples reported in Pettenella *et al.*, 2006). Support and expertise are especially needed for their initial development. As the support by different rural or regional policy measures is often limited to the initial phases, a well-defined network can assure the survival and continued development of the initiatives after the start-up phase. The analysed case studies show that public administrations can play a key role in network creation and development. Public agencies can support networks by providing funding (i.e. through partnerships, sponsorships, etc.), technical assistance and communication services and they can also act as mediators among various partners in conflict resolution and in cost and benefit sharing.

Because of its niche product nature, the NWFP&S sector is often very cross-sectoral in nature and cross-sectoral networks are needed in order to successfully carry out the innovation activities in the sector. This calls for cross-sectoral or horizontal policy support and measures at both regional and national levels. To enhance the innovation activities in the NWFP&S sector, effective policies (especially based on a cross-sectoral approach or on the integration between different sectors) should be planned to help the diffusion of network structures among the NWFP&S SMEs that will allow a concrete local development for rural and forest areas. However, at the same time national level tools should not be forgotten in order to build effective thematic initiatives. All the key ideas mentioned in the Community strategic guidelines for rural development (Box 10.1) fit perfectly with the described network strategies to be applied in rural areas for the development of NWFP&S markets.

Notes

[1] The Leader approach is based on the idea of encouraging the implementation of integrated, high-quality, bottom-up and original strategies for local sustainable development. It has a strong focus on partnership and networks of exchange of experience among different rural actors.

[2] PGI certification (see EC Reg. 2081/92) concerns the name of an area that is used for identifying an agricultural product or a NWFP with a particular quality, tradition or reputation and whose production and/or processing take place in that specific area.

[3] Information and data concerning this example come mainly from authors' interviews and from the First Level degree thesis by Francesco Loreggian.

[4] Information and data concerning this example come mainly from the authors' interviews with the project coordinator (Mr Esa Säkkinen) and from the Final Report of the LUPPO project.

[5] Information and data concerning this case study is based on interviews with the people participating in the establishment of the network, final report and other reports of the Luontoyrittäjyyden imago ja verkostoituminen (Image and Networking of nature-based entrepreneurship project), material from the Association board meetings and the web pages of the Finnish Nature-Based Entrepreneurship Association: www.luontoyrittaja.fi.

References

Aarne, M., Hänninen, R., Kallio, M., Kärna, J., Karppinen, H., Ollonqvist, P., Packalen, K., Rimmler, T., Toppinen, A., Kajanus, M., Matilainen, A., Rutanen, J., Kurki, S., Peltoniemi, J. and Saarinen, J. (2005) Country Report: Finland. In: Jáger, L. (ed.) COST E30 Economic integration of urban consumers' demand and rural forestry production. Forest sector entrepreneurship in Europe: Country studies. *Acta Silvatica & Lignaria Hungarica*. Special Edition 2005.

Dewar, D.R. and Dutton, J.E. (1986) The adoption of radical and incremental innovations: an empirical analysis. *Management Science* 32, 1422–1433.

FAO (1999) Towards a harmonized definition of non-wood forest products. *Unasylva* 50, 63–64. Also available at: http://www.fao.org/docrep/x2450e/x2450e0d.htm#faoforestry

Harju-Autti, A. and Ryymin, J. (2007) *Matkailun yleisosa. Toimialaraportti 10/2007.* Ministry of Trade and Industry, Helsinki.

Hirsch, F., Korotkov, A. and Wilnhammer, M. (2007) Private forest ownership in Europe. *Unasylva* 58, 23–25.

Kangas, K. (2001) Commercial wild berry picking as a source of income in northern and eastern Finland. *Journal of Forest Economics* 7, 35–68.

Luonnontuotealan teemaryhmä (2001) Luonnontuotteiden talteenoton ja käytön edistämisohjelma, Maaseutupolitiikan yhteistyöryhmän julkaisuja 5/2001.

Luontoyrittämisen toimintaohjelma (1998) Maa- ja metsätalousministeriön asettaman työryhmän raportti. Maaseutupolitiikan työryhmän julkaisu 1/1998.

Luostarinen, M. and Vanhamäki, P. (2001) Ecological entrepreneurship: a study on the views of consumers, entrepreneurs and members of parliament on developing the ecological entrepreneurship in Finland. MTT Agrifood Research.

Matilainen, A. and Aro, M. (2002) Luontoyrittäjyyteen liittyvät mielikuvat. Helsingin yliopiston Maaseudun tutkimus- ja koulutuskeskus, sarja B, 24.

Matilainen, A., Suutari, T., Kattelus, P., Zimmerbauer, K. and Poranen, T. (2005) *The Role of Information Technology in Mediating External Information to the Rural Micro Enterprises.* National Literature Review. Reports 4. University of Helsinki, Ruralia Institute, Helsinki.

Ministerial Conference on the Protection of Forests in Europe (2007) *State of Europe's forests 2007. The MCPFE report on sustainable forest management in Europe.* Ministerial Conference on the Protection of Forests in Europe, Liaison Unit Warsaw, Warsaw.

Ministry of Agriculture and Forestry of Finland (2002) Työryhmämuistio 2002: 6. Metsämarja- ja sienialan erityistoimet –työryhmämuistio.

Pettenella, D., Ciccarese, L., Dragoi, S., Hedegus, A., Hingston, A., Klohn, S., Matilainen, A., Posavec, S. and Thorfinsson, T. (2006) NWFP&S marketing: lessons learned from case studies in Europe. In: Niskanen, A. (ed.) Issues Affecting Enterprise Development in the Forest Sector in Europe. University of Joensuu, Faculty of Forestry, Research Notes 169.

Rametsteiner, E., Weiss, G. and Kubeczko, K. (2005) Innovation and Entrepreneurship in Forestry in Central Europe. European Forest Institute Research Report 19. Brill, Leiden.

Rutanen, J. and Luostarinen, M. (2000) Luontoyrittäjyys Suomessa. Alueelliset verkostot: luontoyrittäminen, matkailu- ja maaseutuklusteri –hankkeen loppuraportti, Maatalouden tutkimuskeskuksen julkaisuja. Sarja B 23.

Trochim, W.M.K. (1989) Concept mapping: soft science or hard art? *Evaluation and Program Planning* 12, 87–110.

UNECE/FAO (2009) *Forest Products Annual Market Review 2008–2009.* UNECE/FAP Timber Committee, Geneva.

Varamäki, E. and Vesalainen, J. (2003) Modelling different types of multilateral co-operation between SMEs. *Entrepreneurship and Regional Development* 15, 27–47.

Yin, K.R. (2009) *Case Study Research. Design and Methods.* Applied Social Research Methods Series, Volume 5. Sage, Thousand Oaks, California.

11 The Role of Cooperation in Enhancing Innovation in Nature-based Tourism Services

Anne Matilainen,* Gerhard Weiss, Zuzana Sarvašová, Diana Feliciano, Carmen Nastase and Monika Prede

Abstract

Nature-based tourism is a rapidly growing industry sector providing new sources of livelihood for rural areas to diversify the traditional economies. Although innovativeness is an important element in the competitiveness of companies, the level of innovations in nature tourism and recreation services has not been reported as very high. The innovations in nature-based tourism typically occur not as a result of specific innovation systems, but rather 'between' the existing ones, and as a result of spontaneous, project-oriented cooperation of various public and private actors. These actors play a significant role in creating new ideas and cooperation is essential for their successful development. However, combining multiple actors to cross-sectoral networks is a challenge, particularly when small and micro-companies are concerned. By using case studies across Europe, this chapter aims to illustrate what kind of cooperation types exist between nature-based tourism enterprises and critical stakeholder groups affecting the innovation process, and how the cooperation is managed by innovation carriers. In addition, the chapter aims to illustrate what kind of influence different policies have had in this cooperation. The empirical data consist of 12 case studies from six western and eastern European countries (Austria, Estonia, Finland, Romania, Scotland and Slovakia). The results highlight the important role of informal cooperation in nature-based tourism innovation processes. This type of cooperation is supported mainly through horizontal rural and regional development policies. The results emphasize that the policy instruments with a bottom-up horizontal approach have an important role in enhancing innovation in the nature-based tourism sector.

11.1 Introduction

Nature-based tourism is a growing business sector providing sources of livelihood to the rural areas, diversifying the traditional economics. It can broadly be defined as tourism with main activities related to nature (Saarinen, 2001). Nature tourism is very territorial as such. A tourist usually has to come to a certain place or region to 'consume' the product. The income generating from nature tourism typically remains in the rural regions. In addition, the sector is labour intensive (Honkala, 2001; Saarinen, 2003). All these characteristics make it especially interesting concerning rural development aspirations.

Tourism is currently one of the most rapidly growing business sectors. Nature-based

* Corresponding author.

tourism has been the tourism subsector with highest growth rates (Franklin, 2003; Balmford *et al.*, 2009) and the growth has been estimated to continue in the near future with increasing respect for pure, authentic nature by the consumers (Ryymin, 2006). For example, in Finland the growth rate of the turnover of nature safari companies rose up to 8.5% during 2003–2004 and up to 6.0% in smaller nature tourism companies (Ryymin, 2006). At the same time, the growth of traditional sawmill industry was practically non-existent (Hänninen and Toppinen, 2004).

Regardless of the rapid growth, the level of innovations (both product and process innovations) in nature tourism and recreation services has not been reported as very high (Nybakk *et al.*, 2005; Rametsteiner *et al.*, 2005). This raises the question of whether the possibilities and opportunities have been recognized and used to their full potential. The competition in nature tourism is continuously increasing, especially concerning niche customer groups. Innovativeness is an important element in the competitiveness of companies and has been seen as one indicator predicting the future development of the sector (Rametsteiner *et al.*, 2005).

In creating new ideas and opportunities, the role of key actors and cooperation partners is essential. Those actors, who have knowledge of and access to natural resources, play a very important role. In the European context, nature-based tourism typically utilizes forests in different forms and, in many cases, the used forest areas are not solely owned by the entrepreneurs. This brings out not only the question of property rights, but also that of combining the interests of different user groups of forest areas (Matilainen and Lähdesmäki, 2009). Those actors who have 'access to customers' in the marketing channels have an equally important role.

The capability of combining these different types of actors for cross-sectoral networks and cooperation is essential and provides a big challenge to the sector, especially at small and micro-company level (Lunnan *et al.*, 2005; Luostarinen, 2005; Rametsteiner *et al.*, 2005). In order to support innovation activity in the nature-based tourism sector, the important questions are: how to simultaneously guarantee access to the needed natural resource (in this study, forest areas) and to the customers; how to organize the business activities effectively; and how to combine the interests of different stakeholder groups in order to develop the nature-based tourism sector successfully as part of rural economies.

To overcome these problems, small and micro-size companies have to find successful networking and cooperation strategies (Virkkala, 2006). Typically, rural nature-based tourism companies form complex cooperation relationships to ensure their activities. They continuously develop this kind of cooperation, without which they could not survive. Especially in rural regions, the social sustainability of business activities plays an important role for the success of the company (Lähdesmäki, 2005).

This chapter aims to illustrate what kind of cooperation types exist between nature-based tourism companies and the key stakeholder groups affecting the innovation process, business activities and social sustainability of the companies, and how it is managed by the innovation carrier. In addition, the chapter aims to analyse what kind of influence the different policies have in this type of cooperation.

11.2 Theoretical Approach

11.2.1 Innovation research describing the innovation system of nature-based tourism companies

Innovation research has often studied innovation processes in large firms that pursue explicit innovation strategies and research and development departments. Scholars, however, have learned that innovation is a process taking place not only within firms, but also between firms and amongst firms and many other actors. Besides the various types of private and public actors, the institutional framework is important for the innovation processes. The 'systems of innovation' approach (Lundvall, 1992; Nelson, 1993; Edquist, 1997) defines innovation as a result of systems that consist of actors and institutional settings. Actors include, besides and as important as the firm(s), authorities, interest groups, consultancy agencies and research and education institutes. Institutional settings are understood as formal and informal rules and norms, e.g. public policies or the innovation and interaction culture in certain regions or sectors. Innovation systems are often understood, in a narrow sense, as established, enduring systems that are explicitly and strategically oriented towards creating innovations in a national economy (national innovation system), in a sector (sectoral innovation system) or in a region (regional innovation system).

Innovations in nature-based tourism typically occur not as a result of specific innovation systems but rather 'between' the existing ones, and as a result of more spontaneous, project-oriented cooperation of various actors (Kubeczko *et al.*, 2006). Alternatively, they can be described under a wider notion of innovation systems (Asheim and Isaksen, 2002). Such kinds of innovation processes are particularly important in regions with weak economic and institutional structures and are crucial for the economic development of such regions. This is often the case in remote rural regions. These kinds of processes are studied under the headings of regional development, regional governance and learning regions.

Some representatives of the 'learning regions' approach focus on the role of social capital and trust, formal and informal interfirm networks, and the process of interactive learning (Asheim, 1996; Morgan, 1997). The important resource for innovation and economic development is the capacity of people, organizations, networks and regions to learn. Authors often look at network-organized innovation projects. In these studies, the basic features of innovation systems are used, but they are more broadly interpreted and applied to any cooperation or network of actors in innovation projects (innovation systems in a broader sense). This approach fits well the innovation processes in the nature-based tourism sector and highlights the importance of successful cooperation. According to Saxena (2005), sustainable tourism products are territorially embedded in ongoing social networks and relationships, and tourism destinations can be seen as 'regions' that support processes of innovation as a key source of competitive advantage.

11.2.2 The role of stakeholders in forming successful cooperation

The critical networks and cooperation partnerships for innovation systems in a broader sense can be seen to be formed from different types of stakeholders relating to the innovation initiatives. A stakeholder can be defined as any group or individual that can affect or is affected by the achievement of a corporation's purpose (Freeman, 1984). The stakeholders have influence on the company's possibilities to implement the innovation process, whether they are business partners, local people, resource holders, administrative bodies, etc. Common to them all is that they have an interest in whether the innovation initiative is carried out, and they also can have their own agenda towards it. The crucial issue is how all the key stakeholders are taken into consideration so that successful networks and partnerships can be formed and sustainability of the business activities can be guaranteed.

The stakeholders can be divided into primary stakeholders, who have a formal, official or contractual relationship with the company, and secondary stakeholders, who represent the rest of the groups interested in the business environment, such as local people, forest owners, etc. (Carroll, 1993; Clarkson, 1995; Näsi, 1995). The influence both of these stakeholder groups have on companies' activities can be direct or indirect. Frooman (1999) has divided the stakeholder influence between a focal company and a stakeholder group based on the resource dependence. If the company's dependence on the stakeholders' resource (e.g. in case of nature tourism, forest land) is high, more often direct influence mechanisms are used in the cooperation and interactions between the company and stakeholders. In cases when the dependence is low and stakeholders do not control the critical resources for company's operations, the indirect influence methods via other stakeholders are used (Frooman, 1999; Sharma and Henriques, 2005). Nevertheless, this does not mean that stakeholders using indirect influence methods would necessarily be less influential than those using direct influence methods (Keskinarkaus and Matilainen, 2009).

Successful cooperation with stakeholder groups has been found to have clear connections with the company's business performance and development (Besser, 1999). According to Näsi (1995), in the long run, the company must operate in such a way that the stakeholder groups are satisfied: 'The more dissatisfied the main stakeholders are, the more certain is that the company's activities will cease'. Bryson (2004) also highlights that it is important to find ways to satisfy the key stakeholders at least minimally according to their own criteria for satisfaction. This brings out the need for entrepreneurs to be able to understand the stakeholders' point of view concerning the innovation process. Sometimes the stakeholders' primary agenda is difficult to identify. The failure to understand the often unforeseen hidden power and influence of stakeholders has led to countless project and business failures (Nutt, 2002; Bourne and Walker, 2005).

In rural areas, it seems that typically there are more expectations from the stakeholder groups towards the companies than in urban areas (Lähdesmäki, 2005). The stakeholder groups are unique for each company and its actions based on, e.g. location, line of business and customer base. They are also numerous and form complex networks (Neville and Menguc, 2006). In many cases, it is impossible to fully satisfy all the stakeholder groups. Therefore, it is important to find the key stakeholders for each case (Bryson, 2004). According to Mitchell *et al.* (1997), the critical attributes in defining key stakeholders are the power of the stakeholder, the legitimacy of the stakeholder concerning the stake and the urgency with which the stakeholder requires attention to his or her claims.

In this chapter, the approaches mentioned above are combined to a certain extent. The stakeholders without whose acceptance or cooperation the company's innovation process would not have been possible or the business activities could not continue successfully are considered key stakeholders concerning the innovation process, regardless of whether the stakeholders can be seen primarily or secondary, or whether their influence is direct or indirect.

11.2.3 Policies supporting innovation in the nature tourism sector

Policy documents from the EU, at national and regional levels, highlight the need to improve the competitiveness of rural regions in Europe. Enhancing innovation is one of the key elements in improving competitiveness. At the operational level, various policy instruments and actors are related to the innovation environment of rural small and medium enterprises (SMEs), and a huge amount of money is used to achieve these goals. Regarding nature-tourism, several policies, both sectoral and horizontal, are linked to the innovation processes. At the European level, one of the most important is the rural development policy, which also provides practical measures for nature tourism development. Other related policies can be seen, such as forest policy and environmental policy, as these influence the use of nature resources. In addition, in some countries the innovation policy as a separate policy sector provides some support to tourism and nature tourism sectors (Rametsteiner *et al.*, 2010).

It has been questioned whether these policies and support instruments actually target the right problems and provide assistance in overcoming them. Also, a very segmented policy environment typically shows its face in a very complex way to the companies on a practical level (Matilainen *et al.*, 2007). In addition, the regional priorities have a strong role in directing the elements of a national innovation system at a regional level (Kautonen, 2006). The situation is especially complex for companies operating in the interface of different sectors, like nature-based tourism, or which are small or micro-sized conglomerates.

Many European countries have strong sectoral forest policies, which often focus mainly on timber production-related industries (Rametsteiner *et al.*, 2010). However, forests as natural resources for different activities cannot be taken separately from other rural areas or rural development. It has been stated that, in order to maintain forest-related economies, the innovation support related to the sector has to include coordination and integration with different related policies, e.g. a rural development policy and a regional policy (Verbij and Schanz, 2002).

According to Lafferty (2004) and Briassoulis (2005), the attempts at coordination and integration of policies have been led by two basic approaches: vertical (intrasectoral) and horizontal (intersectoral). The vertical approach describes coordination and integration of policies at European, national, regional and local level within one sector. The horizontal approach, on the other hand, features coordination and integration of policies between the sectors at one political level. In reality, those two approaches are combined and it is not possible to clearly distinguish the direction of coordination and integration of policies (Persson, 2002). Both vertical and horizontal policies are combined when the approach of regional governance is taken to the political discourses. Regional governance also connects the political discourses to rural development. It is seen as an intersectoral cooperation, encompassing both a horizontal and a vertical dimension in the form of regional partnerships and networks like those between the regions and higher political levels (Frey, 2002; Adamaschek and Pröhl, 2003; Benz and Fürst, 2003; Scherer, 2006; Böcher, 2008; Giessen, 2008).

11.3 Material and Methods

11.3.1 Methodological approach

In this study, a descriptive perspective was taken for examining the cooperation relationships between the nature-based entrepreneurs and key stakeholder groups. A qualitative approach was chosen, which is a justified choice in order to understand phenomena about which little is yet known (Strauss and Corbin, 1990). The aim is a rather inductive analysis (Strauss and Corbin, 1990). For the study of innovation processes, a case study approach was chosen. The cases used are designed as innovation cases at an enterprise level. The innovation carrier can be either a private or a public actor. A case study is an appropriate research strategy to investigate contemporary phenomena within their real-life context, especially when the boundaries between the phenomena and the context are not clearly evident (Perry, 1998; Yin, 2003), which is typical in an innovation process. The case studies can be descriptive, explanatory or exploratory in nature (Yin, 2003). In this chapter, the case studies are mainly descriptive and explanatory.

The empirical data consist of 12 case studies representing six different European countries (Austria, Estonia, Finland, Romania, Scotland and Slovakia) providing a representative collection of cases, each presenting different institutional settings with regard to access to forest land, the innovation support system and forest ownership. The sampling of the interviewees in each case was made by a purposive sampling in order to ensure manageable and informative data (Patton, 2002). The case data were collected by the authors of this chapter using joint semi-structured thematic interview guidelines, which allowed flexible conversations to take place but still ensured that all the main issues were discussed (Patton, 2002). The themes were chosen to cover the critical aspects relating to cooperation networks of forest-based nature tourism companies, especially focusing on mapping out the critical stakeholder groups and their management.

The interviews were conducted between 2004 and 2009. In most cases, the innovation carrier was visited by the case author. The data collection methods comprised personal face-to-face, telephone and e-mail interviews with core actors of the innovation project. In addition, written sources such as internal or official project documentation, press releases, newspaper articles, information on websites and brochures were used.

In the analysis, first the critical key stakeholder groups for each case were identified by the case authors based on the data (Table 11.1). After that, the data on the cooperation relationships were analysed by using analyst-constructing typologies in order to point out cooperation typologies. In analyst-constructing typologies patterns, categories and themes are looked for from the data and, based on these, typologies are formed (Patton, 2002). For this purpose, a common analysing framework was created. Typologies are built on ideal types or illustrative endpoints rather than complete and discrete sets of categories providing one simple form for presenting the qualitative comparisons (Patton, 2002). Unlike classification systems, typologies do not provide decision rules for classifying. Instead, typologies usually identify multiple ideal types, each representing a unique combination of the attributes that are believed to determine the relevant outcome (Doty and Glick, 1994). In other words, the typologies and their characteristics emerge from the data during the analysis instead of being decided beforehand. Also, since typologies present ideal types, the cases can have elements from several different typologies. The case descriptions were cross-checked by co-authors in order to ensure the quality of the analysis and to avoid the risk of creating typologies too much influenced by the analyser (Patton, 2002).

The policy influence was analysed in the third analysing phase by looking for indications of the direct policy influence in the innovation cases and especially in establishing or maintaining the cooperation relationships. The influence was analysed based on in what form it occurred (financial, institutional, informational, other), whether the influence was positive or negative and which policy was in question. The analysis was made by the case study authors and cross-checked by the co-authors. The summary of cases is presented in Table 11.1.

11.3.2 Case studies

All the cases, though representing different nature tourism activities, represent new forms of how to utilize forest areas for tourism purposes. Common to all cases is the importance of several key stakeholder groups without whose support the activities could not have been established or maintained. The critical cooperation relationships are presented in Table 11.1.

'Almliesl' – Marketing of forest cottages for tourists, Austria

The carrier of the innovation is a regional unit of the Austrian Federal Forests. The innovation was to renovate and lease 12 traditional forest houses and hunting cottages to tourists. In the first phase, the project experienced severe troubles. In the second phase, the project was reorganized and marketing was delegated to a tourism agency, offering quality country cottages in the Austrian mountains under the brand 'Almliesl', which brought high demand for the cottages. Because of the success, there are plans to expand the tourism activities in the future. In this case, it was evident that active cooperation with different stakeholders, including company staff, neighbours, business partners and authorities, was crucial for success. Ignoring these groups in the first phase led to business failure. The new cooperation policy in the second phase included, for instance, an informal relationship with the relevant authorities as well as including the neighbours in the business activities. There was no direct influence from innovation policies, except for the earlier start-up of the Almliesl brand. The case company did not actively seek support from institutional actors either. The policy influence stems from the restrictive land use law that limits options for development of buildings scattered in the landscape.

Canopy walkway Sauwald, Austria

The initiator of this innovation was a private farmer aiming to diversify his business activities by establishing in his forest a

Table 11.1. Summary of the case studies and critical cooperation relationships.

Country	The case	The identified key stakeholder groups for the innovation	Benefits gained from the cooperation
Austria	'Almliesl' – Marketing of forest cottages for tourists	(1) MTS Almliesl tourism agency and other local tourism partners	(1) Professional marketing outsourcing the costs and risks
		(2) Local people	(2) Acceptance for the activities
		(3) Public authorities	(3) Acceptance for the activities
Austria	Canopy walkway Sauwald	(1) Firms involved in construction of the canopy walkway	(1) Professional work
		(2) Members of the society Baumkronenweg	(2) Acceptance for the activities, wider knowhow base
		(3) Public administration	(3) Acceptance/official approval for the activities
		(4) Funding bodies	(4) Funding for the investments
Estonia	Leigo Tourism Farm	(1) Local farmers	(1) Ensuring extra services or workforce in bigger events, some complementary services
		(2) Local community government	(2) Acceptance and official approval for the activities
		(3) Otepää tourism office	(3) Professional marketing
		(4) Other public institutions in Valga county (state regional forest department, environmental units)	(4) Acceptance for the activities, informational support
		(5) Local SMEs	(5) Complementary services, acceptance for the activities
		(6) State business and regional development agency Enterprise Estonia	(6) Financial support for technical equipment
		(7) Other state institutions (ministries)	(7) Financial support for landscape management and maintenance
		(8) Regional Leader group	(8) Financial support for new initiatives and cultural events
		(9) Private sponsors	(9) Financial support
		(10) Professional music collectives	(10) Complementary products
		(11) General public	(11) Acceptance for the activities
Estonia	Woods Crone Family Park	(1) Family Park members (ten farms + one individual entrepreneur)	(1) Marketing support, complementary products and services
		(2) Local municipality	(2) Informational support, acceptance for the activities
		(3) Other local SMEs	(3) Complementary services
		(4) Regional state development agencies in Valga county	(4) Financial and informational support for the investments
		(5) Local and Regional Tourism Offices	(5) Professional marketing, information of the customers
		(6) Karula Nature Park	(6) Complementary services

Continued

Table 11.1. Continued.

Country	The case	The identified key stakeholder groups for the innovation	Benefits gained from the cooperation
		(7) Local hotel-spa in Võru county	(7) Complementary products and services
		(8) State institutions (Ministry of Interior)	(8) Financial support
Finland	Private hunting enterprise	(1) Local hunting clubs	(1) Access to hunting grounds, subcontracting services
		(2) Land owners	(2) Access to hunting grounds
		(3) Selling agencies	(3) Professional marketing and selling
		(4) Other SMEs	(4) Complementary services
		(5) Local people	(5) Acceptance for the activities
		(6) General public	(6) Acceptance for the activities
Finland	Horseback riding tours	(1) Private land owners	(1) Access to the forests
		(2) Metsähallitus	(2) Access to the state forests
		(3) Other SMEs	(3) Complementary services
		(4) Hunting club	(4) Access to infrastructure (hunting cabin)
		(5) Regional developers	(5) Knowledge of development instruments
Romania	Dorna Adventures, private nature tourism company	(1) Public administration (municipality tourism officials)	(1) External funding, knowledge on customers
		(2) Other private SMEs	(2) Complementary services
		(3) Volunteers, trainers, guides	(3) Volunteer guides
Romania	The Calimani National Park	(1) National Administration of Forests (Suceva and TG Mures branch offices)	(1) Financing, administrative guidelines
		(2) Forest Research and Management Institute	(2) Knowledge for environmental management
		(3) The Association of Forest Owners	(3) Joint development projects (funding, infrastructure, knowledge)
		(4) Forest Group Josenii Bârgâului	(4) Joint development projects (funding, infrastructure, knowledge)
		(5) Local SMEs	(5) Complementary services, acceptance for the activities

Scotland	Huntly peregrine wild watch/ Wildlife interpretation centre	(1) Scottish Agricultural College (SAC) (2) Gartly Primary School (3) North East Raptors Study Group	(1) Volunteers for the wildlife centre (2) Dissemination channel, acceptance for the activities (3) Knowledge and free population monitoring
Scotland	Mountain biking	(1) Forestry Commission (2) Local tourism consortium (3) Local farmers (4) Local community council	(1) Access to the forests (2) Complementary services (3) Acceptance for the activities, complementary services (4) Acceptance for the activities
Slovakia	Forest tourism in Velky Klíž forests	(1) Local forest owners (2) Local associations (like hunting clubs) (3) ALEA (association focusing on marketing on web, organizing exhibitions of tourism, etc.) (4) Local joiners (5) Local municipality	(1) Access to the forests, complementary services (2) Acceptance for the activities, infrastructure (e.g. meeting places) (3) Professional marketing (4) Timber constructions for the activities (5) Acceptance for the activities
Slovakia	Vydrovská valley joint nature tourism	(1) Lesy SR state forest enterprise Cierny Balog (2) CHZ (local company operated narrow-gauge railway) (3) Local municipality (4) Other enterprises	(1) Access to the forests, equipments for the activities (recreational) (2) Additional services (transport) (3) Financial support (4) Complementary products and services

SMEs, small and medium enterprises.

canopy walkway and a forest restaurant. The canopy walkway initiative was strongly influenced by Regional Development Policy as it was carried out as EU Leader+ project. The development strategy of the Leader+ region had already included the idea of a forest canopy walkway, proposed by the regional development office. The 8-ha recreational forest with the canopy walkway offers 250 m of wooden bridges reaching up to the tree tops at 22 m, a forest restaurant, a children's playground and various nature interpretation stations. In the first season the project attracted over 100,000 visitors and the project employed 12 people. The canopy walkway itself is managed by the society 'Baumkronenweg', formed of a number of local people with the initiator as president. The forest restaurant is managed by the farmer's family. The networking in the case focused on including influential actors of the society, e.g. the mayor, a school director and a bank director, to support the innovation, rather than accumulating the professional know-how from the fields of tourism or marketing. Success factors were the entrepreneurial attitude of the farmer developing the idea, who invested private money, the strategic forming of the society Baumkronenweg and good networking beyond, as well as the Leader+ programme under which about half of the total costs were financed and which provided consultancy for the regional development strategy and project development.

Leigo Tourism Farm, Estonia

Leigo Tourism Farm is a private company started in the middle of the 1980s. In the 1990s, because of the collapsing agriculture industry in Estonia, the owner changed his agricultural activity to tourism. The farm has 27 double rooms, two suites and 60 additional lodgings. The tourism product development is based on the innovative use and design of landscape (including plans for 28 artificial lakes, of which 14 are already implemented) for different purposes combining natural beauty and diversity with traditional rural lifestyle, arts, music and extreme sports. The tourism products and services vary from organizing wedding parties to cultural events and military sports. The main attraction of the farm, of the Otepää region and of the whole of south Estonia is the internationally recognized Leigo Lakemusic Festival. The farm provides eight to ten full-time jobs, including four family members. Also, part-time staff of up to 240 persons are used when needed. There have been both positive and negative policy influences in the case. The main financial support has been provided by rural and regional development programmes and ministries. Also, national tourism offices have provided help for marketing. The building and planning legislations, on the other hand, have caused obstacles in the implementation of the innovation and there were problems in utilizing the development support from national agencies because of the time-consuming bureaucracy in the application process.

The Woods Crone Family Park, Estonia

The Woods Crone Family Park is the first tourism village in Estonia, founded by a young couple in 1990. They started with nature tourism based on local natural resources. By 2009, the Family Park had 11 informal cooperation partners, ten local farms and one individual entrepreneur, all sharing one common trademark: The Woods Crone Family Park. Since *c.*70% of the area is covered by forests, it provides the main theme and resources for the products and services: nature tours in forests and meadows, climbing trees, Estonian cultural traditions, folk medicine and forest heritage. In addition, the company makes decorations from dried plants and offers herbal products. All business is based on common ethic and spiritual values and on deep patriotic emotional bonds between the partner organizations. The company has successfully combined the expertise of different partners in order to create wide sustainable product provision. The park has gained institutional, informational and financial support during the development process. For example, regional development agencies have provided supporting policy measures based on rural and regional policy programmes.

In addition, the local and regional tourism offices have provided help for marketing activities, influenced by the state tourism development programmes.

Hunting in eastern Finland

The private company Finnhunt Oy organizes moose and small game hunting activities in private and state forest areas in eastern Finland. The company has managed to transform moose hunting, which is a traditional leisure activity, holding passionate interest from different stakeholder groups like recreational hunters, local hunting clubs and the general public, into a commercial activity in an innovative way. Also, the company has been successful in finding target groups for their products by close cooperation with central European sales organizations. One of the most important factors has been the close cooperation with local hunting clubs and landowners from the very beginning in order to maintain the social sustainability of the activities. At the beginning of the activities, the Rural and Regional Development Policies had direct positive influence by providing both informational and financial support in the form of a development project in order to access the markets and make contacts with the selling agencies. Hunting legislation, on the other hand, has caused challenges to the activities because of the hunting licence policy, according to which the licences should be primarily allocated for recreational, not commercial, use.

Horseback riding tours in Finland

The private company Kainuun vaellustalli organizes horseback riding tours with Icelandic horses in Finland, mainly in private forest areas. The base of operations is the home farm of one of the partners of the company. The company organizes daily tours throughout the year. It has managed to find an additional source of income for a rural region as well as utilizing the customer base of bigger tourism companies, like local hotels and sport estates, in a remote region. Also, it has managed to negotiate successfully with several private non-industrial forest owners (up to 100) to be able to establish riding routes wide enough for their activities. At the beginning of the activities, the regional development agency had a significant role in providing both knowledge and financial support, e.g. for establishing the cooperation with the private and public forest owners in order to build the tracks needed for the company's activities. Also, the organization provided help in detailing the business plan and other business-management-related issues.

Dorna Adventure, Romania

Dorna Adventure is a private company initiated by two partners who started by providing boat rides on the Bistrita River and expanded later to mountain climbing, paintball, horseback riding tours and courses and mountain biking. All activities are developed in nature, and the firm utilizes only equipment that does not endanger the environment. The innovativeness of Dorna Adventure was to offer something unique in the forest area and they developed prerequisites for that. With regard to the opportunities for the enterprise in the long run, it is essential to collaborate with the private and public actors even more than they have done in the past. In addition to the economic benefits from close cooperation with other companies, the enterprise is confronted with pollution of the river that destroys the beauty of the place. The initiative has gained support from regional development policy measures in forming both public and private partnerships, even though in all cooperation relationships no direct policy influence was observed, e.g. in the cooperation with other SMEs or volunteers.

The Calimani National Park, Romania

The Calimani National Park is a public park focusing on the protection and conservation of natural elements, also providing resources for scientific, educational, recreational and touristic visits. Its establishment in 2004 provided many opportunities for developing new innovative nature tourism and recreation activities in the area. There are currently several activities offered for tourists, such as hiking, mountain biking, equestrian tourism in the Calimani Mountains, photo safaris, junior ranger camps, cross-country skiing and bird-watching tours. To maintain and develop the innovation, it is very important to sustain fluent cooperation between the National Park and the interest groups. The policies supporting sustainable development in general have influenced at a strategic level the establishment of the national park and therefore supported the innovation case. However, in the cooperation relationships relating to the product development, no direct policy support was detected.

Huntly peregrine wild watch, Scotland

The wildlife interpretation centre provides wild watch opportunities and is managed and implemented by the Forestry Commission on its own woodland. Several animal species can be seen in the area, but the main attraction is the peregrine's nest visited by a pair of peregrines every year. Cameras film the peregrines feeding the chicks and their other activities around the nest. The Huntly peregrine wild watch centre is open during the peregrine immigration time from mid-January to mid-September and it is free of charge for the visitors. The site (2 ha) belongs to the Forestry Commission, a government department responsible for the protection and expansion of Britain's forests and woodlands. The project is considered innovative because it uses new technology to show wildlife activities live. The interaction of permanent staff and volunteers with the visitors makes the centre special. The centre was awarded a prize for customer care. In terms of the role of public policies, as the Forestry Commission is a public body, there are no policies impeding the activity. It was mentioned that the Scottish Forestry Strategy fits well with objectives of the project.

Mountain biking, Scotland

A private company in the Tweed Valley in Scotland provides mountain biking and organizes courses related to it in the forests, mainly owned by the Forestry Commission. Most of the customers are focused on the mountain biking activity, not on the natural resources, even though the forest plays a major role in the attractiveness of the services by providing a unique environment for the tracks. The company works within a network of businesses in the area, which has created an innovative hospitality scheme to improve the forest-based mountain bike tourism activities. In the network there are 16 businesses offering services like bike courses and accommodation packages, which all are created within the Mountain Biking Hospitality Scheme. The accommodation packages, including bike storage, maintenance, cleaning facilities and technical assistance, are targeted especially at mountain bikers. Scottish forestry policy supports sustainable development and therefore it can be seen to have had a positive influence on this innovation case. However, the influence has been more at a strategic than an operational level. Other policies critical for the success of the activity have been the Land Reform (Scotland) Act 2003 and the Scottish Outdoor Access Code. According to these policies, access to land is free by default without asking permission to use forest areas for mountain biking activities, apart from when events are organized.

Forest tourism in Velky Klíž forests, Slovakia

The Urbarium (special type of shared ownership) of the village Velky Klíž associates about 600 owners of agricultural and forest land with a total area of 786 ha. The most important challenges for innovations were efforts to diversify their own production activities, ensure additional income for the members of the Urbarium Velky Klíž, and enhance the development of the municipality and the region with the use of natural and cultural potential of the municipality. The facilities were built up to serve as accommodation for guests and provide background for other services being offered by Urbarium Velky Klíž. Nowadays they provide visitors in urbarial forests various recreational services, e.g. accommodation in the forester's house, nine round trails, forest guides, sports and a playground for children. Regarding the innovation case, at the beginning of the activities, a development project was funded by the Special Accession Programme for Agriculture and Rural Development (SAPARD) programme (rural development policy). This project initiated the activities and brought all actors together from different levels and sectors for cooperation. There was a significant positive impact in the form of financial and informational support for establishing the cooperation.

Vydrovská valley, Slovakia

Vydrovská valley is a new touristic destination located in Čierny Balog, Middle Slovakia. The village of Čierny Balog is one of the largest villages in Slovakia, with over 400 years of rich history, culture and traditions. The valley includes several tourist attractions related to forestry, e.g. the narrow-gauge Čiernohronská railway (ČHŽ), a forestry open-air museum and the birth place of Jozef Dekrett Matejovie, the founder of sustainable forest management in Slovakia. All the new activities are based on the Vydra (Rural Development Activity), which aims at cooperation with other similarly oriented non-governmental organizations to contribute to the sustainable development of the region. The project and trademark Vydrovská valley was formed based on this cooperation. In the innovation case, at least two different kinds of direct positive support by the policy measures can be detected. The ERDF-funded Sectoral Operational Programme provided the financial support and, in addition, the Local Action Group (LAG) provided both institutional and informational support for the innovation case.

11.4 The Results

11.4.1 Cooperation with stakeholders

When studying the cooperation between the innovation carrier and the key stakeholder groups, the cooperation types were further analysed based on four characteristics, which emerged from the data to represent the typical characteristics of the cooperation:

- **Formality of the cooperation relationship** (formal–informal): the relationship was divided into formal and informal relationships. In the formal relationship, written contracts were typically issued and even some compensation can occur, while the informal relationships were based on verbal informal agreements or discussions.
- **Types of communication** (official–unofficial): in this study, the communication between the innovation carrier and stakeholder groups was analysed based on: (i) its regularity; and (ii) the fora in which it was primarily conducted. Based on these attributes, the communication was divided into official and unofficial. In official communication, the communication is regular and happens in formal business meetings and/or agreements. In some cases, minutes are even taken and distributed to the participants. The unofficial communication, even though it can be very vivid, happens on a non-regular basis, such as random chats or discussions. Also, the role of personal relationships between the innovation carrier and the stakeholder group representatives is significant.
- **Form of cooperation** (unisectoral–cross-sectoral): the cooperation was analysed further by using the concept pair unisectoral and cross-sectoral cooperation. Unisectoral cooperation happens within one industry sector. Cross-sectoral cooperation is seen as cooperation between the different sectors, i.e. the innovation carrier represents a different sector than the stakeholder group and the cooperation occurs between more than one industry sector. The tourism sector has been considered as one sector including, e.g. catering and accommodation.
- **Regionality of the cooperation** (local cooperation–wider cooperation): based on the geographical level of cooperation, the cooperation occurring within regional partnerships on local level and

in local networks has been identified as local cooperation. On the other hand, the partnerships between the region and higher levels like national actors etc. has been referred to as wider cooperation.

By using the four characteristics mentioned above, two distinct cooperation types were found. These were evident in all cases, even though the cases represented different nature tourism activities in different institutional settings. The first cooperation type can be called **structured cooperation**. This type of cooperation was established and maintained based on business activities typically providing monetary benefits to both parties. These were also used as justification and positive arguments for cooperation and 'selling the innovation' to the stakeholder group in question. The relationship is often a business relationship with written agreements and contractual rights. In almost all vertical cooperation relationships, the cooperation was 'structured' in nature, since as part of the production chain the stakeholders were usually subcontractors or selling and marketing organizations/actors. However, the structured approach was also used in horizontal cooperation relationships, especially when other local SMEs were in question. Because of its formality, structured cooperation occurred both with local stakeholders and those outside of the region. The communication was typically official based on the business actions between the innovation carrier and stakeholder group. The personal relationships were important in enhancing cooperation, but they were not highlighted or seen as a prerequisite for its establishment.

The second cooperation type visible in the case studies can be called **unstructured cooperation**. Typically, unstructured cooperation occurred in horizontal and cross-sectoral cooperation relationships. The communication could be either very vivid or relatively random, but it was typically very informal. The role of personal relationships between the innovation carrier and the stakeholder groups was highlighted. In cases where the personal relationship was non-existent, local mediators were occasionally used.

In unstructured cooperation relationships, the innovation was justified and argued with 'benefits to the whole region' in an 'improvement of regional economies' discourse by the innovation carrier. Also, more general level values like 'nature conservation', 'nature education' and 'increased knowledge on forests' were used as positive arguments. Regardless, whether the innovation process was carried out by a private company aiming to maximize its benefits or a public actor generating public goods, this approach was used especially in cooperation relationships between the innovation carrier and local or regional stakeholder groups. The unstructured cooperation model was not utilized regarding stakeholder groups outside the region in the cases. The cases illustrate the crucial role of local stakeholders as well as unstructured cooperation in maintaining successful cooperation in forest-based nature tourism innovations.

In unstructured cooperation relationships, the compensation for the stakeholder group for their work, land, etc. was not necessarily paid. The innovation carrier expected the stakeholder groups to contribute for general good and 'benefit to the whole region', even though the direct benefits would be allocated primarily to the innovation carrier. An interesting point concerning unstructured cooperation was that in cases where the privately owned forest land was not owned or administrated by the innovation carrier, in order to gain the access to required forest area, all the relationships in the cases were informal.

In some cases, the stakeholder groups with whom the unstructured cooperation type typically prevails, the cooperation was altered by the innovation carrier into a structured form. By selecting this type of cooperation, the social sustainability of the innovation carrier's activities was increased significantly. Typically, the main goal for the cooperation was to gain acceptance for the activities, even though other benefits could entail as well. This was the situation, e.g. in the cases of hunting enterprise in Finland (cooperation between the local hunting club and the tourism company was formalized) and forest tourism in Velky Klíz

Forests in Slovakia (the cooperation between the local forest owners and the innovation carrier was formalized), or in Canopy walkway Sauwald Austria (the cooperation between the members of the Baumkronenweg was formalized into a society). In some cases like the mountain biking in Scotland, there were also indications that the local farmers were invited to take part in the business activities in order to divide the benefits, but the cooperation has not reached the structured form yet. By recognizing the local stakeholder group at least partly as a 'business partner' and allocating benefits to them (even as a token), the local acceptance for the activities was improved. The activities were seen in practice to 'benefit the whole area'.

When looking at the innovation processes in the cases, the role of one key person, the innovation carrier or the innovator, seems essential. The innovator's personal capacities, social position and networks seem to be highlighted especially in unstructured relationships, but are also important in structured cooperation relationships. For example, in the case of Klíz Forests in Slovakia, the personal relationships of the innovator with all key stakeholders were probably the biggest fostering factor in the innovation process, even bigger than the financial support. The same can be detected from the case of Leigo Tourism Farm in Estonia.

In addition to the two mentioned cooperation types, with certain key stakeholder groups the innovation carriers did not have any kind of cooperation relationship at all, even though the stakeholder group was identified as critical to the innovation success. In some cases, the importance of these stakeholder groups was perhaps not recognized properly by the innovation carrier, like in the case of renting forest cottages in Austria, but also in some cases these stakeholder groups were seen to be too massive or difficult to manage for the innovation carrier even to try to maintain a proper cooperation relationship with them. Therefore, they ignored the stakeholder group in question. This was the situation, e.g. related to the general public concerning hunting tourism in Finland. The entrepreneur realized the importance of the public as a key stakeholder group in influencing the business environment, but had not come up with any actual active way to cooperate with it.

11.4.2 Policy influence

The influence of different policies in the innovation cases in establishing and maintaining the different kinds of cooperation relationships was also analysed. No significant policy influence was discovered in approximately half of critical cooperation relationships, these cooperation relationships being equally structured and unstructured in nature. In cooperation relationships in which the policies had clear influence, both negative and positive influences were found (Table 11.2). Typically, the positive influence occurred when rural and regional policies influenced the innovation. On the other hand, in some cases legislation caused negative influence on the innovation cases. However, it must also be noticed that the legislation enabling the innovation was mentioned, such as in the case of mountain biking in Scotland. Sometimes the cooperation relationships were established in the cases in order to overcome the limitations the legislation caused to the innovation process, such as in the cases of hunting tourism in Finland or Leigo Tourism Farm in Estonia.

The most relevant policies supporting the cooperation relationships are rural and regional development programmes and tools based on them. In many cases, the policy influence was direct, providing financial support, but many times the influence was also more informational or institutional in nature. For example, the rural policy measures influenced the establishment of local enterprise and development centres, as in the case of horseback riding tours in Finland, which provided knowledge and other support to the innovation case. The Leader+ programme and the local action groups as part of regional policy measures in some cases (Canopy Walkway, Austria and Vydrovska Valley, Slovakia) were mentioned as positive promoters of

Table 11.2. The policy sectors and programmes directly influencing the innovation cases.

Type of cooperation	Policy sector/programme influencing the innovation process
Structured cooperation	+ Rural development policies (financial, informational support), including, e.g. SAPARD (Special Accession Programme for Agriculture and Rural Development) programme + The Programme of Community Aid to the Countries of Central and Eastern Europe (Phare) 2000–2006 + Regional development policies (financial, institutional) including Leader+ programme + Sustainable development policies (strategic influence) + National tourism policies and programmes – Hunting legislation – Environment protection policies – Building and planning legislation + The Land Reform (Scotland) Act 2003 and the Scottish Outdoor Access Code
Unstructured cooperation	+ Regional development policies (financial, institutional) including Leader+ programme + Forest policy (strategic level) + Rural development policies (financial, informational) including, e.g. the SAPARD programme + Phare programme 2000–2006 + The Land Reform (Scotland) Act 2003 and the Scottish Outdoor Access Code

+, positive influence; –, negative influence.

the innovation. In addition, national or regional tourism organizations were involved in the innovation case (e.g. Leigo Tourism Farm in Estonia), which highlights the role of policies enabling these organizational units. It was interesting to notice, though, that the forest policy as such did not play a significant role in enhancing the critical cooperation for innovation. In some cases, however, it was mentioned to have positive influence at a strategy level, as in the case of mountain biking in Scotland or Calimani National Park in Romania, but no actual measures related to it were found from these cases. Also the state agencies committed to innovation development had, at the end, a fairly moderate role in the establishing the critical cooperation to enhance the innovation process in the nature-based tourism sector.

The rural and regional policies had a positive influence on both structured and unstructured cooperation types. Typically, the policy influence was more significant at the beginning of the cooperation relationship.

11.5 Discussion and Conclusions

Both structured and unstructured cooperation types were found, in establishing and maintaining the critical cooperation for each innovation case. In general, the structured approach was applied to cooperation with business partners and the unstructured approach was used in establishing and maintaining the local-level stakeholder cooperation. *The results clearly highlight the important role of informal unstructured cooperation and cooperation networks in nature-based tourism innovation processes.* Even though the cases represent different institutional settings and entrepreneurial environments, in all cases the role of informal, unstructured, typically local-level cooperation was vital for the sustainable innovation activities. These kinds of relationships are important regarding very different types of critical stakeholders, including authorities, neighbours or interest groups. Even if the stakeholder groups are not formally involved in the business activities, they might put the project at risk if good relations are not maintained. The unstructured cooperation networks were vital in both the cases in which the innovation carrier was business focused and those in which the innovation carrier did not seek direct profit from the activities.

Even though the forest owners can be seen, according to literature classification (Mitchell *et al.*, 1997), as the key stakeholder group of nature-based tourism, holding the vital natural resource for the innovation activities and having direct influence mechanism to them, in the cases where the forest area was not owned or managed by the innovation carrier the cooperation was informal and no actual compensation was prevailing. The forest owners were not typically seen as business partners in the innovation processes, even though in some cases there were indications in this direction.

In areas where there is a high pressure to use forests owned by those other than the innovation carrier for nature tourism and recreation activities, this approach raises interesting questions concerning the forest owners' role as resource provider. It should be considered to what extent the innovation carriers can count on the 'good will' of the forest owners and the local social pressure to enable the innovation processes. In northern Europe, the long tradition of free access to forests also influences the cooperation. This historic custom may have raised the forest owners' tolerance towards utilization of their forests by other people (Matilainen and Lähdesmäki, 2009). This could influence the selection of the cooperation type between the resource holder and the innovation carrier, and an informal relationship is still seen as adequate.

However, the *results also indicate that in some cases the business success and the social sustainability of the activities can be strengthened when the cooperation with the local stakeholder groups is transferred to a structured relationship*. In this way, the benefits of the innovation activities will be distributed more widely in the network and the area, and the formal recognition of the stakeholder group indicates the appreciation by the innovation carrier towards successful cooperation. This reduces the risks and provides more predictability regarding the future activities (Matilainen and Lähdesmäki, 2009). Formalizing the cooperation does not necessarily bring any main business benefits to the innovation carrier other than acceptance for the activities. However, this seems to be successful strategy in cases where it is vital to gain the key stakeholders' acceptance.

The policies form part of the institutional settings and therefore also a part of the innovation environment of nature-based tourism companies (Weber, 2002; Fornahl and Brenner, 2003). Based on the results, the policies seemed to have a similar kind of influence to both structured and unstructured cooperation. It was anticipated that the policy instruments would fit better to support structured rather than unstructured relationships because of the formal nature of many support measures. However, according to the results, unstructured cooperation seems to be well supported by existing measures. Nevertheless, it must be noticed that most of the cooperation relationships had features of both cooperation types. Also, in making the conclusions, the limitations of the data must be kept in mind. The case-study approach is more suitable for exploring the phenomena than making generalizations.

The innovators' personal characteristics can sometimes be the most significant driver for the innovation, especially in the innovation processes where unstructured cooperation relationships prevail. In order to enhance the innovation processes, it is vital to recognize these people and find ways to support them with specific services or through capacity building. These people may not have the skills and knowledge to be able to utilize the support and policy measures fully aimed at enhancing innovation.

Innovative ideas typically include risks. The policy measures, e.g. project-based funding, should be flexible enough to accept a certain level of risk in the development activities. This principle is not always realized at a practical level though. For example the Leader programme, which is one of the EU initiatives for rural development, was originally planned to be risk-funding in nature. However, in many cases, the local and regional authorities have tried to limit the risks of the initiatives by different regulations, e.g. concerning the acceptance of the initiatives, maybe even too much. The truly new innovative ideas may not have room to develop or the bureaucracy is too heavy for small local actors to take part in the programme. There is a possibility that this 'carefulness' sometimes causes obstacles to the innovation processes rather than supporting it.

According to the results, it seems that *rural and regional policies are more relevant in supporting innovation in the nature-based tourism sector than are forest or innovation policies.* In most of the cases, the cooperation networks in the nature-based tourism sector are complex and cross-sectoral, which is the reason why *horizontal or regional policy measures provide more suitable tools for those activities.* However, at the same time, it should be noticed that in many cases no direct policy influence was detected relating to the critical cooperation in the innovation cases. Negative policy influence was not very significant in cases in general, even though there were indications that legislation in some cases created cooperation, the main aim of which was to overcome the obstacles set by the legislation.

This importance of the horizontal policy measures highlights a challenge related to various innovation support schemes at EU, national and regional levels. How do we create more flexible measures enabling cooperation within the complex networks needed for innovation and allowing better integration of the forest sector to the horizontal policies? It must be noted that both horizontal and vertical policy approaches are needed. The question is how to create the necessary integration between these two approaches. The flexibility in the policy support measures also requires functional local-level activities as a joint effort of different actors. This is especially important when considering the innovation environment of nature-based tourism companies, where innovation often occurs 'between existing innovation systems' (Kubeczko *et al.*, 2006). Regional governance approaches to innovation research as well as innovation support provide a promising way to analyse and foster the innovation environment in the nature-based tourism sector. Its further elaboration might also bring new practical level tools to understand and develop innovation support for nature-based tourism and related landscape-based or ecosystem services.

Acknowledgement

This work was supported by the Slovak Research and Development Agency under the contract No. APVV-0692-07.

References

Adamaschek, B. and Pröhl, M. (eds) (2003) Regionen erfolgreich steuern. *Regional Governance – von der kommunalen zur regionalen Strategie.* Verlag Bertelsmann Stiftung, Gütersloh.

Asheim, B. (1996) Industrial districts as 'learning regions': A condition for prosperity? *European Planning Studies* 4, 379–400.

Asheim, B. and Isaksen, A. (2002) Regional innovation systems: the integration of local 'sticky' and global 'ubiquitous' knowledge. *The Journal of Technology Transfer* 27, 77–86.

Balmford, A., Beresford, J., Green, J., Naidoo, R., Walpole, M. and Manica, A. (2009) A global perspective on trends in nature-based tourism. *PLoS Biology* 7, 1–6.

Benz, A. and Fürst, D. (2003) Region – Regional Governance – Regionalentwicklung. In: Adamaschek, B. and Pröhl, M. (eds) *Regionen erfolgreich steuern. Regional Governance – von der kommunalen zur regionalen Strategie*, 2nd edn. Verlag Bertelsmann Stiftung, Gütersloh, pp. 11–66.

Besser, T.L. (1999) Community involvement and the perception of success among small business operators in small towns. *Journal of Small Business Management* 37, 16–29.

Böcher, M. (2008) Regional governance and rural development in Germany: the implementation of Leader+. *Sociologia Ruralis* 48, 372–388.

Bourne, L. and Walker, D.H.T. (2005) Visualising and mapping stakeholder influence. *Management Decision* 43, 649–660.

Briassoulis, H. (2005) Complex environmental problems and the quest for policy integration. In: Briassoulis, H. (ed.) *Policy Integration for Complex Environmental Problems; the Example of Mediterranean Desertification.* Ashgate, London, pp. 1–49.

Bryson, J.M. (2004) What to do when stakeholders matter? *Public Management Review* 6, 21–53.

Carroll, A.B. (1993) *Business and Society: Ethics and Stakeholder Management*, 2nd edn. South-Western Publishing, Cincinnati, Ohio.

Clarkson, M.E. (1995) A stakeholder framework for analysing and evaluating corporate social performance. *Academy of Management Review* 20, 92–117.

Doty, H.D. and Glick, W.H. (1994) Typologies as a unique form of theory building: toward improved understanding and modelling. *Academy of Management Review* 19, 230–251.

Edquist, C. (ed.) (1997) *Systems of Innovation. Technologies, Institutions and Organizations.* Pinter, London.

Fornahl, D. and Brenner, T. (2003) *Cooperation, Networks and Institutions in Regional Innovation Systems.* Edward Elgar, Cheltenham, UK.

Franklin, A. (2003) *Tourism. An Introduction.* Sage, London.

Freeman, R.E. (1984) *Strategic Management: a Stakeholder Approach.* Pitman Publishing, Boston, Massachusetts.

Frey, R.L. (2002) Regional Governance. Input paper in the Avenir Suisse Workshop, 15 October 2002, Basel.

Frooman, J. (1999) Stakeholder influence strategies. *Academy of Management Review* 24, 191–205.

Giessen, L. (2008) Forstwirtschaft zurückhaltend bei Programmen zur integrierten ländlichen Entwicklung: Analyse der politischen Ursachen. *Forstarchiv* 79,116–121.

Hänninen, R. and Toppinen, A. (2004) Saha- ja vaneriteollisuuden vienti ja tuotanto. In: Hänninen, R. (ed.) *Metsäsektorin suhdannekatsaus 2003–2004.* Metsäntutkimuslaitos, Helsinki.

Honkala, A.M. (2001) *Oulangan kansallispuiston kävijätutkimus 2000.* Naturpolis Kuusamo koulutus- ja kehittämispalvelut. Tutkimuksia 2/2001.

Kautonen, M. (2006) The regional innovation system bottom-up: a Finnish perspective. A firm-level study with theoretical and methodological reflections. Electronic dissertation, Acta Electronica Universitatis Tamperensis. Available at http://acta.uta.fi/pdf/951-44-6702-7.pdf

Keskinarkaus, S. and Matilainen, A. (2009) The permit hunters' opinions towards professional hunting tourism and the current hunting license policy as a perquisite of social sustainability. The XXIII European Society for Rural Sociology Congress 17–21 August 2009,Vaasa, Finland.

Kubeczko, K., Rametsteiner, E. and Weiss, G. (2006) The role of sectoral and regional innovation systems in supporting innovations in forestry. *Forest Policy and Economics* 8, 704–715.

Lafferty, W.M. (2004) From environmental protection to sustainable development: the challenge of decoupling trough sectoral integration. In: Lafferty, W.M. (ed.) *Governance for Sustainable Development: the Challenge of Adapting Form to Function.* Edward Elgar, Cheltenham, UK, pp. 191–220.

Lähdesmäki, M. (2005) *Yrityksen yhteiskuntavastuun käsite pienyrityksissä – kilpailuvaltti vai kilpailun este.* Raportteja 7/2005. Helsingin yliopisto, Ruralia-instituutti.

Lundvall, B.-Å. (1992) (ed.) *National Systems of Innovation. Towards a Theory of Innovation and Interactive Learning.* Pinter Publishers, London.

Lunnan, A., Nybakk, E. and Vennesland, B. (2005) Innovation in non-timber products and services in Norwegian forestry. Past experiences and future development. In: Innes, J.L., Edwards, I.K. and Wilford, D.J. (eds) *Forests in Balance: Linking Tradition and Technology.* XXII IUFRO World Congress, 8–13 August, Brisbane. Abstracts. The International Forestry Review.

Luostarinen, M. (2005) Ekologinen klusteri ja innovaatiopolitiikka. Ecological cluster and innovation policy. *Maa- ja Elintarviketalous* 70.

Matilainen, A. and Lähdesmäki, M. (2009) *Nature-based entrepreneurship in private forests – the preconditions for the sustainable co-operation between private forest owners and entrepreneurs.* Reports 48/2009. University of Helsinki, Ruralia Institute.

Matilainen, A., Kajanus, M., Niskanen, A., Mynttinen, S., Ollonqvist, P. and Vihinen, H. (2007) Innovations in forest-based value chain firms. The Finnish approach to the discussion. In: Rametsteiner, E. (ed.) *Proceedings of the 1st COST Action E51 joint MC and WG meeting*, 12–14 October 2006, Grosspetersdorf, Austria. Department of Economics and Social Sciences. University of Natural Resources and Applied Life Sciences Vienna, pp. 88–100.

Mitchell, R.K., Agle, B.R. and Wood, D.J. (1997) Towards a theory of stakeholder identification and salience: defining the principle of who and what really counts. *The Academy of Management Review* 22, 853–866.
Morgan, K. (1997) The learning region: institutions, innovation and regional renewal. *Regional Studies* 31, 491–504.
Näsi, J. (1995) What is stakeholder thinking? A snapshot of a social theory of firm. In: Näsi, J. (ed.) *Understanding Stakeholder Thinking.* LSR-Julkaisut Oy, Helsinki, pp. 19–32.
Nelson, R. (ed.) (1993) *National Systems of Innovation.* Oxford University Press, Oxford.
Neville, B.A. and Menguc, B. (2006) Stakeholder multiplicity: towards an understanding of the interactions between stakeholders. *Journal of Business Ethics* 66, 377–391.
Nutt, P. (2002) *Why Decisions Fail: Avoiding the Blunders and Traps that Lead to Debacles.* Berrett-Koehler Publishers Inc., San Francisco, California.
Nybakk, E., Vennesland, B. and Lunnan, A. (2005) Innovation in non-timber products and services in Norwegian forestry. Manuscript, Norwegian Forest Research Institute, Ås.
Patton, M.Q. (2002) *Qualitative Research & Evaluation Methods*, 3rd edn. Sage, Thousand Oaks, California.
Perry, C. (1998) Processes of a case study methodology for postgraduate research in marketing. *European Journal of Marketing* 32, 785–802.
Persson, A. (2002) *Environmental Policy Integration: an Introduction.* PINTS – Policy Integration for Sustainability, Stockholm Environment Institute, Stockholm.
Rametsteiner, E., Weiss, G. and Kubeczko, K. (2005) *Innovation and Entrepreneurship in Forestry in Central Europe.* European Forest Institute Research Report 19. Brill, Leiden.
Rametsteiner, E., Weiss, G., Ollonqvist, P. and Slee, B. (eds) (2010) *COST Action E51 Policy Integration and Coordination: the case of innovation and the forest sector in Europe.* OPOCE, Brussels.
Ryymin, J. (2006) *Matkailun ohjelmapalvelu.* Toimialaraportti 14/2006. KTM ja TE-keskus. Helsinki.
Saarinen, J. (2001) Luontomatkailu osana metsien käyttöä. In: Kangas, J. and Kokko, A. (eds) *Metsän eri käyttömuotojen arvottaminen ja yhteensovittaminen.* Metsäntutkimuslaitoksen tiedonantoja, pp. 129–133.
Saarinen, J. (2003) The regional economics of tourism in northern Finland: the socio-economic implications of recent tourism development and future possibilities for regional development. *Scandinavian Journal of Hospitality and Tourism* 3, 91–113.
Saxena, G. (2005) Relationships, networks and the learning regions: case evidence from the Peak District National Park. *Tourism Management* 26, 277–289.
Scherer, R. (2006) *Regionale Innovationskoalitionen. Bedeutung und Erfolgsfaktoren von regionalen Governance-Systemen.* Haupt Verlag, Bern.
Sharma, S. and Henriques, I. (2005) Stakeholder influences on sustainability practices in the Canadian forest products industry. *Strategic Management Journal* 26, 159–180.
Strauss, A. and Corbin, J. (1990) *Basics of Qualitative Research. Grounded Theory Procedures and Techniques.* Sage, Newbury Park, California.
Verbij, E. and Schanz, H. (2002) Inter-sectoral co-ordination: state of the art and beyond. In: Tikkanen, I., Glück, P. and Pajuoja, H. (eds) *Cross-sectoral policy impacts on forests.* EFI Proceedings 46, 91–101.
Virkkala, S. (2006) What is the role of peripheral areas in a knowledge economy? A study of the innovation processes and networks of rural firms. Paper presented at the Conference 'Innovation Pressure', Tampere, Finland, 15–17 March 2006. Available at http://www.proact2006.fi/index.phtml?menu_id¼16&lang¼1
Weber, M.K. (2002) Innovation networks and the transformation of large socio-economic systems: the case of combined heat and power technology. In: Pyka, A. and Küppers, G. (eds) *Innovation Networks: Theory and Practice.* Edward Elgar, Cheltenham, UK.
Yin, R.K. (2003) *Case study research. Design and methods*, 4th edn. Applied Social Research Methods Series 5. Sage, Thousand Oaks, California.

12 Innovations in Wood-based Enterprises, Value Chains and Networks: an Introduction

Pekka Ollonqvist

Abstract

Linear intra-firm innovation process paradigms do not correctly identify the technological and innovation infrastructure development that is important to small and medium-sized enterprises (SMEs) in traditional industries. Instead, the recent industrial dynamic literature identifies the competitive business infrastructure of traditional industry enterprises by bringing forward the potentials available through actor interactions and knowledge transmission in sectoral and regional innovation systems. The view on innovations as systemic process outcomes has identified proximity in knowledge distribution and tacit knowledge transformation as supportive elements, but path-dependent and business culture routines as impeding for innovating SMEs. Wood-related value chains and SMEs are dependent on external knowledge and competence services in their innovation processes. The current broad-based innovation policy has expanded the old view of proximity benefits that was based on unintended knowledge spillovers and other positive externalities. The innovation-induced activities under consideration in this introductory chapter try to wrap up the specific entrepreneurial fields of interest in timber construction, furniture industries, wood-based bio-energy and timber transportation relevant for successful innovations. There are specific examples of each of those industries in this chapter. Traded firm-to-firm knowledge relations and formal capability creation networks, discussed in a separate chapter, have expanded in innovation-related knowledge and competence service markets. The content and extent of the cluster concept are relevant in the regional or local context because of the modes of transactional contacts among the enterprises and the knowledge and competence suppliers

12.1 Background and Objectives

12.1.1 Low- and medium-tech in traditional industries

This chapter aims to discuss the creation of competitive advantages (CAs) among wood-related micro-, small and medium-sized enterprises (SMEs).[1] The potentials for CAs can be identified in various stages of the parallel wood-related vertical value chains concerned. The innovation-induced activities are the specific CAs under consideration in this introductory chapter trying to wrap up the specific entrepreneurial fields of interest: timber construction, furniture industries, wood-based bio-energy and forest transportation are discussed in the specific sections. These fields of business provide new entrepreneurial opportunities, but existing competition makes them also challenging. Innovation has, in the literature, a widely applied definition as a new or substantially improved product (good or service or their combination) in the market, or introduction of substantially improved processes within the company during recent years (Community Innovation Survey, CIS, by Eurostat). Innovation activities in traditional industries classified into low or medium technologies (LMT) in innovation statistics typically comprise incremental product design and quality improvement, changes in organization and management routines and marketing, but also diffusion of existing knowledge from competitors and business partners (Malerba, 1992). The latter is related to their low level of direct science, technology and innovation (STI) investment activities. The major innovation knowledge among traditional industry SMEs covers practical, engineering, design, marketing, logistics, production, sales and distribution activities, i.e. doing, using and interacting (DUI) activities (Asheim, 2009). Traditional industry enterprises pursue incremental innovations by adopting high-tech through best-practice solutions and the achieved economic benefits mainly come from gradual development more than from innovations (Fagerberg, 2005).[2] Empirical findings support the importance of ability to innovate as a key factor for the competitiveness of SMEs in traditional industries (Asheim *et al.*, 2003).

Traditional industries have provided biased policy signals in the prior innovation policy that was based on formal research and development (R&D) outlay measures. There are at least two major sources for the biased conclusions: (i) the utilization of high-tech equipment; and (ii) the technology life cycle. The process technologies used in LMT industries are developed within outside sectors that are frequently classified as being high-tech. The current new way of identifying innovation activities as interactive processes inside the enterprises, and even within learning activities, has improved innovation policy incidences (Lundvall, 1988). The conceptualization of innovativeness has been standardized by the European Union (EU) CIS and forms the basis of the OECD third Oslo Innovation Manual. The R&D intensity of an industry is a temporary state that is determined by the life cycle of the industry concerned (Bender, 2006). The stage of industry technology(-ies) is cyclical and typically low-tech during the mature stage (Klepper, 1997). The low-tech stage of an industry is connected to the cessation of possibilities for newly formed innovations, but also to the fair interests among the enterprises managing with dominant technologies.

Regional and sectoral clusters provide arenas for systemic innovation processes to organize business activities or discover new channels or methods in marketing (Robertson *et al.*, 2008; Nybakk, 2009). The radical specification on innovation can be used only in cases where the impacts do not change the basis of the competition.[3] Innovating enterprises carry out interactive processes with public and private institutions, and knowledge-intensive business service (KIBS) providers towards tradable products and commodities. Innovations currently include creative activities that contribute to diversity and therefore generate profits (Hirsch-Kreinsen, 2006). The new approach identifying technology diffusion as a public promotion base has changed only gradually in parallel with the intake of systemic innovation policy view (Ollonqvist and Rimmler, 2010).

12.1.2 Systemic innovation approach

The demand for innovation-related knowledge among the traditional industry enterprises, especially among SMEs, has been raised in the innovation policy arena as an objective to create and maintain the supply of their innovation-related knowledge and competencies. Non-science-based knowledge and capabilities that underpin their needs are especially important (Hirsch-Kreinsen, 2006). The positions of innovative milieu and informal knowledge transmitting institutions have also been empirically identified in the research traditions of industrial economics and economic geography, respectively (Asheim *et al.*, 2003; Bender, 2006).

The systemic approach considers innovation processes as socially and territorially embedded interactive learning processes between knowledge providers and innovating enterprises (Lundvall, 1988; Asheim, 2009). Linkages with external actors are crucial for network relationships, interactive learning processes and distributed knowledge bases (Heanue, 2009). The knowledge transformation can be distributed through externalization, dissemination, internalization and socialization of learning activities (Smith, 2002).

The current systemic innovation approach supports the organized (traded) knowledge transfer processes that involve organized multiple actors and encompass vertical interactions with either upstream or downstream sources of new knowledge. It also acknowledges the horizontal collaboration with other enterprises in the industry. The institutionalized systems for interactive innovation processes are necessary and can be arranged either on a regional or sectoral basis (Rametsteiner *et al.*, 2005).

The regional innovation system (RIS) has a territory-oriented policy that focuses on regional or subregional innovation processes and is not conceptually different from national or sectoral systems (Boekholt and Thuriaux, 1999). The inferior position of R&D activities in the work agenda of traditional industry entrepreneurs is not only related to the low risk endurance and long expected repayment periods, but also to bounded knowledge about innovation opportunities and insufficient entrepreneurial capacities towards innovation processes.

12.1.3 Harvested and harvestable wood as innovation potential

Wood as a raw material used in existing first-degree industries (sawing and wood plate industries as well as pulp and paper making) covers only part of the harvestable wood material available in forests. There are material potentials and opportunities for new products and business opportunities. Innovative wood material uses can substitute or complement current ones by providing opportunities for higher aggregate benefits.

There are three parallel vertical value chains, construction, furniture and bio-energy, which are discussed and analysed in the subsequent chapters of this book. The interconnected relationships of the enterprises in those value chains are described in Fig. 12.1. There are two 'cluster' interfaces to express the existing business culture boundaries between industrial clusters. Path dependence and cluster-specific business culture tend to be common among traditional industry enterprises and especially among SMEs, but also on those applied in the clusters in the downflow and upflow along the value chains.

Entrepreneurs in forest transportation command the technical equipment, but innovative new business solutions shall be needed. Timber-based solutions have, in the residential and furniture industries, lost market shares to more competitive materials. There is a need to search for new timber-using innovative solutions. The availability of innovative wood-based bio-energy processes and solutions has opened up competitive complementary and substituting wood fibre demand options in parallel with the current modes of use.

The position of innovation activities in enterprise are worth discussing in a wood-related value chain context. The businesses analysed in the COST E 51 wood value chain subgroups are identified in Fig. 12.1.

12.1.4 Objectives of the chapter

This chapter discusses features related to innovation activities among SMEs in wood-based vertical value chains with specific interest on the suppliers and customers and their contracts throughout the value chain. The focus in this chapter is on the knowledge dissemination among SMEs to complement their knowledge base and innovation capabilities. Policy, in this context, concerns broad-based competence creation promotion through innovation policy and also other complementary policies. The resources and capabilities as well as innovative entrepreneurship in low-tech industries are discussed in Section 12.2.

Rethinking innovations as integrated elements in the entrepreneurial activities implies a new management agenda. Innovation Enabling

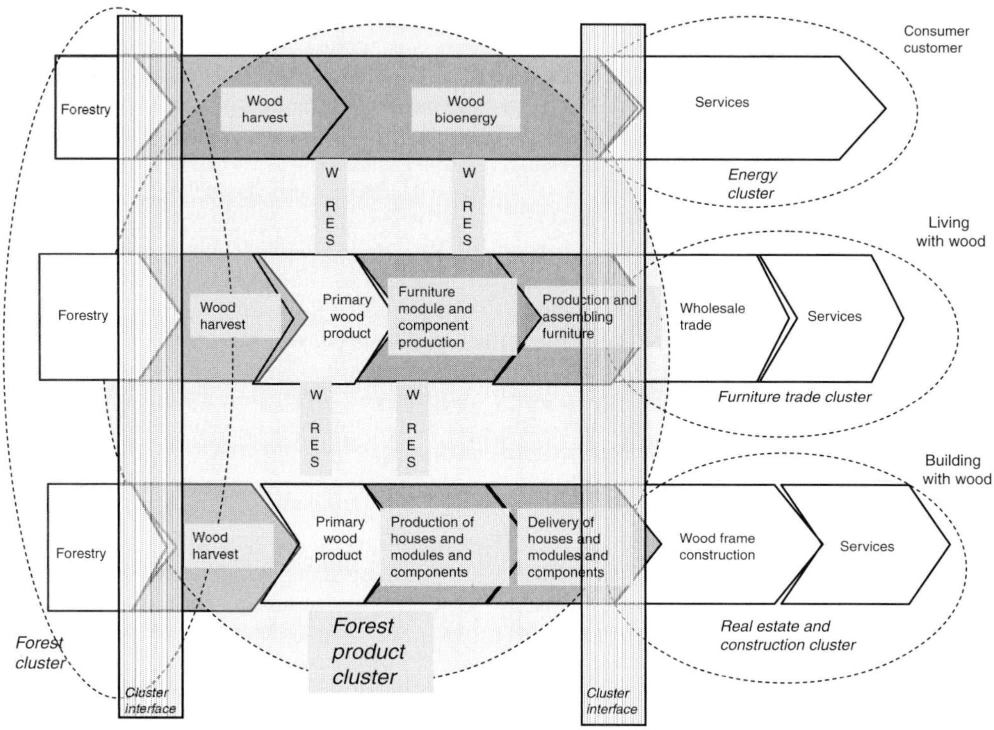

Fig. 12.1. Use of wood products in consumer segments related to real estate and construction, furniture and energy clusters and the vertical value chains. There are cluster interfaces between forest and forest product clusters and consumption and forest product clusters respectively (source: Ollonqvist, 2009).

Capabilities (IEC) among SMEs are discussed in Chapter 3 and also in Section 12.5.

Section 12.4 discusses the traded knowledge transfer and organized collective learning in rural infrastructure. The question of market balance between knowledge and competence demand and their organized supply are discussed in Section 12.4.3. The traditional industry-specific innovation policy needs and challenges, and innovation integration into development policies, are contained in Section 12.5.

12.2 Technical Maturity and Innovation Activities

12.2.1 Competencies and capabilities

Innovation is based on the generation, diffusion and use of new knowledge, but also the combination of old and new knowledge along all stages of the process. The generation and implementation of new ideas, processes or products aimed at increasing the competitiveness of enterprises imply capabilities to identify available factors owned or controlled by the enterprise but needed for complementary capabilities. Amit and Schoemaker (1993) identify management-related uncertainty, business infrastructure complexity and intra-organizational conflict as critical for innovation. Most innovation processes in enterprises imply external knowledge and competencies derivable, directly or indirectly, from several sources, and innovators need effective mechanisms for the knowledge diffusion (see PILOT; Bender, 2006). Innovation capacity constitutes intra-firm knowledge and capabilities but also potentials achievable in the innovation infrastructure. Absorptive capacity, an ability to recognize, assimilate and apply the potentials from new external knowledge, is important. The dimensions, acquisition, assimilation, transformation and exploitation of external knowledge, as specified by Zahra and George (2002), have also been empirically identified (Asheim *et al.*, 2003).

Wood industries provide an example that is characterized by saturated market demand, low return on invested capital (ROI) and meagre growth of turnover in individual enterprises. The mental capacity demand among entrepreneurs for day-to-day battle against competitors is outstanding (Stendahl, 2009). Thereby the temporary competitive advantages achieved through incremental innovations tend to be preferred by entrepreneurs. The requirements for external resources and capabilities seem to have excluded the rationality of radical innovation processes among wood industry SMEs and wood product market segments. The outstanding position of cost leadership and consequent production outsourcing in the US furniture industries is an acute example of these challenges (see Ollonqvist *et al.*, Chapter 14, this volume).

Technology maturity tends to decrease entry barriers in industries, but also search interests towards radically new solutions. These features make it important for traditional industry SMEs to target sustainable competitive advantages (SCAs) with their innovation processes. Korhonen and Niemelä (2005) have identified three basic managerial patterns to innovative capability use in wood industry enterprises. First, the proactive development of new capabilities to understand the upcoming changes in the competitive infrastructure. The wood product industry competes for transferable individual skills (technical trades, accounting) with other industries especially in remote areas. Second, the reactive utilization of existing capabilities is valid in new market contexts, e.g. by learning from prior successful export activities. Third, the possession and identification of immobile and non-substitutable resources and capabilities help to maintain and upgrade their value over time.

12.2.2 Innovative entrepreneurship in traditional industries

In most cases, traditional industry SMEs are manufacturers of standard products, components or modules. They operate as subcontractors to international major customers with limited possibilities to create other features but technical productivity (Hirsch-Kreinsen, 2006). Their position keeps their potentials easy to imitate, thus impeding their innovation activities. Local networking is one of the major supporting factors for gaining CAs. These kinds of solutions are possible in local clusters with flexible relations. Industrial dynamic literature identifies enterprises in their competitive business infrastructure (Malerba, 1992). Competition in traditional industries brings forward sectoral (SIS) and regional (RIS) innovation systems as arenas to identify interaction potentials among different actors (enterprises, universities, polytechnics, research institutes, financial organizations and KIBS). The value chain configurations analysed in the chapters on bio-energy (Nybakk *et al.*, Chapter 15) and wood harvesting (Bouriaud *et al.*, Chapter 16) in this volume provide examples of production processes among the SMEs applying high-tech production technologies. The substitution of labour with highly advanced process technologies in simple product manufacturing is rational. The innovations in the intake of new technologies are radical to a single investing enterprise at the time of investment but represent gradual development in the industries delivering high-tech equipment. These investments are also frequently demanding to the investing enterprise chief executive officer (CEO) (see research findings in PILOT; Bender, 2006).

Service orientation is one way to make market-induced product innovations open up flexible sales opportunities in a highly flexible manner. Wood product industry SMEs in high-income countries put increasing efforts into organizational and marketing innovation issues. Inventory management by substituting customer inventories (builders' merchant stores) is one of the complementary services in primary wood product business-to-business (sawmill strategies, see Stendahl, 2009). Increased focus on market segmentation enhances short-term CA potentials. Custom-graded sawn wood supplied to furniture enterprises that once provided CAs has become dominant and a mature business model. This involves both investments in product design capabilities for functional and technical product segmentation and flexible reactivity to special customer needs. Rethinking innovations as a continuous integrated SME manager activity implies new management agenda and thinking. The first fundamental decisions among the CEOs towards this agenda are: (i) where and to what extent the key person resources are allocated there; and (ii) how to create the managerial practices to build up a temporary innovation process team comprising key intra-enterprise knowledge with complementary competencies and capacities in each case.

12.3 Innovations for Customer Responses and Quality Generation

12.3.1 Knowledge base for innovation activities

The development of global competition is behind the innovation policy tasks addressed to traditional industries in the EU. Product differentiation, successful specialization to niche markets and product innovations inside successful value chain architectures have substituted prior CAs in traditional industries (Hirsch-Kreinsen, 2006). Innovation-relevant knowledge is acquired and generated within the enterprises but external knowledge sources are also relevant for them. The necessary knowledge base involves many intra-enterprise actors representing different specialities of trade and technology (Smith, 2002).

The industrial and innovation policies that are based on the local proximity paradigm concerning knowledge and competence transmission among traditional industry SMEs have had both supporting and impeding impacts (Forsman, 2009). The creation of local value networks was advocated with the advantages from interlinks by repeated transactions and flexible governance structures in vertical and lateral directions (Porter, 1999). The cluster concept, gradually adopted into regional and industrial policy debates and widely used as a public support allocation premise, needs redefinition to match current development challenges. The addressed potential proximity advantages and local agglomeration economies suffer increasingly from imitation and duplication. A few cluster operations relevant for traditional industry SMEs and cluster output can currently be traded across locations even on a global scale (see the survey in McDonald and Belussi, 2002). The latter poses a major challenge for the existing and potential mini-cluster to maintain innovation as a support environment through functional, social and geographical proximity (see Rimmler *et al.*, Chapter 7, this volume). Value chain efficiency, end-customer orientation and total quality generation imply value chain configurations that pass through traditional cluster boundaries.

The codified and formalized knowledge and related competencies for use can be complemented in traditional industry SMEs by: (i) value chain connected partnering; (ii) working with knowledge and competence creating and distributing organizations; and (iii) other competence-related actors. These factors, which stimulate innovation activities, are external. Consultants also provide an important source of information (Aslesen and Isaksen, 2007). Innovation-relevant tacit knowledge, covering DUI elements, is important during implementation (Asheim, 2009). Information from marketing people, reflecting responses from the customers and users, gives important signals towards innovation process generation. The customers, frequently from user-innovator communities, lead to the appearance of user-purchasers who want to buy the products or services of the lead users rather than building the customization themselves. The objective of lowering variable costs will tend to drive prices down to expand the market (Baldwin *et al.*, 2006).

Successes in innovation processes are not solely related to the competencies available in the enterprise but cover the capabilities among the managers to acquire, transform and adapt new technical knowledge. The choices of knowledge sources can be attributed to different capabilities among the enterprises and their network partners (consumers, universities, etc.) in creating and utilizing (exploration and exploitation) respective innovation-enhancing knowledge (see Ukrainski and Kajanus, Chapter 4, this volume).

12.3.2 Innovative capabilities: enterprise demand and complementary knowledge supply

The concept of IEC traces back to the research findings from European traditional industries (see Bender, 2006). IEC address competence-building requirements composed of transforming and configuration capabilities. Traditional industry SMEs normally have knowledge and demand from KIBS, polytechnics and other knowledge-generating and distributing institutions. Transformative capabilities allow the transformation of general knowledge (R&D outputs) into the modes usable in a locally specific context. Configuration capabilities focus on the enduring abilities to synthesize new configurations of established knowledge, artefacts and actors (see PILOT; Bender, 2006). Organizational competencies cover abilities to use distributed sources of relevant knowledge and competence through linking actors together with available relevant knowledge, technology and competence. Design competencies are needed when identifying customer needs and consumer preferences and cover the process to create the added customer value demands in an output.

Cohen and Levinthal (1990) proposed the absorptive capacity as a decisive innovation ability element. Following Teece *et al.* (1997), this ability can be conceived as the dynamic capability to cultivate and develop the knowledge base strategically and mobilize individual knowledge elements towards generating innovative solutions in the long run. The operational dimension of IEC refers to the routines in the value-creating transformation process, by which an enterprise converts resources and knowledge into the customer values of goods or services.

Rosenberg and Nelson (1994) indicated that the content and source of knowledge varies greatly across industry sectors. Scientific knowledge is the force driving technological change in some sectors, whereas there are industry sectors in which formal R&D applications or DUI knowledge play a major role in innovations.

Sophisticated suppliers of technical components, machinery, tools or software are important for the diffusion of new process technologies. External relationships and collaboration, therefore, are essential for tapping external sources of innovation. The creation of cross-sectoral linkages could become an important means to foster the transfer of information and communication technology (ICT) and other state-of-the-art technologies into use (Hirsch-Kreinsen, 2006; Ollonqvist and Rimmler, 2010).

12.4 Innovation-supporting Infrastructures

12.4.1 Individual and collective knowledge creation support

Average traditional industry SMEs cannot afford key experts in their permanent staff and their major innovation activities are separate, discrete tasks. Those features make enterprises dependent on the complementary knowledge and competence supply, and the right identification of their competence demand provides important signals to policy makers. The positive knowledge spillovers and access to externalities are the most traditional and also the most cited benefits from the local innovation infrastructure. The extent and intensity of informal knowledge spillovers and externalities within industries and across industries have been the key industrial policy targets for technology transfer (see survey in Tödtling *et al.*, 2009). Empirical evidence suggests that policy support should strengthen the innovative milieus among traditional industries. This should be based on proximity and embedded relations (Asheim *et al.*, 2003; and PILOT; Bender, 2006). This traditional approach utilizes the proposition of atomized SMEs in local clusters. There are also empirical findings from traditional local cluster arrangements, characterized by voluntary collective learning. Skive Carpenters' Guild operating in the Salling furniture cluster in Denmark is a well known example of this mode (Asheim *et al.*, 2003).

New broad-based innovation policy agendas have advocated trade-based knowledge and capability service supply. They can be based on individual or network activities and organized through SIS and RIS. The enhanced demand also explains the emergence of regional and sectoral KIBS supply. In addition to the knowledge services, knowledge-intensive service activities (KISA) network modes provide benefits for SMEs by reducing uncertainties that relate to development of innovation base. The innovative milieus, established by public institutions, can be based on regional or rural development programmes to enhance collaboration between R&D institutions (universities and polytechnics), KIBS and SMEs (Asheim and Isaksen, 1996). The emergence of regional knowledge and competence services available for the furniture cluster (see Box 12.1) is an example.

12.4.2 Rurality and local innovation support

Economic geography research disciplines provide relevant starting points to characterize innovation systems through the emergence of innovation infrastructures (exogenous or institutional push versus organic growth). The taxonomy of Asheim and Cooke (1999) identifies exogenously pushed and endogenously grown innovative networks. Exogenously pushed innovative networks are based on the establishment of supportive organizations and institutes (e.g. polytechnics, R&D institutes, technopoles or science parks). Exogenous creation tends to emerge in circumstances that are characterized by large enterprises leading product development and process technology development. The creation of multi-product bioenergy knowledge systems with full size field implementation in Austria has these kinds of development features (Box 12.2). The created innovative milieu was supported by regional or rural development programmes to enhance collaboration between R&D institutions (universities and polytechnics), KIBS and SMEs (see other findings, Asheim *et al.*, 2003).

Endogenously grown innovative networks are based upon existing regional or local industrial districts. There the innovation activities tend to be based on the DUI traditions utilizing interacting and learning connections among the local enterprises. These activities in the minimum case cover active competition for skilled labour in the Marshallian district sense. Active cooperation can be horizontal or vertical. The initial innovative network activities among the local enterprises can continue self-serving, thus substituting the needs of public initials (McDonald and Belussi, 2002). Innovative enterprises can even consider public support activities too time consuming and their support so limited that they continue with their own financing, thus excluding direct public interventions. Endogenously grown innovative infrastructures provide opportunities for acquiring skilled people, but it can also become a challenge to lose intellectual resources to competitors. Rural enterprises tend to be in an inferior position concerning the supplementary knowledge and competence supply with respect to their urban counterparts. The organizational thinness, empirically identified by Tödtling and Trippl (2005), weakens the regional and local support on innovation activities. The latter is a challenge to the development of technology knowledge or network distribution activities in the regions. The identification of network initials and the emergence of a local innovation system are valuable when

Box 12.1. Enhanced innovation capability supply for the Southern Ostrobothnia furniture industry. (Box prepared by A. Pirc, A. Viitala, P. Ollonqvist and T. Rimmler.)

Innovation support activities for furniture SMEs have been centred into a regional Research and Development and Innovation (RDI) centre, Nikkarikeskus, in Jurva from 2004. The centre, providing and distributing marketing and promoting know-how for Ostrobothnian furniture manufacturers and also design and technology services, was created through a bottom-up approach. The planning process and the project plan for the centre, financed by the local Suupohja Leader programme, provided an alternative to the prior top-down RDI national programmes. The development centre houses a regional coordination on furniture expertise with university and polytechnic partnerships to enhance the supply of knowledge for innovation-enabling capabilities among the SMEs, but also to provide tailored services towards new CAs in the region. The supply provided by the centre can be divided into: (i) business concept creation; (ii) business incubation; and (iii) culture and concept development.

(i) Business concept creation was organized as a part of a new business concept section of PuuOske* and implemented through the training programmes (TAKO).†

(ii) The centre provides start-up and spin-off as well as other business incubating services through entrepreneurship training and knowledge specialist services. These tailored programmes extend to 6–12 months. The centre building and business incubator (FAARTTI) was organized in Nikkarikeskus.

(iii) The culture and concept development through the design project connects furniture industry enterprises and students at Seinäjoki University of Applied Sciences aiming to cooperate towards R&D projects. This activity associates students into enterprise-specific research projects, thus allowing them opportunities to show their skills during their diploma research. These projects help students in their career interests if the associate furniture enterprises recognize their talents and employ them. Seinäjoki University of Applied Sciences is getting some support money from TEKES for developing new design projects but most of the projects have been financed by the involved furniture enterprises.

The current supply of knowledge and innovation supporting services has grown and includes sponsors in addition to local Leader and EU goal programmes and national development programmes:

- Intelligent Furniture and Space Concept Development Platform Project: www.alykaluste.fi.
- Habit 5xm – Consumer profiler: www.habit5xm.org.
- Experimental Design Lab.
- Partner in Centre of Expertise Programme of Living business: www.oske.net.

There are also adult education centre activities and a furniture branch technology village in the Suupohja subregion.

*PuuOske was a nationally networked Centre of Expertise for Wood Products (WP CoE) during 1999–2006. WP CoE cooperated with the Wood Wisdom Research Programme and PuuSuomi (WoodFinland) network creation action programme under national coordination of wood focus (Paajanen *et al.*, 2007).

†Tako was coorganized in 2001 and 2003 for the future key personnel in furniture industry enterprises by Levón-Institute (University of Vaasa), Seinäjoki Polytechnics and Finland's Entrepreneurship Institute. The business concept creation was thereafter continued by the University of Vaasa in with Politecnico Di Milan.

considering the potentials among the individual innovative enterprises. The gradual development of business networks is discussed in Box 12.3.

12.4.3 Innovation knowledge demand and supply balance

Demand and supply of innovation relevant knowledge and competencies among traditional industries are different at levels of aggregation: project, enterprise, value network and technical infrastructure. Continuous interaction between different stages of the innovation process from research to marketing and enhancing partnerships between industry, science and the governmental sector are needed, as well as inter-enterprise collaboration along value chains. Cultural, institutional and social structures are important frame

Box 12.2. Wood-based bio-energy value networks: biomass power plant in Güssing Austria. (Box prepared by A. Bauer, BOKU Vienna, and G. Weiss, EFICEEC/BOKU Vienna.)

Güssing is a small city in the Burgenland province. In 1990, the region was one of the poorest in Austria because of its remote location in eastern Austria at the border of the so-called 'Eastern Bloc'. Industry and manufacturers were almost missing and, because of the lack of jobs, 70% of the population had to commute weekly to Vienna or other regions. As a consequence, the migration rate was high and the infrastructure poor.

Because of this situation, the municipality took the decision to become self-sufficient in its energy supply. An energy concept was developed and the first biomass-based heating plants installed. EU funding was successfully used for further innovative energy plants and the European Centre for Renewable Energy Güssing (EEE) was founded as a focal point for developing innovations in renewable energies. The flow chart of the combined heat and power plant in Güssing is presented in Fig. 12.2. The innovative process, combining heat and power production simultaneously, is based on steam gasified in a dual fluidized bed reactor. The EEE conducts research, carries out demonstration projects and runs commercial operations. The centre cooperates with various businesses and research organizations on a variety of projects.

Today, 25 plants exist in the region on the basis of renewable energies, including: bio-diesel, solar energy, one of the largest biomass district heating nets in Austria and new technical concepts for biomass power plants. Furthermore, the centre offers further education and training and energy-related services, and runs bio-energy-related tourism projects. The many visits from interested audiences are professionally managed as a business in itself. In effect, a bio-energy cluster was created: 50 new related enterprises developed in the region, directly and indirectly offering more than 1000 new jobs. The successful example of the city of Güssing led to the goal of the whole region of Southern Burgenland becoming energy self-sufficient.

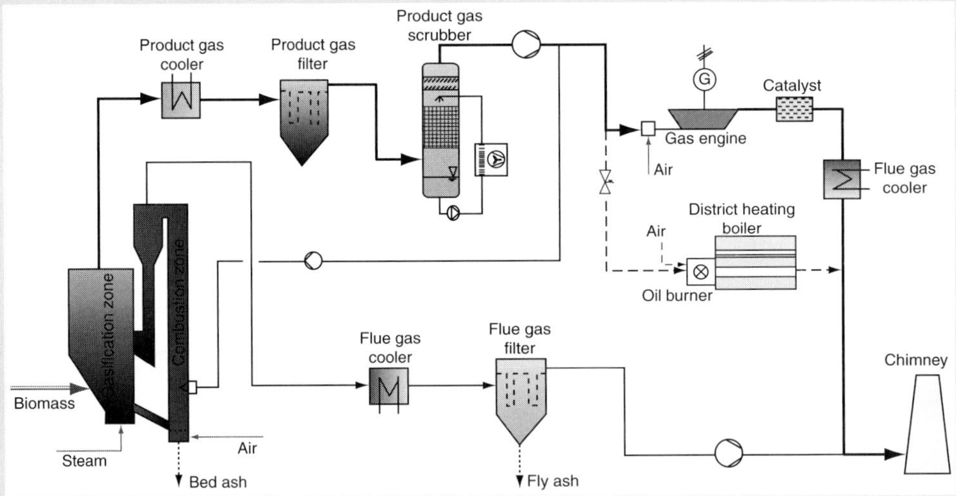

Fig. 12.2. Flow chart of the combined heat and power plant in Güssing. (Source: Bolhàr-Nordenkampf *et al.*, 2007).

conditions for the flow of non-codified information among enterprises. Therefore, strengthened linkages between different actors of the innovation system, in general, are of policy concern. The latter extends over cluster boundaries along value chains (Fig. 12.1). The challenges to supply knowledge and strengthening capabilities in system integration are the most difficult, implying enhanced abilities to

Box 12.3. Regional and rural policy support on timber construction and sustainable development – timber construction module industrial park and micro-cluster, North Satakunta, Finland. (Box prepared by P. Ollonqvist, H. Perko and T. Rimmler.)

The regional centre of seven municipalities (four from rural heartland areas and three from sparsely populated rural areas) agreed on the Pohjois Satakunta entrepreneurship development strategy in 2005. The adopted regional strategy compensated for the secondary position in this region in the Satakunta county wood product business and technology strategies. Strategic positioning of the regional development company (Pohjois-Satakunnan Kehittämiskeskus) provided a tacit knowledge base and actively supported public financing for micro-cluster creation. The bottom-up strategy improved commitment among the related entrepreneurs.

The micro-cluster, in the form of a timber construction module industry park and subcontractor network formation, was promoted and managed by the wood processing project from 2002 in the regional Centre of Excellence (CoE) with business knowledge development, materials technology and distant technology as their core interest areas. Those CoE competencies provided cross-fertilization to a wood processing project and network promotion. The micro-cluster and the policies relevant for the cluster creation are presented in Fig. 12.3. The creation of the micro-cluster benefited from the tacit knowledge potentials among the locomotive park enterprises through: (i) factory line organization and management experience for pre-fabricated on-site construction modules and elements; (ii) polyurethane isolation technology experience in large-scale production; and (iii) rural joinery and carpentry traditions among the networked micro-enterprises.

The business knowledge creation and other documented knowledge utilization were provided by the regional CoE through their special knowledge but also through regional implementation of WoodFinland (PuuSuomi) business network activation top-down programme 1999–2005. The latter programme's implementation benefited from the bottom-up approach accentuated by the enterprises in Satakunta. The regional CoE provided promotion and assistance to adopt external knowledge partners and establish business networks. The community coordinated a rural heartland programme during 1997–2002 provided by EU 5b, and ALMA financing thus assisted the wood park formation and regional micro-woodworking network creation. The creation of micro-cluster expanded the utilization of positive scale economies through the regional business networks and paved the way for the establishment of the timber construction module industrial park. The timber construction module industrial park financing was mainly based on national regional development financing. The formation of regional SME networks and subcontracting network formation were financed through ALMA and Leader+.

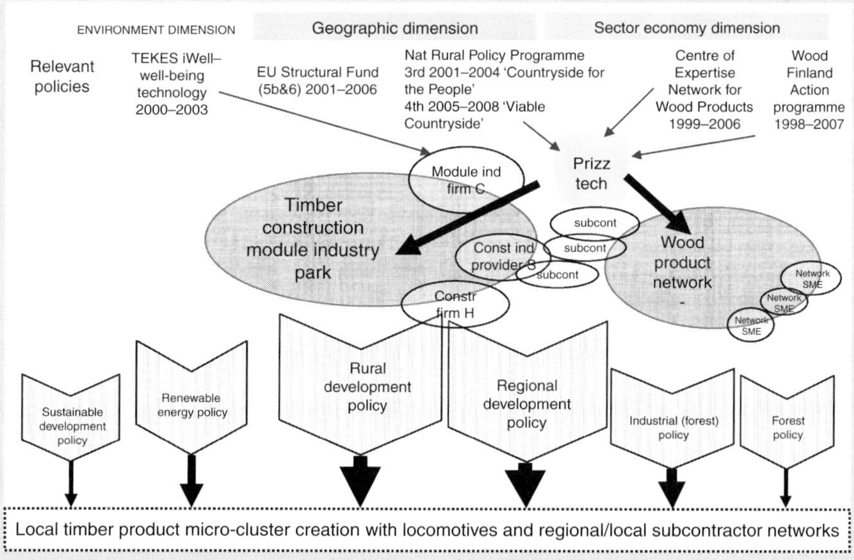

Fig. 12.3. The micro-cluster and policies relevant to cluster creation.

utilize knowledge created outside the local knowledge centre. Timber construction enterprises face a knowledge deficit to integrate their business into modern construction systems based on interactive 3D modelling and products, components and modules available in that mode for architects and construct planners (see Box 12.4 and Nord *et al.*, Chapter 13, this volume).

12.5 Discussion and Conclusions

Traditional industry SMEs would benefit from the broad-based innovation policy where domains of industrial policy are integrated with sector and regional development policies and coordinated with research and education policy. Deep understanding among the policy makers is needed to make these coordination and integration activities valuable.

Rapid progress in globalized business networks has induced strong regional and local structure changes among traditional industries. Local SMEs have benefited from these changes through enhanced labour supply qualification and joint knowledge transfer. Competence development circles, common in this context, provide participation arenas for the entrepreneurs that are characterized by low formalities in the cooperation commitments. The emergence of jointly acting project groups, joint ventures and joint units could be supported with integrated innovation and rural policy efforts (see Finnish and Swedish furniture industry examples in this chapter).

Industrial district-specific policy activities (with regional traditional industry clusters as a mode) frequently have a place-specific social and cultural background. The regional or local cluster research has accounted for the knowledge spillovers and other positive external economies among the local SMEs. These immaterial benefits rely on person-to-person relationships and are fostered by trust, reciprocity and commonly shared values and norms. They are discussed by Rimmler *et al.* in Chapter 7 of this volume. Features supporting positive conditions for SMEs also cover options to promote cooperation.

The traditional industry entrepreneurs typically prefer process innovations because of their process technology educational background. Investments in new production technology, however, require training and technical knowledge accumulation in the enterprise. The emergence of regional and local KIBS supply has helped influential entrepreneurs with the high-tech knowledge requirements over the path-dependent narrow local milieu that can become an impediment for innovations (Tödtling and Trippl, 2005). There have been significant benefits from the technology development related to ICT and material processes. There is often a need to arrange these knowledge transformations on a professional supply basis in the local market. These service needs were identified in bio-energy and timber transportation subgroups (see chapters in this volume). The creation of innovative entrepreneurship through polytechnic and research institute spin-offs is a major issue in the current innovation policy regime. Innovative entrepreneurs with a strong scientific knowledge background are considered the major agents for future production and employment growth.

Structural regional reorganization can create high transaction costs when the global network subcontractors have afterwards lost their contracts. The latter can also become true in the future if the enterprises do not have alternative business plans available or capabilities to renew through innovation processes. The empirical results show that ensuring knowledge creation band distribution by establishing a regional knowledge centre is not enough. SMEs need knowledge services to enhance their innovation-related strategic, internal and external capabilities. The services and capabilities needed among SMEs must be able to be implemented hand-in-hand with their daily business. This potential demand is a challenge but also an opportunity for the local polytechnics and other knowledge transmitting actors.

Box 12.4. Multi-storey timber-frame residential construction in Sweden: national and local cooperation. (Box prepared by T. Nord.)

The Swedish Building Regulations were revised in 1994 from being prescriptive to functional. The revision made possible the introduction of new materials, systems and processes, e.g. timber-frame systems in larger structures. Increasing from a market share of 0% in 1994 to a present 10–15% has been the result of activities on the national and regional/local levels involving actors from the superstructure as well as infrastructure areas.

The development in the project-based construction industry shows a high degree of path dependency. Resistance to change characterizes the superstructure of the system, i.e. clients, regulators and innovation brokers, who shape and follow the construction industry culture and standards (Hobday, 1998; Winch, 1998). The production and knowledge infrastructure of timber-framed structures were concentrated to single-family house manufacturers predominantly located in the southern part of Sweden and to structural wood products producers, i.e. sawmills, because of the ban on large structures for more than a century.

National level activities and progress:
1998–2001 – A public–private financed research programme aimed at wood products and construction initiated basic research around timber construction and established timber construction research at two universities.

2004 – An initiative to a national strategy to enhance timber usage in construction, which called for joint activities between public and private organizations.

2004–2012 – Public financing to 18 growth areas, including the wood manufacturing industry, to support further (regional) competence platforms from the former research programme.

2005–2008 – The short-term goals for the strategy implementation from a public–private group predominantly facilitating regional and local authority as well as collecting and disseminating knowledge. A number of RDI construction projects were realized.

2006–2013 – A public–private research programme, based on the National Innovation Programme, aiming to initiate additional timber construction knowledge and competence dissemination in the forest sector. Two geographic regional/local level activities and progress for larger timber construction development in Sweden. They have different but complementary knowledge and industry infrastructures creating a positive climate for growth.

Northern Sweden: a city has become a major contributor of innovative timber projects. The local authorities have used the new regulations to commission and build timber-based structures, and arranged seminars, workshops and information meetings around timber construction. The local and regional timber-based enterprises have established long-term collaboration with the regional university and the public research organization as innovation brokers and KIBS. The collaboration and brokerage have moved from basic research of product and system properties to a production and organizational process creating market opportunities. With successful development of the innovation infrastructure in the region, further support from regional authorities in cooperation with private financing has taken the development further.

Southern Sweden: with a strong timber-framed housing industry, a university with a wood technology division, a strong wood mechanical industry and local authorities searching and being guided by environmental friendly solutions and decisions. The latter has resulted in a strategy of one city becoming a 'green' city and with one residential area with a high share of timber-framed housing. The strategic development has been a top-down process in cooperation with the university and external construction clients interested in timber housing. The city has provided land and a legal structure, the university together with other research organizations research projects, and external clients with the financial means to develop larger residential timber structures.

The two regional/local districts have provided a fertile base for innovative behaviour of the timber construction industry from their attitude to changing traditional behaviour in the construction industry, cooperating with research organizations and supporting enterprise growth and development.

Notes

[1] Medium-sized enterprises are defined as employing fewer than 250 persons and having an annual turnover of less than €50m. Small enterprises are defined as employing fewer than 50 persons and having an annual turnover of less than €10m. Micro-enterprises are defined as employing fewer than ten persons and having an annual turnover of less than €2m.

[2] Incremental innovation is understood to exploit existing forms or technologies, thus improving something that is already available or reconfiguring a technology to serve another purpose (Lindqvist and Ghazi, 2005).

[3] Innovation is defined as radical if one or more of the following characteristics are satisfied:

- an entirely new set of performance features;
- improvements in known performance features of five times or greater; and
- a 30% or greater reduction in cost (Leifer *et al.*, 2000).

References

Amit, R. and Schoemaker, P. (1993) Strategic assets and organizational rent. *Strategic Management Journal* 14, 33–46.

Asheim, B. (2009) The formation of competitive regional innovation systems. Presentation at COST E51 Meeting, Zagreb, 15–16 January.

Asheim, B. and Cooke, P. (1999) Local learning and interactive innovation networks in a global economy. In: Malecki, E.J. and Oinas, P. (eds) *Making Connections: Technological Learning and Regional Economic Change.* Ashgate, Aldershot, UK.

Asheim, B. and Isaksen, A. (1996) Location, agglomeration and innovation: towards regional innovation systems in Norway? STEP Report R 13. Oslo.

Asheim, B., Coenen, L. and Svensson-Henning, M. (eds) (2003) Nordic SMEs and regional innovation systems. Final Report. Department of Social and Economic Geography, Lund University, Lund.

Aslesen, H. and Isaksen, A. (2007) New perspectives on knowledge-intensive services and innovation. *Geografiska Annaler* 89B(S1), 45–58.

Baldwin, C., Hienerth, C. and von Hippel, E. (2006) How user innovations become commercial products: a theoretical investigation and case study. *Research Policy* 35, 1291–1313.

Bender, G. (2006) PILOT Final Report. Peculiarities and Relevance of Non-Research-Intensive Industries in the Knowledge-Based Economy. Available at: http://www.pilot-project.org

Boekholt, P. and Thuriaux, P. (1999) Public policies to facilitate clusters: background, rationale and policy practices in international perspective. In: *Boosting Innovation: The Cluster Approach, 1999 OECD Proceedings*, Chapter 16. Paris.

Bolhàr-Nordenkampf, M., Rauch, R., Bosch, K., Aichernig, C. and Hofbauer, H. (2007) Biomass CHP Plant Güssing – using gasification for power generation. members.aon@biomasse/thailand.pdf

Cohen, W. and Levinthal, D. (1990) Absorptive capacity: a new perspective on learning and innovation. *Administrative Science Quarterly* 35, 128–152.

Fagerberg, J. (2005) Innovation: a guide to the literature. In: Fagerberg, J., Mowery, D. and Nelson, R. (eds) *The Oxford Handbook of Innovation.* Oxford Handbooks, Oxford.

Forsman, H. (2009) Improving innovation capabilities of small enterprises: cluster strategy as a tool. *International Journal of Innovation Management* 13, 221–243.

Heanue, K. (2009) Aspects of the furniture industry in Ireland. PhD thesis, Dublin City University Business School, Dublin.

Hirsch-Kreinsen, H. (2006) Low technology – a forgotten sector in innovation policy. Paper presented at the International ProACT Conference, 15–17 March, 2006, Tampere, Finland.

Hobday, M. (1998) Product complexity, innovation and industrial organization. *Research Policy* 26, 689–710.

Klepper, S. (1997) Industry life cycles. *Industrial and Corporate Change* 6, 145–182.

Korhonen, S. and Niemelä, J. (2005) A conceptual analysis of capabilities: identifying and classifying sources of competitive advantage in the wood industry. *The Finnish Journal of Business Economics* 1, 11–47.

Leifer, R., McDermott, C., O'Connor, G., Peters, L., Rice, M. and Veryzer, R. (2000) *Radical Innovation: How Mature Companies can Outsmart Upstarts*. Harvard Business School Press, Boston, Massachusetts.

Lindqvist, M. and Ghazi, S. (2005) Disrupt or become disrupted – a framework to identify and evaluate disruptive innovations. Department of Technology Management and Economics, Chalmers University of Technology. Available at: http://www.sadreghazi.com/publications/thesis/MEI_thesis_Disruptive_Innovations.pdf

Lundvall, B. (1988) Innovation as an interactive process: from user producer interaction to national system of innovation. In: Dosi, G., Freeman, C., Nelson, R.R., Silverberg, G. and Soete, L. (eds) *Technical Change and Economic Theory*. Pinter, London, pp. 349–369.

Malerba, F. (1992) Learning by enterprises and incremental technical change. *Economic Journal* 102, 845–859.

McDonald, F. and Belussi, F. (eds) (2002) Industrial districts: a state of the art review. Project West–East ID 'Industrial districts' Re-location Processes: Identifying Policies in the Perspective of the European Union Enlargement.

Nybakk, E. (2009) Innovation and entrepreneurship in small firms: the influence of entrepreneurial attitudes, external relationships and learning orientation. PhD thesis, 2009/07, Norwegian University of Life Sciences.

Ollonqvist, P. (2009) Integrating innovation and development policies for the forest sector. Phase II – WG 2 Tasks. Presentation at COST E51 Meeting, Zagreb, 15–16 January.

Ollonqvist, P. and Rimmler, T. (2010) Policy integration supporting innovation policy design for SMEs. In: Rametsteiner, E., Weiss, G., Ollonqvist, P. and Slee, B. (eds) *Policy Integration and Coordination: the Case of Innovation and the Forest Sector in Europe*, pp. 27–40. Available at: http://www.boku.ac.at/coste51/

Paajanen, T., Patokoski, R. and Saarenmaa, L. (2007) Tekemiseen tietämistä – tietämiseen tekemistä – Puutuotealan osaamiskeskuksen toimintakertomus 1999–2006. Tampere, Finland.

Porter, M. (1999) Clusters and the new economics of competition. *Harvard Business Review* 76, 77–90.

Rametsteiner, E., Weiss, G. and Kubeczko, K. (2005) Innovation and entrepreneurship in forestry in central Europe. EFI Research Report Nr 1egic9. Boston, Massachusetts.

Robertson, P., Jacobson, D. and Langlois, R. (2008) Innovation Processes and Industrial Districts. Working Paper 2008-03, University of Connecticut, Storrs, Connecticut.

Rosenberg, N. and Nelson, R. (1994) American universities and technical advance in industry. *Research Policy* 23, 323–348.

Smith, K. (2002) What is the 'knowledge economy?' Knowledge-intensive industries and distributed knowledge bases. UNU INTEC Discussion 6/2002.

Stendahl, M. (2009) Product development in the wood industry – breaking Gresham's Law. *Acta Universitas Agriculturae Sueciae* 2009:3. Uppsala.

Teece, D., Pisano, G. and Shuen, A. (1997) Dynamic capabilities and strategic management. *Strategic Management* 18, 509–533.

Tödtling, F. and Trippl, M. (2005) One size fits all? Towards a differentiated policy approach with respect to regional innovation systems. *Research Policy* 34, 1203–1219.

Tödtling, F., Lehner, P. and Kaufmann, A. (2009) Do different types of innovation rely on specific kinds of knowledge interactions? *Technovation* 29, 59–71.

Winch, G. (1998) Zephyrs of creative destruction: understanding the management of innovation in construction. *Journal of Building Research and Information* 26, 268–279.

Zahra, S. and George, G. (2002) Absorptive capacity: a review, reconceptualization and extension. *Academy of Management Review* 27, 185–203.

13 Role of Policies and National Programmes on Innovations in Timber-frame Construction

Tomas Nord,* Saana Tykkä, Denise McCluskey, Fahrudin Bajric, Laura Bouriaud, Mårten Hugosson, Anders Q. Nyrud, Pekka Ollonqvist, Anders Roos, Kadri Ukrainski and Kristian Bysheim

Abstract

The market share of timber-frame construction has increased in many European countries after the revision of national building regulations and implementation of national development programmes as well as marketing efforts of the forest industry. Our research within COST E51 has looked at the timber construction industry and assessed the innovation process in eight European countries, the role and position of this industry in the construction process and the influence of policies as drivers of knowledge creation. Results indicate that innovative timber construction firms have acted on the changed regulations overcoming the lock-in effects of construction industry standards and culture, and have in some countries been at the forefront to adjust the traditional construction process. These firms have in most cases come from the forest industry sector, as they have had the most to gain. The innovation type and progress differ, indicating that construction is a local activity influenced by national traditions, in both industry behaviour and policy aim. It is recommended that policy makers support the establishment of local and regional timber-based competence centres. Also, public construction clients ought to have bigger leeway to support innovative construction products and processes, as a means to generating the new innovations and knowledge creation and transfer necessary to create a sustainable future society.

13.1 Introduction

Recent EU policy emphasis concerns the necessity of ensuring the sustained competitiveness of European economies (EU Lisbon Strategy, Competitiveness and Innovation Framework Programme, 2006). The construction industry is one particular area of concern, where policy and business analysts consider there to be a need to significantly enhance the industry's effectiveness.

The construction industry is often referred to as traditional, conservative and non-innovative (Pries and Janszen, 1995; Winch, 1998; Manseau and Shields, 2005; Sardén, 2005) with a short-term focus. These characteristics arise from the peculiarities of the production process of the building, which is perceived to comprise a number of subsystems often assembled on-site by temporary project groups hired on a cost-price basis; thus, overall there is little long-term

* Corresponding author.

learning (Koskela, 2000; Dubois and Gadde, 2002; Bergström, 2004; Björnfot, 2006).

In an industry culture where standards and norms have developed over time, and in a context where construction is seen as an economic vehicle for countries' economies as a whole, state and policy makers have seen the need to innovate policy. Such policies ought to drive innovation, as well as come to terms with the general condition of the industry. The EU has therefore moved to harmonize building regulations and codes throughout the member states, e.g. the Construction Product Directive (CPD) (EU Council Directives 89/106, 93/68 and Regulation No 1882/2003).

With the introduction of CPD in 1989, the agreement of the free and open market in Europe and the corresponding harmonization efforts, a number of countries changed from prescriptive to functional-based building regulations. The result has been an increase in innovations in general (Gann and Salter, 2000; Seaden and Manseau, 2001; Miozzo and Dewick, 2002) and among timber construction companies in particular (Goverse *et al.*, 2001; Nord, 2008). With the revised building regulations, new markets for timber structures have emerged, with application areas being extended from components for interior solutions, for infill walls and roof structures also to become a real alternative as structural framing in single- and multi-family housing as well as commercial buildings and bridges (Bregulla *et al.*, 2003).

At a European level, the market share of timber structures in the residential housing sector has grown, albeit from a low level. Strong markets are found in the Nordic countries as well as Scotland, with shares of more than 50% (TMF, 2009, personal communication; UKTFA, 2009). Single-family housing is the single dominant segment.

One distinctive element of the development of timber constructions is a new focus on preparing building elements in factory type environments. This has moved production from on-site to off-site and implies a more industrialized construction process, which is a noteworthy innovation (Lessing *et al.*, 2005; Björnfot, 2006; Apleberger *et al.*, 2007; Nord, 2008; Stehn *et al.*, 2008).

These findings reflect the image that there are no policy barriers to the enhanced use of timber in construction; however, many perception obstacles remain (Bregulla *et al.*, 2003). Construction industry conditions have differed from country to country and many of the conditions have proven difficult to change. The market share of timber-frame housing is much lower in central and southern Europe (CEI-Bois, 2004). Again it is the single-family market where timber is an alternative, and where demand for energy efficiency is spurring timber-framed housing.

Accordingly, when considering growth rates in timber-frame construction, a second image is that building regulations in some countries are still mismatched and overabundant, which leads to considerable administrative burdens and a fragmented, unsustainable construction market. Taken together, it augments the low willingness to change and a path dependency perspective (Bengtson, 2003; COM 860, 2007; Mahapatra and Gustavsson, 2008; Nord, 2008).

When assessing the sources of innovation (Drucker, 1998a,b) in timber frame firms in the construction industry, policy changes are a precursor of change, and the most important sources of innovation are incongruities in the present processes with regard to market and industry needs (Tykkä *et al.*, 2010).

In summary, timber construction is an emerging market resulting from firms acting upon business opportunities that have been enabled by policy changes and re-regulation. Change has been driven by efforts from society and entrepreneurial companies in parallel. Timber frame's share of total construction at European level is still low but there are different and good examples of successes at national level in many countries. It is proposed that examples of successful innovation should be highlighted and described in detail in order to foster a positive image of timber-frame construction development throughout the EU.

13.2 Scope and Limitations

The purpose of this chapter is to describe and assess variations in innovations and in the innovation processes in the timber construction industry, and the role of policies in the innovation development in eight European countries. To realize the scope, the following questions are addressed:

- Who are the characteristic actors in the timber construction industry, and what are their roles?
- What is the role of policies and industry promotion programmes in promoting innovation in the development of timber-frame construction actors?
- How have timber-frame construction actors innovated, and what is the degree of innovation?
- What roles do these innovative actors play in the construction industry today?

The sample countries represent different construction market sizes and developments, governmental structures and forest industry dependence, and thus give a good overview of differences in Europe. The choice of an innovative firm and its development as a case are all considered characteristic of each country but there are, of course, other types as well.

13.3 Policy and the National System of Innovation Framework

EU policy aims to promote sustainable development and manage climate change (EU Lisbon Strategy, Competitiveness and Innovation Framework Programme 2006). Accordingly, to enhance competition and promote more sustainable energy use, building codes have therefore been harmonized throughout EU member states (e.g. EU Council Directives 89/106, 93/68 and Regulation No 1882/2003). These codes place functional requirements, such as fire resistance and thermal insulation, on the building's performance. Moreover, some countries have introduced other policies, which aim to enhance considerably the construction industry's effectiveness by radically changing the industry's processes: instead of constructing on-site, it is advocated that building elements be pre-fabricated in factory-type environments and then assembled on-site (e.g. Postnote 209, 2003).

The policy field has recently changed its view of innovation from a product-based emphasis to the perspective that innovation is a process that is systemic (Lundvall, 1992; Galli and Teubal, 1997; Tödtling and Kaufmann, 2001; Edquist, 2004). In the systemic view, it is considered that the environment where the companies operate also affects innovation (Malerba and Brusoni, 2007). Accordingly, rather than only focusing on firms' internal research and development processes, it is further considered important to also include other actors and institutions, such as markets, political system and infrastructure, as well as the interaction of these actors within the whole system (e.g. Freeman, 1992; Nelson and Rosenberg, 1993; Lundvall and Edquist, 1997).

In this respect, the construction industry is viewed as particularly conservative, with low innovation rates (Winch, 1998; Bregulla *et al.*, 2003; Taylor and Levitt, 2004; Widén, 2006; Visscher and Meijer, 2007). Moreover, there is a general concern that the construction industry's structure may negatively influence innovation and its adoption (Pries and Janszen, 1995; Winch, 1998; Bergström, 2004; Blayse and Manley, 2004; Sardén, 2005). With a view to assessing the environmental factors that contribute to the relatively low rates of innovation in the construction sector, Winch (1998) defined the construction industry's particular structure and types of actors. Winch's model is a development of complex product system (CoPS) theory.

Winch considers that the actors that ought to be influential in promoting innovations are clients, regulators and professional institutions. These three groups of actors comprise the innovation superstructure level (Fig. 13.1) and demand innovation from other actors (trade contractors, specialist consultants and component suppliers) who form the innovation infrastructure level. The final actor in the Winch model is the system integrator. The system integrator

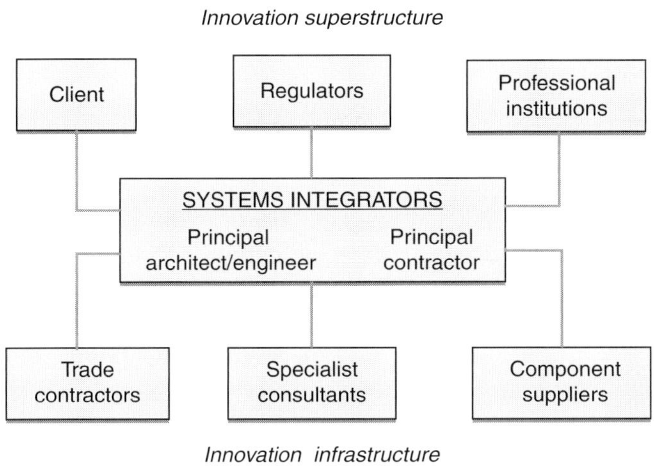

Fig. 13.1. The complex systems view of the construction industry (Winch, 1998).

is a key actor in the system, which ought to connect both subsystems through a role, and integrates a range of technologies as well as a mediating in the sense of matching the client needs, and regulatory requirements, with the infrastructure innovations.

In the construction industry, Winch (1998) considered that the systems integrator role was a particular obstacle to innovation: the defector way of working is that the system integrator is, in practice, two different actors: the principal architect/engineer, most influential at the design stage, and the principal contractor, more influential at the construction stage. Thus, regarding innovation, there is a practical challenge of these two different actors aligning their views and driving innovation amongst other actors (Winch, 1998). With regard to timber-frame construction, it has been found that a general lack of timber engineering skills is also a practical obstacle to innovation (Bregulla *et al.*, 2003; Taylor and Levitt, 2004; Goodier and Gibb, 2007; Roos *et al.*, 2008).

Winch also emphasized another significant structural obstacle to innovation, that of the fragmentation of construction's professional bodies, which weakens their ability to act as honest brokers of innovations, as any given innovation typically threatens the interests of one or another of the professional bodies. Noteworthy obstacles are also that specialist suppliers (trade contractors) are rarely given full technical authority by the principal contractor, and even governmental clients may not have the expertise to assess innovation risk and potential (Winch, 1998). In this respect, with regard to timber-frame construction, it has been found that perceived cost risks are practical obstacles to innovation (Bregulla *et al.*, 2003; Taylor and Levitt, 2004; Goodier and Gibb, 2007; Visscher and Meijer, 2007; Roos *et al.*, 2008).

13.3.1 Policy challenges in national innovation systems

The general policy literature is concerned that any uncertainties are exacerbated by the fact that knowledge is distributed amongst a number of actors and that its exchange depends on the connecting and aligning of multiple actors. These factors lead to uncertainties in the innovation process, which creates a series of problems from a political perspective (Edquist, 1999; Metcalfe, 2007). The central policy dilemma is to ensure that there is a rich knowledge environment from which innovation processes can draw, as well as a set of rules of the game that openly facilitates the formation of innovation systems to solve problems (Metcalfe, 2007).

In general, policies can influence innovation as *science policies*, *technology policies* or *innovation policies* (Lundvall and Borras, 2005). Science policy focuses on scientific knowledge and concerns the funding of public research institutions, the creation of tax incentives to firms and the definition of intellectual property rights. Technology policy aims at the advancement and commercialization of sectoral technical knowledge through public procurement, aid to specific sectors or technologies, and the creation of public–private partnerships or the support of standardization. These two policies are often also regarded as more traditional measures that governments and other authorities may employ in supporting innovation. Finally, innovation policy intervenes on the conditions that influence enterprise behaviour and focuses on the overall innovative performance of the economy through regulation, corporate law and the support of organizational innovation, competition policy or cluster policies at regional level (Metcalfe, 2007).

In the systemic approach, especially in national systems of innovation (NIS), a great deal of weight is placed on national policies and the regulatory environment to create possibilities for and sustain innovation. However, increasing globalization and the defector deregulation processes that follow, and information and communication technology (ITC) development contribute to the rapid change of national economies and create new challenges for policy makers (Lorenz and Lundvall, 2006). Another issue relevant from a NIS perspective is that the EU, which has placed more effort in the harmonizing of regulations, further challenges the concept of *national* systems of innovation. In a certain perspective, similar regulations offer business new markets and business opportunities, making it easier to move cross-nationally. However, as Lorenz and Lundvall (2006) note, benchmarking Europe as a single entity might not always benefit overall innovation generation and could easily limit the innovation policy perspective. Similar regulations are by no means a guarantee that the whole of Europe can and should be benchmarked as 'one'. Variety, as Levinthal (2007) notes, is one critical factor in innovation generation.

13.3.2 Adaptation of the construction national innovation system model

Such a globalizing world, particularly in an EU context, leads to a critique of the implicitly national perspective of the Winch model (Fig. 13.1). To provide a framework for assessing the effectiveness of EU policies in promoting timber-frame construction in the construction sector, the following modifications to the Winch model are proposed.

1. Formal separation of the regulatory actor into two actor categories. The first proposed category is policy-making bodies, which can be either national or transnational. Such policy-making bodies define policy, with national policy-making bodies defining such policies as laws. The second proposed actor category is regulatory bodies, who are actors that have the legal authorities to interpret such policies or laws into, e.g. building codes, and also to enforce compliance with such laws and codes.
2. Modification of the model into a more interactive structure. To denote interaction explicitly, such lines are replaced with interconnecting exchanges.
3. Explicit inclusion of interactions between actors at the innovation superstructure level. Since there are a number of policy fields, e.g. science policies and technology policies, it can also be assumed that the policy makers may interact with non-regulatory actors, such as universities and professional bodies. Although Winch stated that there are interactions, the lines in his model imply rigid lines of command to and from the system integrator. To denote interaction explicitly, such lines are replaced with interconnecting exchanges.

The explicit inclusion of interactions between actors is further motivated on the basis of developments in the business literature, which considers that phenomena such as globalization and changing customer requirements fundamentally require a change in business perspectives from firm-centric to network-centric (Drucker, 1998b; Hamel, 1998; Prahalad and Ramaswamy, 2000). Such a network perspective considers that value is delivered through interactions, and firms can develop their *business model* via innovative ways of combining their own and other actors' resources for service development, production and delivery (Hamel, 1998). This, as well as the fact that a given firm can in practice play several actor roles, implies the need for a fourth change to the Winch model, which is:

4. Inclusion of possible interactions between actors at the innovation infrastructure level, and even the possible interactions of actors at the innovation infrastructure and superstructure levels, with which the system integrator is not directly involved.

This modified construction industry system model is illustrated in Fig. 13.2. While the model implies homogeneous roles, firms as actors can be heterogeneous.

13.3.3 Learning and change

Learning processes are in essence feedback-driven (Levinthal, 2007; Metcalfe, 2007). As a result, the particular context in which a given actor operates influences the feedback received: firms both differ and act as main agents in variety generation through their selection process (Levinthal, 2007) and can also change over time. Forsman (2008), studying innovation in smaller enterprises, has concluded that development work is often a continuum of projects and innovation development is a part of their daily business. It is therefore difficult to differentiate normal business and innovation tasks, and success is thus difficult to assess or appears after long delays (Gopalakrishnan and Damanpour, 1997). Improving learning among smaller enterprises by setting up regional cluster strategies has aimed at improving entrepreneurial capabilities, networking capabilities, managerial capabilities and, finally, organizational capabilities. Assessing the success of such cluster strategies has shown increased front-end adoption of the innovation process and a more beneficial environment for larger, leading firms (Forsman, 2009).

In the construction sector, it is widely accepted that learning occurs through projects – which are often the construction projects themselves (Winch, 1998; Gann, 2003; Widén, 2006). Whereas construction projects follow some general stages, the type

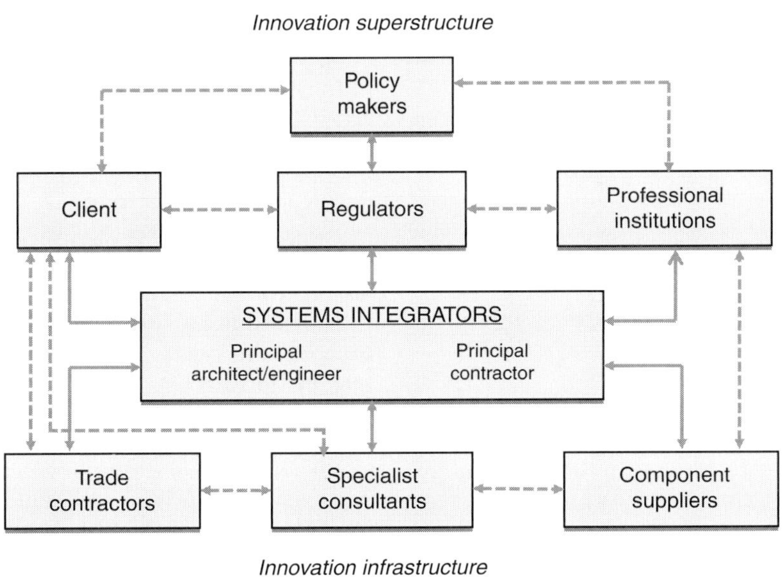

Fig. 13.2. A globalized interactive complex systems industry (modified from Winch, 1998).

of relationship and length of relationship differ. Traditionally, the building process is managed by splitting the project into individual smaller work packages, planned in advance and purchased sequentially. Each package is given to specialists, often selected on the basis of a low price tendering. Given the number of actors and their differing relationships, two different diffusion mechanisms of innovation can typically arise: problem solving in a project, or adaptation stemming from other industries or techniques. This led Winch (1998) to model firms learning in two ways (Fig. 13.3).

13.3.4 Degrees of innovation

Such project activities can be assessed in studies of new product development, where the level of analysis is a given firm's problem-solving and decision-making processes when generating new products for its customers (Gopalakrishnan and Damanpour, 1997). With such a product innovation perspective, it is envisaged that the sawmill sector may innovate its product offerings, from simple timber elements to prefabricated housing elements (Nord, 2005). Implicitly, the construction sector would then adopt such innovations through a diffusion process (Winch, 1998; Widén, 2006).

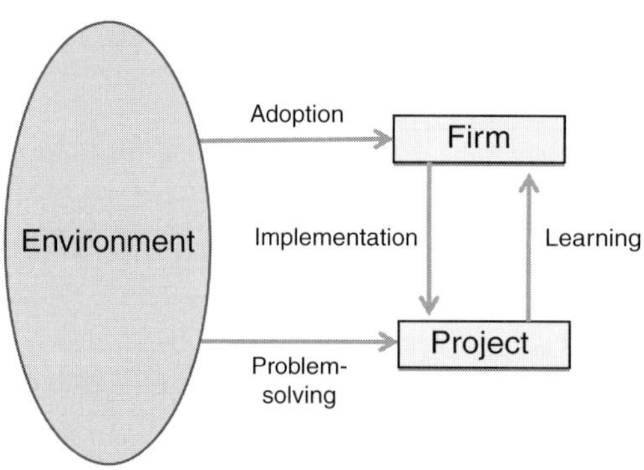

Fig. 13.3. Construction innovation process (adapted from Winch, 1998).

Often, the degree of change to the product would also be assessed and two distinctions noted: *incremental innovations* denote marginal departures from existing practices, where their initiation and implementation require little new knowledge because they are aligned with and re-enforce existing organizational skills and capabilities; and *radical innovations*, which produce fundamental changes in the activities of a firm or industry and represent clear departures from existing practices in all aspects (Gopalakrishnan and Damanpour, 1997).

An implicit aspect of this traditional perspective of innovation is that one single actor has sole control of the innovation. However, in the construction industry, it is implicit that multiple actors need to agree upon an innovation for it to be implemented. The innovation literature, in common with the systemic perspective, has taken consideration of such developments in defining a new dimension to the degree of change: the configuration of roles and tasks amongst firms (Gopalakrishnan and Damanpour, 1997; Bozdogan *et al.*, 1998; Henderson and Clark, 1990; Narayanan, 2001). Innovations in the way that actors allocate roles or tasks amongst themselves are defined as *architectural innovations*, which significantly change the existing configuration of actor roles and responsibilities, while generally tending to use known system practices and technologies.

There can also be innovations in products or services that do not change activities between firms. Defined as *modular innovations*, these are significant changes in products, services or a given actor's organizational practices that do NOT significantly change the existing configuration of the system's actors' roles and responsibilities.

These four types of innovations are illustrated in Fig. 13.4.

It is now generally emphasized that architectural innovations, with industry actor firms developing their offerings by working cooperatively with other firms, can be of significant business potential (Gopalakrishnan and Damanpour, 1997; Bozdogan *et al.*, 1998;

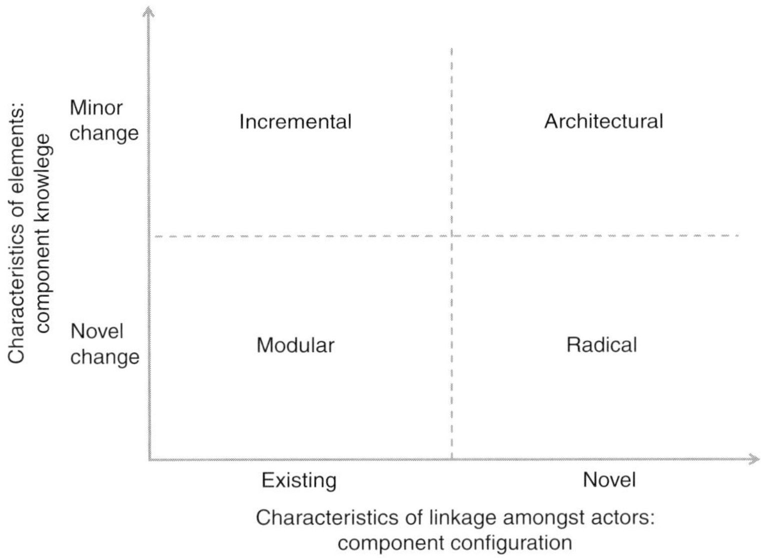

Fig. 13.4. Degrees of change and types of innovation, adapted from Narayanan (2001).

Narayanan, 2001). Such business potential may in fact be greater than that produced by a radical innovation.

13.4 Method

A qualitative case study approach was chosen, on the grounds that the innovation processes need to be studied in their real-life context and with respect to the complexity of such processes (Yin, 1994). The objective was to find patterns of links between the overall construction industry framework, as a CoPS, and innovation processes and internal and external conditions of the particular focal firm and country. The links between these variables were analysed for different representative cases from different geographical and policy contexts.

13.4.1 Criteria for selection of cases

To be able to assess the innovation process and the role of the innovation superstructure and infrastructure of timber-based firms in the construction industry, it was proposed to choose one case for each country for the study, i.e. strategically chosen (Patton, 1990; Silverman, 2000). To obtain appropriate cases, the following criteria were used:

1. Reflect a major business practice in the nation's timber construction industry.
2. Be clearly innovative from a business perspective, i.e. being implemented in practice.
3. Comprise focal actors that were established on a commercial basis.
4. Use timber as a load-bearing element.

13.4.2 Data collection

Using these criteria, case authors selected a case firm. In order to acquire knowledge about the focal firm's business concept and sources of innovation, authors then interviewed people within the firm. Those interviewed all had significant personal experience of the firm and its development, and often held senior positions within the firms, including many owner-managers. To place each case in context, the case authors were also asked to give a brief outline of the key policies in their country that potentially influence the general business situation of the sector in that country.

13.4.3 Data analysis

In the analysis procedure, the authors have looked for common and differing patterns in and between the cases, and explanations thereof. Through moving between theory and empirical results, the analysis procedure adopts an abductive approach (Kirkeby, 1994; Dubois and Gadde, 2002). Such an approach allows for the researcher to be open about ideas and results. This also means that new ideas may appear relevant, and that the course of the research project may change during the analysis procedure, adding findings of importance that were not detected at the start of the analysis (Eisenhardt, 1989).

In describing the cases, the aim was to focus on the policy structure and formulation, along with development of the innovation from an internal case perspective. The analysis procedure followed the case description format first focusing on an internal case analysis by referring to the innovation structure of construction as a CoPS. This was followed by a cross-case analysis for the identified dimensions. The roles of external actors and policies in the realization of the innovation and the learning and competence building of the firm were the result of the analysis.

13.5 Results

The diversity of the European construction industry requires individual descriptions of not only innovative cases but also some general information on the construction environment and the policy innovation superstructure. The following gives a rough picture of the differences and similarities of the eight country cases (Table 13.1). Each case description includes a general business presentation, the learning and change from the innovation and its role in the construction environment.

13.5.1 Austria

The focal Austrian firm

The case company is a micro-sized company with a focus on passive houses and organized as a Limited Company. In its present form, the company was founded in early 2009, but originated in 1994. The company produces flexible passive house buildings using timber and other ecologically friendly material, which differentiates it from other passive house firms in the country.

The company uses only clay plasters and timber, and offers other sustainable energy technology such as solar power panels. The walls and structural elements are produced off-site, and transported to and assembled on the selected locations. The basic concept of prefabrication of large-sized modular parts has appeared to be optimal for the loam/wood/straw passive house construction, such as the increased degree of prefabrication, and has led to reduced costs for further production. Transport activities are by road and by rail according to a predetermined schedule, and the swift assembly of the modular components makes it flexible and rather weather independent.

The company's competence is based on its long-time product development, the flexible and adaptable product, and knowledge of combination of sustainable construction principles with passive house construction standard.

Learning and change

The development of the innovation concept was a long-term product development process, consisting of finding suitable partners and creating a network of excellence, building the knowledge and creating the right concept for the product. In the process of developing the final product (passive house concept), the most important public

Table 13.1. Policy innovation infrastructure and construction environment of case countries.

Country	Construction environment	Policy innovation superstructure
Austria	High per capita construction rates Subsidies by object	Regional, prescriptive building codes Industry development
Estonia	Low construction rates with MFH dominating Industry standards and culture restrict timber framing in MFH despite tradition in SFH	Functional building codes, small changes in construction method Industry-led network creations of enhancing timber in construction
Finland	High per capita construction rate with SFH dominating Regional markets but in MFH a few large contractors dominate Timber frame has high share in single but small in larger building	Functional building codes with conditions Cluster initiatives from national programmes and industry Forest industry development prograGoodier and Gibb, mmes, 1995 and onwards
France	High construction rate evenly split between MFH and SFH Regional markets, but with national contractors Timber frame has low share	Functional building codes allowing timber Prescriptions on air quality, energy efficiency supports timber Regional development aiming at network creation
Norway	Growing construction market with MFH dominating Local markets managed by fragmented structure Timber has high share in SFH and growing in MFH	Functional building codes neutral to material Regional development programmes, some promoting timber Promotion to urban and environmentally sustainable living
Scotland	Housing shortage stimulates construction Building process governed by regulated industry structure Industry under change towards prefabrication and off-site Timber frame dominates	Functional building codes along with promoted construction method enhances timber Network and cluster initiatives support small and medium enterprises
Sweden	Low construction rates after economic and structural changes to lower governmental involvement A few large firms dominate residential construction Timber frame dominates SFH and grows in MFH	Functional codes allowing timber Governmental and industry-led timber promotion and R&D programmes Regional development programmes in some regions to timber construction
Romania	Large growth rates but still huge demand Demand of refurbishment Regional and local markets managed by fragmented structure Timber frame has low share in new build	Functional building codes but small changes in traditional industry Regional and rural programmes supporting networking but difficult for industries to participate

MFH, multi-family housing; SFH, single-family housing.
Sources: EUROCONSTRUCT (2009); INSEE (2009); Romania Statistics (2009); Statistics Finland (2009); Statistics Norway (2009); Statistics Sweden (2009); UKTFA (2009); Rametsteiner *et al.* (2010).

support came through a research and development (R&D) project funded by the Austrian Federal Ministry of Transport, Innovation and Technology. This project, which focused on the development of energy efficient and sustainable buildings, helped the founder to establish a network of professional partners and develop a suitable product concept. The development included architects, universities and different professionals (i.e. an Australian expert to develop termite-resistant buildings). The many years of experience of the present partners in producing timber constructions and building material from clay were used to develop the finalized product, and later the experiences were integrated in the product development before starting the industrial manufacturing of the product.

The innovation and its development were all internal, although the public side had a stronger influence on the innovation development. Policies aiming at construction are increasingly taking an environmental standpoint. One key point for further progress was the release of regulations that limited the use of timber in constructions. In addition, the use of timber for multi-storey housing was later liberated, which may increase the company's development possibilities. In money terms, some subsidies were also received, which were mostly used for additional planning costs for product development.

Presently, the case company's responsibilities are allocated between three different people, who provide the building components for the house production. These people all work in different firms, of which one is responsible for wood construction, one marketing and one clay plaster deliveries. The clay plaster firm's founder started R&D on the use of clay, hemp and timber in the construction industry. The development of the concept took him years, during which his focus moved into modifying and further innovating passive house building standards. He finalized this concept of sustainable passive house context within the Austrian sustainable innovation programme together with his present partners.

The focal firm's role in the construction innovation system

The firm does R&D in the field of ecological passive house construction and for sustainable building technology, and gives consultations to other firms and people with the use of its products and technologies and in different research projects. Because of the newness of the case firm, there are no production or financial data available yet. Officially, the company only employs three persons, but since the manufacturing has been outsourced, the number of employed people can increase up to 40, depending on the construction project. All of the construction projects are done nationwide, and the ready products include both single-family and office houses. The finalized products are designed in cooperation with the customer, and the finalization of the building design is in most of the cases done according to architectural plans.

The company founder has taken part in seminars, meetings and other scientific events organized by branch and research organizations, to gain and share knowledge about sustainable construction methods from the very beginning of work in that field.

13.5.2 Estonia

The focal Estonian firm

The case firm is a family-owned small or medium enterprise (SME) founded in 1988 as a producer of timber-frame wall panels, but since 2003 focused on structural products such as prefabricated timber-frame buildings, and construction elements such as roof trusses and wall elements. The sales turnover has remained slightly below €12 million for the past years with an export share of above 80% (mainly to Norway, Sweden, Germany and Iceland). The growth strategy of the firm has been to focus on larger projects involving row houses, apartment houses and larger house series for different purposes (e.g. kindergartens, office buildings, etc.).

Learning and change

The success on the export markets rendered the firm able to participate in a domestic project of several two- to five-storey residential buildings. The project was realized as cooperation between a main contractor and several subcontractors with process ideas from the architectural design into technical solutions. The project included a number of contracts where the case firm was engaged in the stage of technical design for supplying in-fill external walls in a design and build (D&B) subcontract. This project was the first for the case firm to satisfy the requirements for apartment houses of a supervisory municipal office. Project-developed competences and previous collaborations rendered the company able to undertake the project. In particular, the latter has been a driver for the firm where, for instance, close cooperation with suppliers of windows (as in the external walls, large glass windows were combined with wood surfaces) was beneficial (later on the two firms merged). Furthermore, the firm offered training for its subcontractors to ensure the quality and optimize the production and procurement processes.

The project has been recognized in several ways, from its award-winning architectural design (the ideas have emerged from similar buildings in France and Finland), to the overall performance, as one of the apartment houses was awarded the title of best building of the year. Apart from these recognitions, the project has proved successful also in terms of apartment sales. The following factors contributed to the success:

- alignment of design and production possibilities with the consumers' needs in real time (appropriate planning of the closed area between apartment houses with children's playgrounds, broad selection of apartment designs, etc.);
- appropriate selection of project partners ensuring the high quality of materials used and services performed; and
- supportive market dynamics (the disadvantage of initial high price of apartments stemming from the design and quality applied was diminished as the prices of new apartments on the market were catching up because of the credit-financed demand boost).

The focal firm's role in the construction innovation system

A typical residential project is initiated by a land owner in corporation with a larger real estate or construction firm and realized by a large number of subcontracts with different firms. The leading role in this innovation was carried out by the main construction firm, which was the developer and leading contractor of this project. The role of the focal firm was similar to other projects including the technical design and in-house production of elements. As in any such project, some additional resources were used for coordination with other parties, but relatively more resources were dedicated for complementary changes related to the safety. The larger challenge in this innovation was coping with the higher requirements of fire and noise standards. These are more complicated in the case of multi-storey houses, but additionally governmental authorities supervising this market generally view timber as unsuitable material for multi-storey houses and construction norms can be understood and enforced differently by the municipal officials responsible. This project was the first time for the case firm to satisfy the requirements for apartment houses of the respective municipal office and came from the cooperation within the project. There are still difficult issues in that respect that are opening up further cooperation with a local university helping with R&D in fire and noise conditions.

13.5.3 Finland

The focal Finnish firm

The Finnish case firm is a family owned enterprise Z (established 2001). The founder of Z has a personal background in industrial engineering, connected with construction engineering. The firm started as subcontractors in the single-family segment. An early innovation was to develop a highly standardized factory, which assembled and finished construction modules for use in their single house/detached house subcontracting business. The development included investments in quality management systems and processes. Factory production was based on strength-graded timber product structures and well formulated partner contracts where production logistics and wood component deliveries were concerned.

The annual turnover of Z has increased rapidly from €4.4 million to €30 million in 2004–2009 with employment growing from seven to 66. Z has also been a locomotive of local micro-enterprise networks delivering component solutions to Z. The profit margins in Z's business have remained modest because of the rapid growth and growth-related transaction costs. Low managerial hierarchies and tight connections between marketing, product design and planning staff as well as logistic partners can be identified among the frame conditions to improve competencies and capabilities in Z, in addition to the business innovation discussed.

Learning and change

Firm Z redefined its strategy by developing business in the commercial building segment. This segment comprises nurseries, local schools, homes and care centres for the elderly. The demand in this market segment comes typically from local governmental organizations with a commissioning and owning arrangement. The value creation typically included two different customer channels: customer–construction department and construction department–contractor, with type of contract and the interaction depending on the contract applied.

The firm Z participated in a consortium extending the results of the technology programme by TEKES, iWell – Turning well-being during 2000–2003, concerning the future needs of elderly people in their living areas, into business solutions.

In this project, firm Z learned about the housing needs of the elderly people but also the needs of other actors in producing and maintaining housing services for the elderly. The business innovation of Z was to join the final customers representing the elderly in local governmental organizations to the planning from the very beginning prior to the contacts with the construction department. Based on this learning, and on their own independent assessment of the various actors involving in producing housing and care for the elderly, firm Z innovated in the following ways:

- developed business relationships with actors that were new to firm Z in order to market, plan and produce housing for the elderly;
- developed standardized prefabricated modules, and routines for producing housing modules that were appropriate for constructing housing that was suitable for the needs of the elderly;
- developed routines for transporting modules to the building site, as well as competencies to communicate with other actors to tailor deliveries;
- provided a partner enterprise with specialized architectural services for the construction projects of Z; and
- provided a partner enterprise that has specialized assembling and on-site finishing on the projects of Z.

These housing module solutions used timber as a load-bearing element in construction, with the focal firm considering timber as competitive because of its flexibility when producing in factory-type environments, as well as being lightweight and relative robust when transporting.

Such learning was realized in the Wood Park housing policy project, which was located in the home town of firm Z. This particular project provided firm Z with the financial resources to invest in a new production line to support their business activities in supplying housing for the elderly. The tacit knowledge from the prior construction module industries in the home town of Z could be utilized when the new industry hall was planned, constructed and furnished.

The focal firm's role in the construction innovation system

The firm has moved from being a supplier and subcontractor of modules to an industrial producer and assembling service provider for construction solutions based on tailor-made wood frame construction modules for the special uses of elderly, nurseries and elementary schools. Z, with its partner enterprises, can provide architect and construction engineer design services but also assemble modules as a principal contractor. The relationships with other actors have thus developed from contractors to that of clients and technical consultants.

13.5.4 France

The focal French firm

In service since 1936, the French case is a family-run sawmill, which presently employs ten persons. The case firm is located in the Alsace region near the German border, characterized by strong forest–wood chains supplying raw material for market procurements and German wood industries.

The sawmill produces sawn softwood timber. In 2007, it transformed 12,000 m^3 of logs, including long-length logs up to 15 m. Production is predominantly for timber-frame construction, e.g. battens, common rafters and beams. With a turnover of €1.5 million (2007), the main customers are local carpenters for new construction (two-thirds of turnover) and for renovation of housing (one-third of turnover). In collaboration with an external structural component manufacturer, the sawmill provides additional services to customers, such as cut-to-length structural products and preservation treatment.

Recently, on top of normal product development, the company has been engaged in developing a green-headed solid wood product. It is made of strands of sawn timber glued together when the wood is still green (30% moisture content). This new 'green' invention offers many improvements such as energy savings, better quality and better adaptation abilities. Before entering the market, the product still needs validation of technical requirements. In addition, customer feedback has highlighted some reluctance from carpenters, whose working habits are linked to product specifications such as moisture rate.

Learning and change

The increasing regional competition on both the supply and the demand side has been the driver of the innovation as the company invested to maintain the business and speciality in carpentry.

The driving force on the supply side has been the continuous difficulty of obtaining appropriate raw material. Although Alsace has high annual harvests (1.6 million m^3 in 2007), its wood resources are heterogeneous (42 species). The forests are dominated by large-diameter trees, too large for the sawmill equipment. The dense network of wood SMEs (215 sawmills and wood-working companies), two major French softwood sawmills and installation of new high-capacity sawmills have strengthened the regional competition. A renewal of timber-selling methods of the French National Office of Forests (administrating 80% of regional forests) has additional impact. Former auctions practice allowed sawmills to focus on and target lots of timber in terms of volume and quality, while the new 'supply contracts method' has weakened its position (unlike those of large companies). The company also faces competition on the demand side. New facilities and imported saw timber put pressure on prices and volumes of timber-frame construction. The residential construction demand for product specifications (i.e. physical, mechanical and aesthetic properties) has also increased.

The idea for green-headed wood product came to the manager from a specialized wood journal. The product is presently under technical review, and, once completed, an extension of innovation is possible by replacing existing glues with new glues made from natural materials. The green-headed glued timber offers the possibility of adopting new organizational procedures (i.e. changing the conducts of business), which is still to take place.

The overall development started with laboratory tests of the product, continuing with tests of the product from production line, which will end with full production. Laboratory tests validated the characteristics of the product, providing guarantees on stability and mechanical properties (mechanical resistance with different cuttings, moisture rates, etc.). The next, ongoing, step aims to validate that the industrial production line has no negative impact on mechanical performance of the product. In parallel, a market assessment, analysing demands of external support and the market opportunities, was conducted. Presently, the company is deliberating on whether to integrate the new equipment into the existing production line or to create an entirely new production line. Finally, the sawmill has recruited an engineer.

The focal firm's role in the construction innovation system

The actors internal to the forest–wood chain have various incidences on the product development and on developments in the construction innovation system. The firm acts as a material supplier and relations with the construction industry are purely transactional, with product development being internal or within the forest–wood chain.

Within the forest–wood value chain, the innovation is strongly supported by professional organizations at regional and national levels, as a possibility to overcome regional constraints. Similarly, suppliers see the innovation facilitating transport (e.g. shorter logs) and enlarging the acceptance of different tree species by the company. Support comes in the form of technical validation from the research community and as potential financial support for investments in buildings and equipment from official organizations. The latter is a long process though.

13.5.5 Norway

The focal Norwegian firm

The focal firm is part of one of Norway's larger sawmill and wood manufacturing groups. In 1950, the group started producing module-based housing and shelters. The company has 186 employees and had an annual turnover of NOK 441 million in 2008 (for the period 2005–2008, annual turnover was on average NOK 370 million). The Building Systems division is presently one of three divisions within the company supplying flexible system solutions: interior walls, modular buildings, electrical installations and load-bearing gluelam structures for projects and contract customers primarily in Norway and Sweden. The other divisions have sales to industrial and retail customers, as well as to builders and contractors in Scandinavia and Europe. The firm is increasingly focusing on the market for multi-storey construction. To do this, the firm developed new timber-based construction technologies and has been involved in several governmental initiatives to develop timber-based multi-storey building technologies and urban construction systems.

The concept – modular building for industrial construction – is based on box units with a timber structure manufactured in the firm's own factories. The modules are transported to the building site, and assembled and completed on-site. The concept aims at projects like residential housing, industrial purposes (temporary barracks), townhouses, multi-storey buildings, student accommodation and public buildings, e.g. kindergartens and schools.

Learning and change

The development of the innovation followed a product development process of learning by doing and using integrating company knowledge with market demands. The firm is continuously developing concepts and systems in close collaboration with customers and experts in the fields of architecture, design and construction. Based on their experiences of what 'works', the firm has incrementally developed a lean production process. The focus on lean building design arose from the need for improved efficiency and cost-minimizing construction practices: construction costs (not life cycle cost) tend to be a key determinant for choice of building material and construction technology. Minimizing construction costs is therefore a competitive necessity. The company also benefited from the competencies within the group that had been producing industrial shelters and prefabricated housing, as well as the general knowledge of wood products from the sawmill division.

At present, the most important driver of innovation is related to maintaining cost efficiency in the construction process, but also the ability to meet changes in building codes and demands from customers.

The focal firm's role in the construction innovation system

Formerly a wood products supplier, the firm has developed into a system integrator with direct contracts with clients or with architects. The role and position come from the capability in designing, producing, assembling and coordinating projects. The capability was created from forward integration strategy and based on the resources in wood manufacture, production processing, engineering, technical knowledge of timber components and its properties, and coordination with external consultants and subcontractors. Creating the capability also required taking a larger responsibility when it comes to providing cost-efficient production of industrialized wood construction. The core competence of the building division is industrial construction, focusing on quality, flexibility and short delivery times. The company handles the building process from beginning to end – designing, manufacturing, assembling and complementary works to complete building structures – but the firm does not provide services for interior finishing.

The engineering competence and ability to construct buildings and to develop projects in collaboration with builders also makes it possible to address customer needs and specifications. The company uses its own competence in building in order to overcome the barrier among clients not used to using timber thus reluctant to D&B with timber. The company also uses its competence in building to mitigate the responsibilities of builders and the main contractor in order to reduce the perceived risk of using timber as a construction material.

13.5.6 Scotland

The focal Scottish firm

The Scottish case firm was established as an engineering spin-off from a larger firm that went bankrupt in the early 1990s. At its inception, it was a micro-sized firm with timber engineering design competencies for construction applications: while Scotland lacks a timber-frame housing culture, its stone and concrete buildings have wood-framed roofing structures.

The regulatory environment increasingly placed requirements on traditional house builders to produce at least 25% of their houses using modern methods of construction (MMC) approaches (Postnote 209, 2003). As this regulatory framework came into place, the case firm's owner-engineers recognized that established house builders lacked timber engineering design, construction and assembly competencies. The firm therefore developed a business concept to become a timber-frame supplier to the major house builders in the UK, who organize a network of regional offices that plan and develop the houses.

Learning and change

A key aspect of the case firm's offering to house builders' regional offices is cooperating with architects: the firm transforms the architects' house designs into timber engineering design plans – advising where designs could be adjusted to be more viable. Working at an early stage with architects also means that the firm has house construction plans 'on the shelf' as the construction site managers initiate projects. The firm then: liaises with construction site managers to plan and manage that particular site's production and assembly needs; produces the structural building elements (outer and load-bearing walls, floors, roof frames) in a factory-type environment; ensures just-in-time (JIT) delivery of those elements to the construction site; and, via partners, assembles the prefabricated elements on-site.

The Scottish case firm recognized that their business success depends upon a broad service concept, with stringent approaches to quality and delivery reliability.

The case firm therefore also developed its own production processes to secure both quality and delivery reliability. This work included cooperating with its suppliers about long-term business prognoses and detailed short-term delivery needs, as well as delivery quality and reliability performance. Lean production and JIT approaches are a well-established business practice.

The focal firm's role in the construction innovation system

The case firm also worked extensively with its partner firms that assemble the prefabricated elements at the customers' building sites – providing specialist training on timber-frame specific techniques and ensuring partners approach the work with due quality and reliability diligence.

The case firm initially perceived scepticism to the entire MMC approach amongst their customers' construction site managers and workers. However, as they worked with site managers to plan the work, and consistently delivered in time, the site managers found that build time for multi-family houses was almost halved from 44 weeks to 24 weeks. This overall improvement in construction completion rates was valued, and the entire service was appreciated, particularly during a rapidly growing market that was experiencing labour shortages.

Site managers' confidence in the case firm's track record resulted in business growth: the firm found that they systematically increased their share of business at the regional operations centres they had entered. The case firm now has an expansion strategy of entering regional offices and growing its share of the customers' supply business.

The case firm has self-financed its own growth by 33% per annum for the past decade – including expanding production capacity and expansion into new regions of operation.

13.5.7 Sweden

The focal Swedish firm

The company is a medium-sized privately owned sawmilling company integrating production of sawn timber to value-added production of structural products like gluelam beams and cross-laminated timber (CLT). The development has been from a technical and raw material perspective seeking the optimal yield of raw material, which gradually took them into the timber construction industry. Besides traditional sawmilling activities and markets, the company acts as a subcontractor of a structural framing system mainly in D&B contracts. The turnover has increased from approximately SEK 0.5 billion in the late 1990s to presently more than SEK 1 billion, and maintained a 50/50 mix of sawn timber and value-added products. The growth has come from mergers and acquisitions of local sawmill companies, and organic growth and efficiency measures into new areas.

The company was early to enter the structural products markets, and has been active in branch-induced technical and marketing projects and programmes. The result has led to first mover advantage and close relationships with regional research organizations as well as international presence in marketing campaigns for European organizations.

The innovation is the prefabrication and off-site production of timber-based building systems based on standard and developed timber products like sawn timber, gluelam beams and CLT, resulting in a move from a project to a product and process focus in construction.

Learning and change

Changes in building regulations in Sweden allowing timber structures in larger buildings initiated a chain of activities in which the company was a main participator. The company was involved in a research project with other wood products' companies and a construction company, defining and standardizing CLT boards for usage in construction. The project was financed by regional funds and from involved companies. Direct work was performed by regional research organizations, which previously had worked together with the company establishing a mutual understanding of what was required and possible ways to reach the objective.

Besides engagement in the project, much development was internal with experiences and competencies from sawmilling, gluelam and management. A long tradition in post and beam structures using gluelam had created competence in design, and with a prototype system the company engaged in a project as a frame contractor. Experiences from the initial project were integrated in internal product development and management decisions to position for a frame contractor and preferably D&B contracts. Its knowledge in product development was mixed with the peculiarities of the construction project set-up leading the company to develop direct contacts with a few technical consultants of resources and capabilities not owned by the company itself, e.g. design of water, electricity and ventilation systems. Also, over the years, the company has established relationships with production and assembly companies/units to create a larger production volume as well as regional presence close to projects. The aim has been to maintain core competence in wood product development and still operate in temporary construction projects. From a capability in product development for mainly sawn timber, the company has developed capability in design and production of timber-frame building systems, which also includes the coordination of external consultants and subcontractors. The capability was established as a result of the forward integration strategy, requiring the company to take a larger responsibility and to learn the rules of the game.

The focal firm's role in the construction innovation system

The company has moved from material supplier to frame contractor with its building system. It operates in business to business (BtoB) relations and preferably in D&B with a possibility of coming in early in the decision process affecting the design of the timber frame. Coordination is early mutual agreement with the main contractor and client, and the establishment of degree and art of responsibility in various project issues.

The company was helped by industry-led promotion programmes where the company's system was shown to potential clients. Also, the company participated in seminars, meetings and other events to show the concept. The latter has resulted in direct contacts with main contractors interested in realizing projects with a timber frame. Timber frames being a new solution in larger buildings, clients are more reluctant to choose them and they require a larger responsibility of the frame contractor, i.e. often design, production and assembly. Overcoming suboptimization in design and production, the company looks for early contacts with clients and architects. The building system of the company creates the possibility of reorganizing on-site production and ultimately the entire building process becoming more industrialized. The result is a possibility for new relationships and cooperation.

13.5.8 Romania

The focal Romanian firm

The firm is a medium-size company from the northern part of Romania founded in 1993 and specializing in the production of timber-frame houses. The product portfolio contains prefabricated living and vacation houses, windows and doors, europallet elements, and simple shuttering and shuttering systems. The building system and windows and doors are made from gluelam products.

The main markets are Germany, France and Spain, and only a small share of turnover is from the domestic market. Contracts are found and negotiated by the external partner of the firm and direct with developers. The general drop in the real estate and building sector in Europe may lead to an increased focus on the domestic high-end market. Presently, the firm employs 87 workers compared with 40 employees in 2000. The turnover of the company was €1.9 million in 2008.

The initial owner had good contacts with customers in the German wood products market and, in 1999, the company began the production and export of garden houses to Germany. After 2001, the firm implemented new technologies resulting in increased production capacity and a product diversification into timber-frame living and leisure houses. In 2002, a special production line was purchased from Germany for the production of windows and doors using gluelam timber. The technology has been used to produce panelled houses based on timber frames. The company has also invested in quality management systems.

Learning and change

The innovations come from the distinctive competencies of the firm incorporating the special ability of the owner to tie commercial relationships with partners abroad, to participate in international exhibitions for the timber industry and to look for information on market trends and the latest technologies.

The design and production of timber-based houses came from contacts with customers and, later, partners in the German market. The origin of business partnership is from the early 1990s and included the idea to produce furniture based on the comparative advantage of the Romanian furniture industry. However, the decision of business cooperation turned shortly to the production of small houses for gardens, which was a novelty for the Romanian market. The diversification of the production in the direction of panellized wooden houses has been also driven by demand in the European market, and based on the production and cost efficiency of the prefabricated building system. Since the external partners have opted for diversification and indicated opportunities, the company has followed the trend.

Besides the effect from customers, the acquisition of new technologies was also inspired by information acquired from specialized international trade fares (annual exhibitions for timber industry equipment). This resulted in changing the production layout and the intra-firm value chain to reach an in-house production ratio of 80–90% of regionally produced prefabricated wooden houses.

All along the innovation development, the company did not have any specific public support. The firm benefited from two direct funding schemes, one in the frame of public programme strengthening entrepreneurship and investment in low-income regions, and the second aiming to implement environmentally friendly technologies. The owner believes that they were very important to launch the business, but not to support the innovation.

The focal firm's role in the construction innovation system

Relations with the external market have been via agents (partners) and, at the risk of losing the cooperation, the firm has built on the capabilities to innovate (new products, new technologies and diversification of production). With lack of domestic contact and lack of business capabilities (how to find markets, how to promote, how to sell, etc.), developed up to now via the external partner, the company is at risk. To overcome the absent public support for innovation and the investment in high-value products, the firm became involved in lobbying to change the lack of policy interest for the timber-based building sector and high-value timber-based products. The company founded with 17 other partners the APROCOR (Romanian Association of dwellings buildings with high-speed technologies performing), with the initial aim to promote these types of buildings on the domestic market and to lobby at the political level for the promotion of wooden houses as high-value, ecologically sound products.

The lack of supporting policies for the timber-based industry in Romania hampers the implementation of any innovation in the construction sector, but will have a greater negative impact on the diffusion of the innovation.

13.6 Analysis

13.6.1 Timber construction innovations in different European countries

The analysis includes assessment of type of innovation, role and position of case firms in building process, and policy formation and its impact on promoting innovation in the timber construction industry (Table 13.2). The assessments are based on viewing construction as a complex product industry system, with projects as the dominant source of innovation and learning (Hobday, 1998; Winch, 1998; Gann and Salter, 2000).

Type of innovation

The cases indicate a variety of innovations in the timber construction industry where the maturity of the timber construction industry, policy and regulation influence the position of the industry in the construction value chain. The innovations are described as **incremental**, **modular**, **architectural** and **radical** (Henderson and Clark, 1990; Slaughter, 1998; Narayanan, 2001).

INCREMENTAL. The innovation type exemplified from France is of incremental character as it has little effect on the design of the structural system. The innovation changes the structure and shortens the production process at the firm level. From a relational perspective, the innovation affects the client–supplier relationship of the firm and the structural frame contractor becomes more integrated as the frame contractor may change its internal processes. It can be seen as an initial step towards industrialized production.

MODULAR. The Austrian case switched from the standard use of concrete to timber/clay for structural framing, without changing the overall architecture of the building, which implies a modular change. It is assessed that change is constrained to a modular type because the building of single-family houses in Austria includes a BtoC relationship with an architect as system integrator. This relationship means that the building is defined at the design stage, which requires technical competency but little change to component/subsystem architecture and technology. Also, the innovation as such has little effect on the coordination and relation between other project team members (Bozdogan *et al.*, 1998).

ARCHITECTURAL. The assessment of the Estonian, Finnish, Norwegian and Swedish cases is that these innovations are architectural: the use of a timber-frame structure provided the opportunity for reconfiguring the linkages between actors with the case firms taking production responsibilities from the system integrator in order to optimize production and quality (Gibb, 2001; Höök, 2008). The extent of this change varies between the cases: from the Estonian's modest architectural change of the building process via the Finnish with a new processual way of performing and producing the product for the client; to the Norwegian and Swedish cases where there are noteworthy changes in configuration of subsystems and actor relations. All these innovations created the possibility for off-site production thus also affecting the processes and actor relations at the building site.

RADICAL. The Scottish and Romanian cases indicate radical product innovations as they have a profound effect on component structure and architecture of the end-product as well as relations between actors. The Romanian case indicates such a high radicality from the dominant design that implementation has been difficult and, instead, similarly to the Estonian case, has been more active on the export market. The Scottish case, on the other hand, has had a big impact on the industry development, as it has overturned the former dominant design, although in the UK as a

Table 13.2. Analysis of cases to research questions.

Country	Type of innovation from the construction industry perspective	Type of actor and role in construction process	Role of policies and regulations on innovation	Effect on building process of innovation
Austria	Modular, new products changed assembly procedure but not overall architecture	Specialist supplier acting as system supplier in the low-rise housing market	+Prototype enhancing; +Financial; –Building regulation is a hindrance	Fewer suppliers; BtoC concept; Value-added component production
Estonia	Modular for firm and architectural for construction process, established new relationships and required new technologies	Subcontractor of frame structure	+/–Little effect from policies and no promotion plan from timber industry	Integrated solutions; Network creation and fewer suppliers; Alternative product
Finland	Architectural as timber-based modules changed configuration of subsystem and actor relations	From subcontractor to system integrator in office and other building market	+Establishment of regional competence clusters; –Industry culture affects full implementation	Integrated solution and cooperative relations; New contract arrangement; Fewer suppliers; Developed competences
France	Incremental, concerned only component design	Supplier to frame contractor	+/–In general a support for timber construction but little effect on this particular innovation	Little effect on overall process; Closer ties with frame contractor; Knowledge exchange
Norway	Modular for firm; architectural for construction process as it redesigned product and actor relations	From supplier to system integrator through industrialized production competence	+No prohibiting regulations; +Support programmes of enhancing timber in urban areas	Integrated solution and cooperative relations; New contract arrangement; Alternative product and new competencies
Scotland	Integrated solution of system and practices; radical as innovation overturned former system as dominant design	From specialist consultant to system integrator	+Support of reconfiguring construction enhanced industrialized timber construction	Integrated solution and cooperative relations; Product and process become dominant design
Sweden	Integrated solution; product and process, modular to firm and architectural to construction process; becoming dominant design	Specialist supplier acting as system integrator	+Revised regulations opened market for timber structures; +Regional competence enhancing; +Prototype support	Fewer suppliers and contracts; Additional alternatives for clients
Romania	Radical to firm and construction industry but difficult to obtain acceptance because of newness	Structural frame to contractor of single-family housing	+Financial support; +Competence networking; –Industry standards hinder development, regulation too	New product and process in clash with present process and culture thus difficult to implement

BtoC, business to consumer.

whole the new approach has not dominated because of strong industry culture. The innovation was initially a process innovation but gradually grew into a product affecting the configuration of the end-product as well as individual components. Both these cases require early commitment from the client and in the conceptual design phase, and are based on an active sales effort of the firm and require specific resources and technical knowledge for implementation. The innovations are seen as radical since clients have to step outside the normal architecture of both the final product, the building, and the building process.

Type of actors, and role and position in building process

The timber construction industry is a fairly young industry, as, prior to revision of building regulations in many countries, it was excluded from a number of market segments. It is of interest to assess whether firms in the industry come from the wood products industry or from the construction industry.

TRADE SPECIALIST TO SYSTEM INTEGRATOR. The cases from Austria and Scotland have a similar background to engineering companies moving into the timber-based industry. Their role and position was as specialist consultants and this has grown to become system integrators. They differ in terms of market segments, as the Austrian case is focusing on the single-family housing with a business to consumer (BtoC) perspective and operating together with architects, and the Scottish is focusing more on larger structures and BtoB, with a large degree of cooperation.

MATERIAL SUPPLIER TO SYSTEM SUPPLIER. The Estonian, Finnish and Romanian cases come from the timber-based industry formerly supplying structural products to contractors or more simple products for the end-consumer, but have become active in the residential housing industry from their architectural innovations. The products have filled a demand for quality and accuracy at the site of production and consequently the firms have gradually changed their roles and positions in the building process towards subcontractors or system integrators.

MATERIAL SUPPLIER (WOOD PRODUCTS) TO COMPONENT/SYSTEM SUPPLIER. Finally, the French, Norwegian and Swedish cases come from the wood products industry and started as structural product suppliers to the construction industry. The French case maintains this role and position but, thanks to innovation, is increasing cooperation and product development with frame contractors and smaller contractors. The Norwegian and Swedish cases have taken steps to move towards the clients by offering integrated solutions. They are gradually becoming system integrators operating in a network structure with specialist consultants and trade specialists. All these are operating in a BtoB environment similarly to the Scottish and Estonian cases.

Most of the innovation examples in this study build on the possibilities of prefabrication and modularization that the properties of wood products give. In cases with low knowledge of timber structures, case firms have had to take the leading role in the coordination and implementation process to obtain a position in the market, thus being active early in the design phase. The cooperation sought has come not only from industrial actors but also from public organizations including financing, regulatory and research organizations.

Differences in policies and regulations supporting learning and innovations in timber construction

Policies and regulations can be divided into those that have a direct effect on products and processes used in the actual physical work of producing a building, and those that have an indirect effect on product and process development. Building regulations, industry standards and codes are examples of direct policies and regulations having an effect on product development and building process and its relations. Sweden, Norway, Scotland and Finland have for the past

10–15 years introduced function-based regulations allowing alternative products to enter the market as long as they meet the functions of the product specifications. Specifically for the timber construction industry, the result has been a further engagement of both public and, more so, forest-based organizations to promote the usage of timber. Except for Scotland, the countries are forest rich and have strong forest products industries, and have increasingly moved towards value-added production (Nord, 2005; Hugosson and McCluskey, 2008). In Scotland, the drive has come from rethinking the construction industry and where timber properties of prefabrication have supported the positive development.

In Austria, Romania, Estonia and France, there is a mixture of prescriptive and functional regulations meaning that under some circumstances timber is allowed as a structural frame and not in others. This has prohibited entry in some market segments because of risk and uncertainty of the new products and systems, the lack of knowledge among consultants and clients' subsidiary criteria from authorities.

Policies and regulations with an indirect effect on innovation are related to public and private science and technology and development programmes launched. These are, for example, innovation programmes and rural and regional development programmes. The results of these are the possibility of increasing the knowledge and competence of timber and timber construction, the establishment of networks and alliances of firms to increase competitive advantage (Porter, 1998) and from innovation brokers like universities or research organizations (Winch, 1998).

All European countries have established innovation and development programmes according to EU-wide criteria to improve competitiveness, but the specific implementation process differs between countries because of tradition and legal and industry infrastructure. This is indicated in that countries have established regional and rural development programmes differently using different criteria (see analysis in phase 1 of COST Action E51; Rametsteiner *et al.*, 2010). For the construction industry in general and the timber construction industry in particular, this development is important.

The cases indicate to different degrees the effect of these policies. In Sweden, Finland, Norway and Estonia, the policies have resulted in the creation of regional competence centres where the case companies have been given indirect support via external experts for developing the innovation. The clearest example is perhaps the Finnish, where a regional cluster managing company was the result. This is in line with propositions of Forsman (2009) that, in promoting innovations of SMEs, regional cluster strategies should focus on innovation capability creation by, e.g. providing support for project management and implementation.

The result of policy support indicates differences in learning among innovations firms. Policies aimed at the front end of the innovation process, such as financial support or changes in regulations, require that firms have internal innovations capabilities and are able to include innovations in day-to-day business. For example, the Romanian case obtained financial support in order to realize the technological change and thus initiated the learning process of prefabrication. Policies aimed at networking and knowledge transfer from research organization to firms result in long-term competence building and learning, but require a higher involvement from firms. Examples of these are the innovations in Sweden and Finland with involvement from regional authorities, research organizations and industry. The issue is to convince firms of the merits of participation, which has to be seen in larger market shares and improved profitability.

Based on the results of this study, Fig. 13.5 illustrates the role of policies in establishing a climate for innovative behaviour among the case of the timber construction industry and affecting the traditional building process in each country. The policy support amounts to building regulation changes, R&D programmes and promotion programmes aimed at the forest-based industry (thus also the timber construction industry). These effects are most prominent in

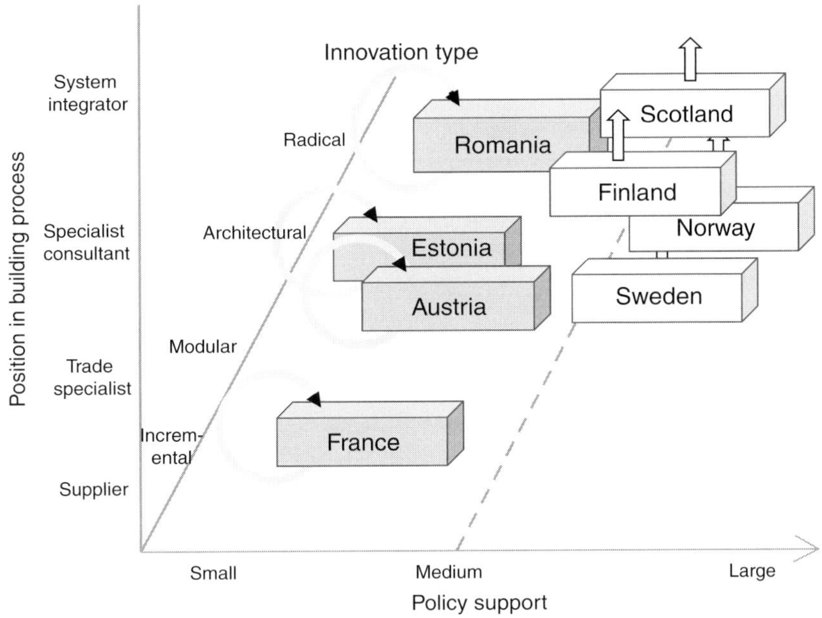

Fig. 13.5. Effect of policies on innovation and position in building process of studied cases (arrows indicate change in position in building process, *y*-axis).

Finland, Norway, Scotland and Sweden, where the cases have moved from different initial positions (supplier, trade specialist) towards a position offering a higher-valued product/service close to being a system integrator (the Swedish case is furthest from that position).

The support for timber construction has been smaller in Austria, Estonia, France and Romania partly because prescriptive building regulations still persist. This has a negative effect on trialability of clients. Furthermore, there are no direct promotion campaigns for timber construction but instead activities related to management and new uses of the forest resource. The Austrian case is interesting since the innovation aims at taking a larger responsibility for the building process even if the industry culture works against it. The R&D programme (Haus der Zukunft) has been positive in this respect. The Romanian case, although radical and moving from one position to another, has little support from policies. Based on this, the cases have more or less maintained their position in the building process.

13.7 Conclusions and Recommendations

The study of CoPS industries is a difficult task, especially when it comes to innovativeness, competence and learning (Hobday, 1998; Winch, 1998). In our study of the timber construction industry and building as a CoPS, we have proposed a development of the model presented by Winch (1998) by indicating the interactions taking place between actors as well as the inclusion of an extra level indicating the specific role of policy makers in the construction industry. Larger markets have appeared for the timber construction industry with revised building regulations, and additional support for growth and competitiveness has come from governmental and industry research and promotion programmes over the years. This has resulted in innovations and changes in the structure and the traditional building process of the construction industry.

The research has shown that changes have taken place on an overarching policy level and on an industry level affecting the innovation process of companies acting in the timber construction market.

Despite the fact that CPD was introduced and amended some 20 years ago, construction is still much of a national activity (Manseau and Shields, 2005). The result of the study supports that view from the large number of different innovations and various roles and positions the cases are taking. Our cases have developed towards a specific market condition and often let go of others (e.g. the Romanian and Estonian cases). This supports the view that each country develops its own industry culture with a high relation to the legal conditions rather than to industry recipes, as put forward by Spender (1989), which is highlighted in studies by, e.g. Winch (2000) and Nord (2008).

Also, the more advanced the innovation is, i.e. architectural or system to the construction industry, the more focused on a specific country it is. There was a difference between countries with revised performance-based regulations compared with those of prescriptive regulations in that the former showed a larger degree of architectural or even radical innovation. In a regulated environment and especially in a CoPS industry, clients are more likely to follow experience and former behaviour, and commission projects based on price (Manseau and Shields, 2005). A regulated environment raises barriers for innovation. In countries with policies and programmes supporting competence growth and knowledge transfer and with promotion, programmes aiming at introducing timber-frame structures have experienced innovations aiming at changing the building process and industry structure.

This study has shown the magnitude of innovations and entrepreneurial behaviour and the different roles national policies have had in this. The conclusion is that the model works on a national level since each country has its own policies and regulations, resulting in different relational schemas in the building process, and thus requires a harmonization of standards and codes if applicable to the EU.

13.7.1 Ways to speed up innovation

Construction often being a local activity, it has been shown that being part of regional policy activities has contributed to timber construction innovations (e.g. Austria, Finland, Sweden). Adaptation, learning and change are related to transfer of knowledge and, for construction, especially tacit knowledge, this is strongly connected to the day-to-day activities, and connected to local areas and regions (cf. Metcalfe, 2007; Forsman, 2009). Speeding up innovation and change in the construction industry would be to support regional and local activities, e.g. by regional project networks, and cluster approaches, but with a national perspective.

13.7.2 Recommendations to the timber-based construction sector

Business in timber construction is different from wood product manufacturing. Timber construction can be characterized as a CoPS involving coordination of value-added products and techniques with teams of specialists operating on a temporary basis, whereas wood product manufacturing is more related to optimal yield of raw material in primary production and low-cost distribution in secondary manufacturing such as windows, doors and stairs, but also structural components and products. The difference requires an umbrella organization to raise the questions of timber construction and monitor progress from its own perspective and not from a wood product manufacturing perspective. There are activities under way as indicated from the study, e.g. Estonia, France and Romania, and perhaps it is more difficult in forest-rich and wood products-strong countries like Norway, Finland and Sweden.

The limited knowledge distribution concerning the potential to use timber solutions in construction has been one obstacle in the wood product industries. The innovation potentials of DUI (Doing, Using and Interacting) activities need to be highlighted and the means of distributing collected information and generating new knowledge in transnational wood product and timber construction networks is of importance. The conditions should be addressed to policy stakeholders and involve organizations with research and education on the curriculum.

The study has highlighted the path-breaking position of new actors towards renewal of an industry (construction) and the degree of innovation. The findings are shown by, e.g. the Austrian case – private person as initiator – and the Swedish case – sawmill as key actor, and where the innovation process has resulted in opportunities to gain market shares. Since many firms are SMEs, the formation of regional networks is required, linking actors of the construction innovation system, the traditional wood product industries to timber-based firms as initiators or with external cluster managers. The development requires improved mutual understanding among partners to be successful.

13.7.3 Recommendations to policy makers

The analysis and conclusions of the study indicate the opportunity created by integrating policy initiatives with integrative actions among policy makers towards enhanced value creative networks and related firms within European timber construction and supporting industries. For a further sustainable development of the built environment, timber construction has a role to play, and one where policies are a way to reach the new objectives and goals. Empirical studies of other industries indicate that innovations take up to two to three decades to diffuse to significant extent (Karshenas and Stoneman, 1995) and in a conservative and slow-to-innovate industry such as construction, it may take even longer. The study indicates some promising ways to shorten this with implications on different actors in the innovation process and supports recommendations from prior studies (Seaden and Manseau, 2001). For policies and policy makers, the following are proposed:

- First, a long-term vision in function-based building regulations, including more active work to harmonize EU-wide building codes and to support introduction of new products and processes, would benefit the industry as a whole. This is because function-based regulations place a higher responsibility and opportunity in the hands of the client and end-consumer, and a possibility for contractors and suppliers to act directly with clients.
- Secondly, further integrative policy actions to promote regional timber construction network initiatives are needed and encouraged. Geographically concentrated networks including firms and R&D organizations should be supported to create a climate for innovation and entrepreneurship. The support could be financial, such as grants for development projects, but also as facilitator of demonstration construction projects for timber construction.
- Thirdly, government organizations and policy makers should make use of their influence as customers to 'pull' changes. Technical development projects sponsored by public financing ought to include an assessment of the obstacles and enablers that current business models and traditional roles and responsibilities create for the adoption and diffusion of innovation. Additional support could include other customers in demanding changes coming from, for example, 'innovative' tendering procedures, information about new business models and innovative technical advisors.
- Finally, since construction is still a local and regional business involving many smaller firms, innovation support should be based from this perspective. For example, projects that provided project management support to the network of different specialists (architects, wood industry workers, energy designers, etc.) were successful in creating new innovative products (e.g. Austria and Scotland). Support-enhancing networking with other construction companies

and suppliers helped to develop business concepts (e.g. Finland) and knowledge transfer and learning. Therefore, the importance of regional impact, national policies and programmes should have a regional and local presence, focusing on access to technology and collaborative activities.

References

Apleberger, I., Jonsson, R. and Åhman, P. (2007) Byggandets industrialisering: en nulägesbeskrivning. Sveriges Byggindustrier, FoU-Väst, Forskningsrapport Rapport 0701, Göteborg.

Bengtson, A. (2003) Framing technological development in a concrete context – the use of wood in the Swedish construction industry. PhD thesis, Department of Business Studies, Uppsala University, Uppsala.

Bergström, M. (2004) Industrialized timber frame housing: managing customization, change, and information. Doctoral thesis, Luleå University of Technology, Luleå.

Björnfot, A. (2006) An exploration of lean thinking for multi-storey timber housing construction – contemporary Swedish practises and future opportunities. Doctoral thesis, Luleå University of Technology, Luleå.

Blayse, A.M. and Manley, K. (2004) Key influences on construction innovation. *Construction Innovation* 4, 143–154.

Bozdogan, K., Deyst, J., Hoult, D. and Lucas, M. (1998) Architectural innovation in product development through early supplier integration. *R&D Management* 28, 163–173.

Bregulla, J.R., Grantham, R., Johansson, H.E. and Enjily, V. (2003) Barriers to the enhanced use of wood in Europe: particular attention to the regulatory barriers. Report prepared by Building Research Establishment (BRE) as a part of the Roadmap 2010 programme of the European Confederation of Woodworking Industries CEI-Bois. Available at: http://www.cei-bois.org/roadmap/reports.html

CEI-Bois (2004) *Roadmap 2010 for the European Woodworking Industries.* European Confederation of Woodworking Industries, Brussels.

COM 860 (2007) Communication from the Commission to the Council, the European Parliament, the European Economic and Social Committee and the Committee of the Regions: A lead market initiative for Europe. European Commission, Brussels.

Drucker, P. (1998a) Management's new paradigms. *Forbes* 162, 152–177.

Drucker, P. (1998b) *Innovation and Entrepreneurship.* Elsevier, Oxford.

Dubois, A. and Gadde, L.-E. (2002) Systematic combining: an abductive approach to case research. *Journal of Business Research* 55, 553–560.

Edquist, C. (1999) *Innovation Policy – A Systemic Approach.* Tema, Linköping University, Linköping.

Edquist, C. (2004) Systems of innovation – a critical review of the state of the art. In: Fagerberg, J., Mowery, D. and Nelson, R. (eds) *Handbook of Innovation.* Oxford University Press, Oxford.

Eisenhardt, K.M. (1989) Building theories from case study research. *Academy of Management Review* 14, 532–550.

EU Council Directive 89/106 (1988) On the approximation of laws, regulations and administrative provisions of the Member States relating to construction products. *Official Journal* L 040, 11 February 1989, 0012–0026.

EU Council Directive 93/68 (1993) Amending 89/106/EEC (construction products). *Official Journal* L 220, 30 August 1993, 0001–0022.

EU Regulation No 1882/2003 (2003) *Official Journal of the European Union.*

EUROCONSTRUCT (2009) European construction market trends to 2010: Country reports. *64th EUROCONSTRUCT Conference*, 22–23 November, Vienna.

Forsman, H. (2008) Business development success in SMEs: a case study approach. *Journal of Small Business and Enterprise Development* 15, 606–622.

Forsman, H. (2009) Improving innovations capabilities of small enterprises: cluster strategy as a tool. *International Journal of Innovation Management* 13, 221–243.

Freeman, C. (1992) The 'national system of innovation' in historical perspective. *Cambridge Journal of Economics* 15, 5–24.

Galli, R. and Teubal, M. (1997) Paradigmatic shift in national innovation systems. In: Edquist, C. (ed.) *Systems of Innovation. Technologies, Institutions and Organizations.* Pinter, London.

Gann, D.M. (2003) Guest editorial: innovation in the built environment. *Construction Management and Economics* 21, 553–555.

Gann, D.M. and Salter, A.J. (2000) Innovation in project-based, service-enhanced firms: the construction of complex products and systems. *Research Policy* 29, 955–972.

Gibb, A.G.F. (2001) Standardization and pre-assembly–distinguishing myth from reality using case study research. *Construction Management and Economics* 19, 307–315.

Goodier, C. and Gibb, A. (2007) Future opportunities for offsite in the UK. *Construction Management and Economics* 25, 585–595.

Gopalakrishnan, S. and Damanpour, F. (1997) A review of innovation research in economics, sociology and technology management. *International Journal of Management Sciences* 25, 15–28.

Goverse, T., Hekkert, M.P., Groenewegen, P., Worrell, E. and Smits, R.E.H.M. (2001) Wood innovation in the residential construction sector: opportunities and constraints. *Resources, Conservation and Recycling* 34, 53–74.

Hamel, G. (1998) The challenge today: changing the rules of the game. *Business Strategy Review* 9, 19–26.

Henderson, R.M. and Clark, K.B. (1990) Architectural innovation: the reconfiguration of existing product technologies and the failure of established firms. *Administrative Science Quarterly* 35, 9–30.

Hobday, M. (1998) Product complexity, innovation and industrial organisation. *Research Policy* 26, 689–710.

Höök, M. (2008) Lean culture in industrialized housing – a study of timber volume element prefabrication. LTU-DT-08/21, Doctoral thesis, Department of Civil, Mining and Environmental Engineering, Luleå University of Technology, Luleå.

Hugosson, M. and McCluskey, D.R. (2008) Strategy transformation in the Swedish sawmill sector. 1990–2005. *Studia Forestalia Suecica* 217, 1–33.

INSEE (2009) La filière forêt-bois: panorama des secteurs d'activité. *Chiffres pour l'Alsace* no. 1, February.

Karshenas, M. and Stoneman, P. (1995) *Handbook of the Economics of Innovation and Technological Change*. Basil Blackwell Ltd, London.

Kirkeby, O. (1994) Abduktion. In: Andersen, H. (ed.) *Videnskabsteori og metodelaere*. Samfundslitteratur, Frederiksberg, pp. 122–152.

Koskela, L. (2000) An exploration towards a production theory and its application to construction. Doctoral thesis, VTT, Espoo.

Lessing, J., Stehn, L. and Ekholm, A. (2005) Industrialised housing: definition and categorization of the concept. In: *13th Annual Conference of the International Group for Lean Construction*, Sydney.

Levinthal, D. (2007) Bringing selection back into our evolutionary theories of innovation. In: Malerba, F. and Brusoni, S. (eds) *Perspectives on Innovation*. Cambridge University Press, Cambridge.

Lorenz, E. and Lundvall, B.-Å. (2006) *How Europe's Economies Learn – Coordinating competing models*. Oxford University Press, Oxford.

Lundvall, B.-Å. (1992) *National Innovation Systems: Towards a Theory of Innovation Interactive Learning*. Pinter Publishing, London.

Lundvall, B.-Å. and Borras, S. (2005) Science, technology and innovation policy. In: Fageberg, J., Mowery, D.C. and Nelson, R.R. (eds) *The Oxford Handbook of Innovation*. Oxford University Press, Oxford, pp. 599–631.

Lundvall, B.-Å. and Edquist, C. (1997) *Systems of Innovation: Technologies, Institutions and Organizations*. Pinter, London.

Mahapatra, K. and Gustavsson, L. (2008) Multi-storey timber buildings: breaking industry path dependency. *Building Research and Information* 36, 638–648.

Malerba, F. and Brusoni, S. (eds) (2007) *Perspectives on Innovation*. Cambridge University Press, Cambridge.

Manseau, A. and Shields, R. (2005) *Building Tomorrow: Innovation in Construction and Engineering*. Ashgate, Aldershot, UK.

Metcalfe, S. (2007) Innovation systems, innovation policy and restless capitalism. In: Malerba, F. and Brusoni, S. (eds) *Perspectives on Innovation*. Cambridge University Press, Cambridge.

Miozzo, M. and Dewick, P. (2002) Building competitive advantage: innovation and corporate governance in European construction. *Research Policy* 31, 989–1008.

Narayanan, V.K. (2001) *Managing Technology and Innovation for Competitive Advantage*. Prentice-Hall, Englewood Cliffs, New Jersey.

Nelson, R. and Rosenberg, N. (1993) *Technical Innovation and National Systems: a Comparative Analysis*. Oxford University Press, Oxford.

Nord, T. (2005) Structure and developments in the solid wood value chain – dominant saw milling strategies and industrialized housing. Licentiate thesis, Luleå University of Technology, Luleå.

Nord, T. (2008) Prefabrication strategies in the timber housing industry – a comparison of Swedish and Austrian markets. PhD thesis, Luleå University of Technology, Sweden.

Patton, M.Q. (1990) *Qualitative Evaluation and Research Methods*, 2nd edn. Sage, Newbury Park, California.

Porter, M.E. (1998) *The Competitive Advantage of Nations.* Free Press, New York.

Postnote No. 209 (2003) *Modern Methods of House Building.* Parliamentary Office of Science & Technology, London.

Prahalad, C.K. and Ramaswamy, V. (2000) Co-opting customer competence. *Harvard Business Review* 78, 79–88.

Pries, F. and Janszen, F. (1995) Innovation in the construction industry: the dominant role of the environment. *Construction Management & Economics* 13, 43.

Rametsteiner, E., Weiss, G., Ollonqvist, P. and Slee, B. (2010) *Policy Integration and Coordination: the Case of Innovation and the Forest Sector in Europe.* COST Action E51, Brussels (in press).

Romania Statistics (2009) Monthly statistics bulletin. Available at: http://www.insse.ro/cms/files_Carhiva_buletine2009/bsci_8.pdf (accessed December 2009).

Roos, A., Woxblom, L. and McCluskey, D.R. (2008) Architects', and building engineers', and stakeholders' perceptions to wood in construction – results from a qualitative study. In: Bergseng, E., Delbeck, G. and Hoen, H.-F. (eds) *Proceedings of the Biennial Meeting of the Scandinavian Society of Forest Economics*, Lom, Norway, April. *Scandinavian Forest Economics* No. 42, 184–194.

Sardén, Y. (2005) Complexity and learning in the timber frame housing. Doctoral thesis, Luleå University of Technology, Luleå.

Seaden, G. and Manseau, A. (2001) Public policy and construction innovation. *Building Research & Information* 29, 182–196.

Silverman, D. (2000) *Doing Qualitative Research: A Practical Handbook.* Sage, London.

Slaughter, K. (1998) Models of construction innovation. *Journal of Construction Engineering and Management* 124, 226–231.

Spender, J.C. (1989) *Industry Recipe: An Enquiry into the Nature and Sources of Managerial Judgement.* Basil Blackwell, New York.

Statistics Finland (2009) Statistical database. Available at: http://www.stat.fi (accessed 26 February 2010).

Statistics Norway (2009) Statistical database. Available at: http://www.stat.no (accessed 25 January 2010).

Statistics Sweden (2009) Statistical database. Available at: http://www.scb.se (accessed 31 January 2010).

Stehn, L., Rask, L.-O., Nygren, I. and Östman, B. (2008) *Byggandet av flervåningshus i trä: erfarenheter efter tre års observationer av träbyggandets utveckling.* Department of Civil and Environmental Engineering, Luleå University of Technology, Luleå.

Taylor, J.E. and Levitt, R.E. (2004) *A New Model for Systemic Innovation Diffusion in Project-based Industries.* CIFE Working Paper #WP086. Stanford University, Stanford, Connecticut.

Tykkä, S., McCluskey, D., Nord, T., Ollonqvist, P., Hugosson, M., Roos, A., Ukrainski, K., Nyrud, Q.A. and Bajric, F. (2010) Development of timber framed firms in the construction sector – is EU policy one source of their innovation? *Forest Policy and Economics* 12, 199–206.

Tödtling, F. and Kaufmann, A. (2001) The role of the region for innovation activities of SMEs. *European Urban and Regional Studies* 8, 203–221.

UKTFA (2009) Market Update 2008. UK Timber Frame Association homepage. Available at: http://www.timber-frame.org (accessed 19 December 2009).

Visscher, H. and Meijer, F. (2007) Dynamics of building regulations in Europe. *International Conference on Sustainable Urban Areas*, ENHR, 25–28 June, Rotterdam.

Widén, K. (2006) Innovation diffusion in the construction sector. PhD thesis, Lund University, Lund.

Winch, G. (1998) Zephyrs of creative destruction: understanding the management of innovation in construction. *Journal of Building Research and Information* 26, 268–279.

Winch, G. (2000) Construction business systems in the European Union. *Building Research & Information* 28, 88–97.

Yin, R. (1994) *Case Study Research: Design and Methods*, 2nd edn. Sage, Beverly Hills, California.

14 Networks and Local Milieus as a Furniture Industry Innovation Platform

Pekka Ollonqvist,* Tomas Nord, Andreja Pirc, Kadri Ukrainski, Vuokko Takala-Schreib, Meelis Teder, Wladyslav Strykowski and Anne Viitala

Abstract

The major European traditional manufacturing industries have been developing in parallel with the globalization of their businesses, generating new tasks for local business clusters and networks. This study identifies various patterns of local furniture industry clusters in five countries – Croatia, Estonia, Finland, Poland and Sweden. The case studies are conducted on two levels: cluster level and firm level. This chapter shows that small and medium-sized enterprises (SMEs) in the furniture sector, which are located close to each other in industrial districts, utilize informal knowledge and joint-interest structures and provide infrastructure knowledge, externalities and spillovers. The new business skills needed in communication technology require organizations to adapt to computerized machinery on the factory floor and meet the expectations of global supply and distribution logistics. Enterprises in the furniture distribution industry search for competitive advantages from customer orientation and mass customization in high-cost countries.

14.1 Introduction

Strong regional clusters characterize the current European furniture industry. The best-known design furniture clusters in northern Italy and Denmark provide a narrow view of the current state of the European furniture industry in the global context. Furniture production in Europe is categorized into low and medium technology (LMT) industries, according to the Organisation for Economic Co-operation and Development (OECD) classification (EUROSTAT Business database, n.d.; Hirsch-Kreinsen, 2006). Limited opportunity to utilize industry-specific, formal research and development (R&D) knowledge explains the strong bottom-up emergence of regional clusters in European furniture production (Maskell and Lorenzen, 2003). Developing a globalized business structure in major, traditional manufacturing industries has also challenged the future of standard furniture producers, who constitute the major share of European furniture manufacturing (Asheim *et al.*, 2003). The furniture industry clusters, frequently featured by numerous SMEs, have relied on tacit knowledge-based developments. However, all major categories of product, process, business architecture and marketing innovations are observed in European furniture clusters. Development opportunities and challenges for such clusters, created through globalized development in down-flow business structures,

* Corresponding author.

can be introduced by identifiable, strong development paths. Those paths are characterized by: (i) the global subcontracting networks mainly in non-upholstered furniture product segments; (ii) multinational European upholstered home furniture and bedroom producers; and (iii) enterprises (or groups of enterprises) offering extensive, comprehensive interior solutions, e.g. in offices, hotels and cruise ships (Heanue, 2009). All three development paths have witnessed consolidations and merging national and international brands both in the European and the US contexts (Maskell, 1996; Acharya *et al.*, 2009). Office-furniture industry enterprises base their competitiveness on their ability to provide comprehensive delivery solutions and demand segment-focused interior products. The office furniture segments and industries are omitted in this research.

Interdisciplinary knowledge transfer activities and restructured delivery and distribution channels have provided bases for systemic innovation activities among SMEs in LMT industries (Antonelli, 2005). There is empirical evidence on informal knowledge transactions in the form of positive knowledge spillovers and other externalities, but also from organized collective networks. The latter, identified in carpenters' guilds,[1] has implied close proximity and strong mutual trust among the joint entrepreneurs. The formal contract-based knowledge transactions have gradually become more common, thus complementing and even partly subsidizing informal modes of knowledge and competence transfer. This has been important for the innovation activities of furniture sector enterprises that build their competitive advantages (CAs) on uniqueness and tacit knowledge. Participation in the current global business sector implies new, interdisciplinary documented knowledge, as well as the tacit knowledge embedded within enterprises and enterprise networks. The debate about whether new knowledge is accessed through informal systems or through formal contracts is relevant in the furniture industry (Heanue, 2009). The local furniture cluster infrastructures in the participating countries of COST E 51 Action are among the points of interest in this chapter.

Knowledge flows within local clusters can be divided into disembodied and embodied modes. Disembodied flows involve the use of knowledge transmitted through scientific and technical literature, consultancy, education systems and movement of personnel (Smith, 2002). Much embodied knowledge is incorporated into technical equipment, which, in the furniture industry, is external to the local cluster (Heanue, 2009). Low barriers to entry and easy appropriation characterize the competitive infrastructure in major furniture industries (Drayse, 2008). These tend to reduce interest in radical innovation processes among single enterprises. Large furniture producers that manufacture coated panel board and other standardized wood products are apt to outsource production to gain positive scale economies through low-cost mass production. European furniture clusters provide infrastructures to SMEs utilizing design, mass customization and high customer responsiveness as the major sources of CA. The industrial districts provide lively, contract-based knowledge transfers, but untraded knowledge externalities and spillovers are also still supported. However, the need for new knowledge has enlarged their demand for traded knowledge and competencies (Heanue, 2009). The creation of global furniture production networks has facilitated the development of high technological solutions. Computerized machinery on the factory floor and standardized global supply and distribution logistics/channels have released the furniture industry from the boundaries of their local clusters, but have also wiped out many sources of CAs that had previously existed.

We know that furniture industry enterprises use information technology in innovation processes, and the role of collective learning in knowledge transfer has been of great interest for their development (Drayse, 2008).

The position of regional innovation system (RIS) and coordinated regional and rural policy actions provide support benefits for the interactive learning of regional, tacit knowledge (Lorenzen, 1998). Nybakk (2009) has identified four levels of relevance when considering the needs of innovation policy actions for LMT enterprises: project level, organizational level, inter-organizational level and society level. Global networking and value creation, based on subcontracting, have downgraded project-level issues in innovative development. Increasingly, innovation projects constitute network activities with the contributions based on documented or tacit knowledge competencies. There the contributions from individuals are frequently difficult to identify. Policy makers have difficulty identifying knowledge needs, and thus support the creation of optimal learning systems. Learning is essential also for the policy makers, who have a strategic role to play. They establish regional innovation support systems, which bring together economic and institutional development. Developing a mix of policies is central to promoting localized learning. Different policy areas are needed to take full account of the systemic and embedded nature of the industry, as well as to promote organic elements of learning (Heanue, 2009). The lead enterprises, frequently participants in global production systems, are relevant to furniture industry clusters when transmitting the acquired codified knowledge for use among the local enterprises (Heanue, 2009).

Recent innovation activities in the furniture industry can thus be traced to the creation of new value-network architectures, new marketing solutions and reorganized value creation. Partnership skills in global business networks and compatibility with the computer-assisted product (CAD) and the business planning (CAM) system of global enterprises contribute to the major sources of CAs for the SMEs in the furniture industry. These skills have now been transformed into competitiveness parameters, namely regional tacit knowledge and strong local chambers of commerce.

14.2 Research Objectives and Case Studies

This chapter looks for inter-linkages between the types of CAs and innovative and local clustering and networking in furniture-industry clusters. The case studies consist of wood-based furniture industry SMEs in five participating countries – Croatia, Estonia, Finland, Poland and Sweden. Local policy actions on learning, which change or build regional learning, is part of our interest. The case studies are conducted on cluster and enterprise levels, respectively. This research structure enables us to examine individual innovations at enterprise level while also looking at connections with regard to respective cluster activities. This allows us to search for possible linkages in different national settings regarding innovation and production systems. The mix of countries covers a large variety of examples in terms of size, different clustering intensity and also very different experiences in cluster cooperation. The differences in national policy initiatives as well as implementation frameworks can also be discussed.

Section 14.3 provides an overview of the structure of the European furniture industry. Section 14.4 examines the new innovation framework of the furniture industry enterprises based on customers' views and configuration capacities. This information is generated by applying functional- and production-based business taxonomies to identify case enterprises. Furniture industry enterprises and their innovation activities are also discussed in Section 14.4 within the framework of functional and product categorization. Section 14.5 characterizes the main types of innovation activities relevant to different types of furniture SMEs (including the intake of new external high technological business applications), implying capabilities within complex technological processes. Section 14.5 also discusses value-chain configuration, business architecture and marketing in new innovative solutions. In Section 14.6, local milieus and business networks in regional furniture clusters are evaluated with country-specific reports, along with the role of policy. Section 14.7 examines the country reports of national furniture clusters and industries as well as summarizes the evaluation of innovation, CA and clustering patterns of furniture firms. The conclusions are presented in Section 14.8.

14.3 The Furniture Industry in Europe

14.3.1 Globalized markets and competition

The furniture industry holds an important position in the EU-27 area economy, providing employment to 1.4 million people, and generated a turnover of €130 billion in 2007 (Acharya *et al.*, 2009; CEI-Bois Memorandum, 2009).

Furniture manufacturing belongs to important traditional industries in some European countries but covers less than 4% of the aggregate gross value of manufacturing industries in European Union (EU) countries (UEA, 2006). This disproportion is understandable because of the national concentrations that are typical in furniture industries. Some of the major furniture clusters with their interconnections to up-flow in the value chain are expressed in Fig. 14.1.

The EU furniture industry has a high level of regional clustering. The four largest furniture producers (Germany, Italy, France and Spain) cover over 50% of the total EU production (Fig. 14.2).

Strong regional furniture clusters are valid also in Poland, the UK, Denmark and Sweden. In our survey, regional concentration (as clusters) is valid also in small countries, like Croatia, Estonia and Finland. SMEs dominate the furniture sector when assessed by the number of enterprises it contains. The countries in our sample comprise about 7% of the total production in the EU.

Furniture industry business can be divided into major product segments: upholstered, non-upholstered, office furniture and interior carpentry. These segments have their own specific value chains, sources of CAs, standardized production processes and business networks. The functional dimensions of enterprise taxonomy (integrated producers but also business network locomotives, their partners and subcontractors) are beneficial. European upholstery producers, as well as some of their component-producing partners, have been able to preserve their CAs under strong global

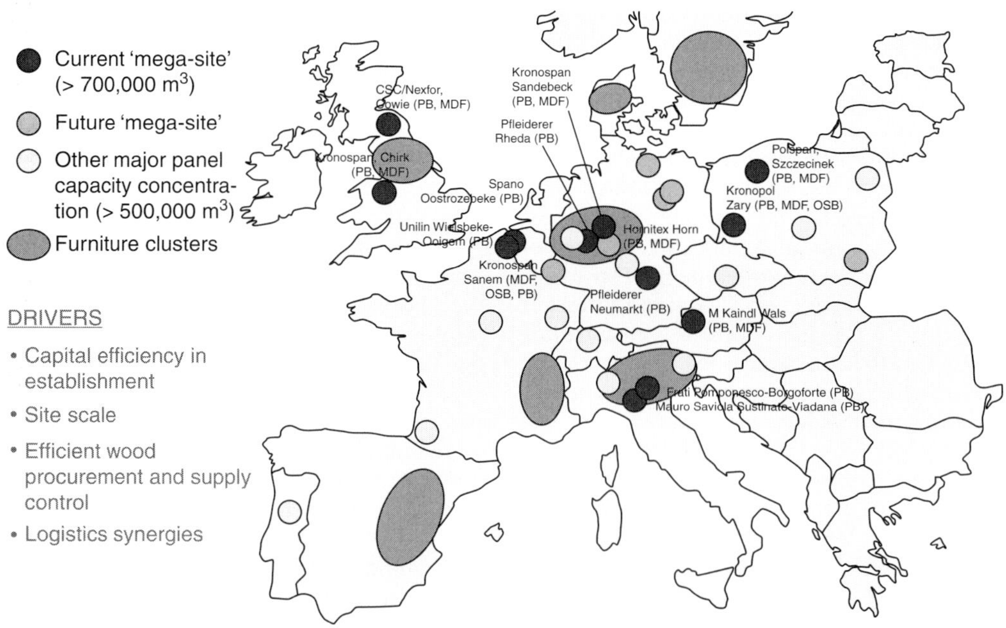

Fig. 14.1. A map of wood-based panel 'mega-sites' in Europe linked with furniture clusters (source: Buehlmann *et al.*, 2003).

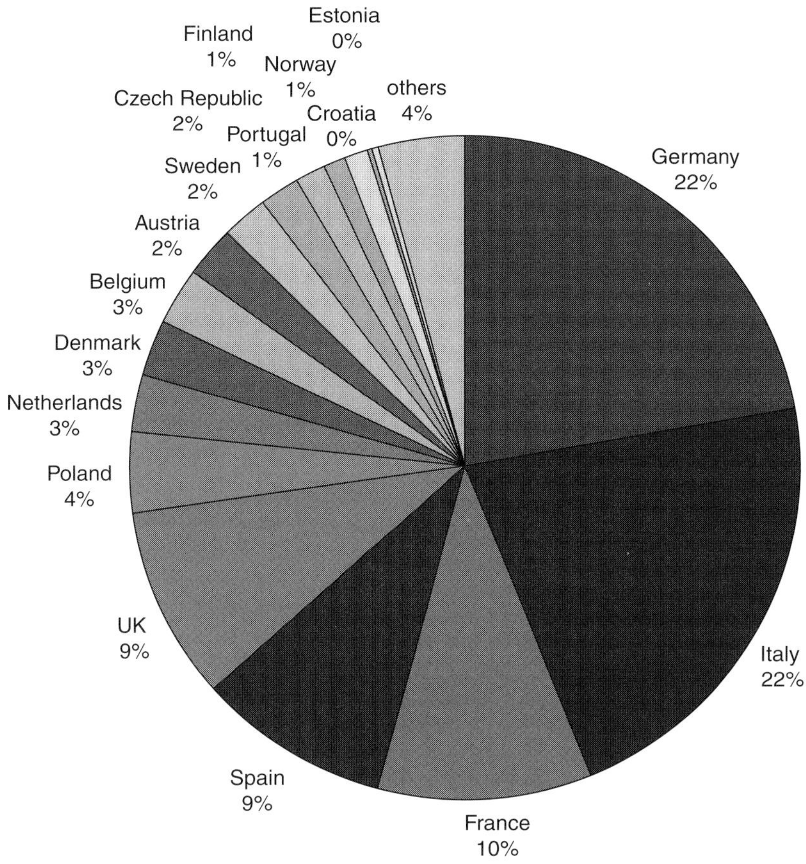

Fig. 14.2. Shares (%) of national furniture industries in Europe (source: UEA, 2006).

competition. This is mainly because they have tailored their products to meet the needs of their customers, and there are limited opportunities to achieve positive economies of scale in upholstered furniture production. Enterprises in these product groups face mostly regional competitive pressures. The upholstered furniture segment has typically remained domestic and been dominated by local SME clusters.

The producers of standard wood plate furniture (case goods) can, and frequently do, benefit from the design segments, whereas the production of standard products has been outsourced extensively to low-cost countries (Drayse, 2008). SMEs, like component producers and finished-furniture assemblers, manage to survive in non-urban industrial districts that provide potential links to the EU industry because of their general flexibility and high adaptability to market changes. Furniture industry enterprises in vertical and horizontal networks can provide support for start-up and spin-off businesses if they benefit cluster development (Maskell and Lorenzen, 2003). Fair investment capabilities and difficulties in obtaining risk financing, together with the limitations in the knowledge base, are the major impeding features among SMEs. Furniture-industry SMEs utilize craftsmanship and tacit knowledge provisions to create CAs but face challenges coming from easy imitation and low barriers to entry (Asheim *et al.*, 2003; Bender, 2006).

Those European furniture enterprises that operate in countries with high wages and high unit costs have gradually outsourced their non-upholstery production activities. Outsourcing originally started through component subcontracting but later spread to cover products and whole product segments. The search for cost advantages has supported increased segmentation towards high customer orientation and shortened product life cycles.

14.3.2 European furniture industry clusters

Much of the entrepreneurship in furniture production relies on SMEs through process industrialization, which has now substituted high-skilled carpentry with industrialized processes (Maskell, 1996; Maskell and Lorenzen, 2003; Buehlmann and Schuler, 2009). The dominance of domestic market production and business characterized this industry until the late 1990s, although design furniture production and export trade are major exceptions. Furniture industry enterprises in the EU face strong competition when efficient production process technologies are available in the commercial markets. Easy access and global transferability using new technical knowledge increasingly challenge enterprises in high-cost countries and generate needs for CAs to remain in business. Organizational competitiveness and survival depend on a firm's ability to develop new or innovative products and services (Drew, 1997).

Domestic market dominance, which used to support regional furniture clusters and their individual furniture enterprises in EU countries and in the USA, have gradually disappeared (Drayse, 2008; UEA, 2008; Buehlmann and Schuler, 2009; Heanue, 2009). It was also noted by Grushecky *et al.* (2006) that continued exploration of new markets and export market opportunities can add more value to current products and are essential for furniture industry enterprises.

Regional furniture clusters, with their historically developed structural features, have supported growth among individual enterprises (and their groups) with specific product orientation (Acharya *et al.*, 2009). Growth modes have been dependent on: (i) the competitive and industrial structure; and (ii) the value-chain orientation among major enterprises. These enterprise concentrations have had easy access to key raw materials, skilled employees, or both. Cluster enterprises frequently form local production networks with subcontractors and partners and they also form horizontal cooperation between operations on the same level of the production chain. Many of these networks act like a unit of a large enterprise. A cluster can comprise an industrial district with institutions promoting and maintaining cooperation between chief executive officers (CEOs) and their employee organizations. Clusters can provide the potential for cooperation in order to develop capacity and RISs. The latter implies the need for a high level of mutual trust and effective learning processes supported by knowledge and skills supplied by industrial service centres, technology centres and training centres (Isaksen, 1996; European Commission, 2002).

The development of local clustering and related inter-firm cooperation has at least three rationalities: first, the heterogeneity of knowledge and competencies in the form of tacit knowledge and competencies. They allow fast adaptation to market changes. Furniture demand is characterized by taste and demand segment varieties, which provide growth opportunities for export-based furniture clusters. Second, enterprises can benefit from localized learning because of their close proximity, which ensures easy knowledge transfer, but at the same time supports coherence within the regional clusters. Third, the absorptive capacity of enterprises supports improved learning in the clusters because they share value systems and have close, professional working relationships at CEO and expert levels.

14.4 Product and Function Types among Furniture Industry Enterprises

Many categories characterize business strategies among furniture industry enterprises.

Brege *et al.* (2001) have defined nine that apply to the major product categories: product orientation (mass producer, customer oriented), the main customer segment (home, office, bedroom and interior) and business concept categories (service orientation, standard manufacturing and process specialization). Value-chain positioning is especially important in furniture industries. The furniture enterprise taxonomy with value-chain positioning developed by Brege *et al.* (2001) covers nine basic categories:

- The enterprises among *production-oriented volume producers* and *component and module subcontractors* both aim to use positive economies of scale as the basis for their CAs. These enterprises, applying automation and standardized production processes, frequently contract or are partner producers of international furniture traders. Independent domestic or export trade is a minor business strategy and they rarely have specific branding.
- The enterprises among *traditional home furniture producers* and *traditional office/public furniture producers* frequently share family or owner/manager tenure and are characterized by the extensive use of tradition and craftsmanship as the sources of their CAs. Traditional furniture producers typically benefit and utilize CAs from local clusters and aim to achieve high product quality and intra-enterprise model design as the sources of their CAs.
- *Design home furniture producers* and *design public space furniture producers* share independent external designers (or design enterprises) as essential service providers for their CAs. The name or even brand of the designer is identifiable in the products of these subgroups. Product design and brand legacy as sources of CAs are heavily based on their reputation as the designer/craftsmanship processors (creating image-, or legacy-based CAs). In many countries, the design home furniture producers have a skewed size distribution, with some large international businesses and many SMEs. High quality, associated with design, is common and high-level designers are also complemented with parallel knowledge-intensive business service (KIBS) supply.
- *Bedroom facility producers* have a variety of enterprises, from international enterprises to niche-oriented producers. Strong branding seems to characterize market competition and in export markets among the niche producers. The latter makes the position among SMEs difficult and also explains the high frequency of consolidations in Europe.
- *Interior wood-furniture producers and joiners* cover many SMEs, which are mainly customer oriented with narrow product segments. Such enterprises are frequently interior-solution providers with attached complementary subcontractors in the value-chain down-flow.
- The enterprises among *office-facility producers* cover large European companies that have grown through consolidations. They are able to provide a complete assortment of alternative products for big projects (whole office buildings) and focus products (like seat producers) with their own delivery systems. The majority of enterprises involve intra-enterprise international production. They operate at an international level more often than that of their counterparts from the other industry subsectors.

These categories are complemented with a more general business concept. *Service-oriented enterprises* control their process chain and logistics thereby acquiring strong market orientation and flexibility in their operations. The regular adaptation to customer needs explains the frequent service-orientation for upholstery furniture producers. They are only restricted by the unit cost conditions: how much a customer is willing to pay for a specialized service. Non-upholstery furniture enterprises can benefit by applying this approach and utilizing economies of scale (Buehlmann and Schuler, 2009). *Subcontracting enterprises* have narrow capability to create an independent business strategy or other network connection towards partnering. *Standard manufacturers* acquire their enterprise-specific CAs and benefit from the economies of scope by handling specific product materials in a systematic manner. Access to high-quality material has been one of the major sources of their CA. Standard manufacturers can improve their innovation enabling capabilities (IECs) by horizontal networking. The potential for CA and growth tends to be connected to the embedded knowledge within local clusters. Their position allows only limited options for process innovations if their product design is arranged externally. Modes of production processes have large batches, high set-up times and low interchangeability of parts in different product lines. Standard manufacturers are dependent on traditional distribution channels where competition from mass merchant houses and large retailer networks is strong. *Offshore production* characterizes non-upholstery wooden residential furniture businesses that allow the utilization of low wage and social costs in addition to low ocean freight costs. Findings from the USA also confirm much of the European evidence (Buehlmann and Schuler, 2009). Buehlman and Schuler (2009) have suggested that furniture manufacturers will outsource more work to specialized entities, allowing them to focus on managing a supply chain, and building and improving efficient distribution operations. Process specialists typically command production-line knowledge and are experienced to ensure failure-free production flows in complex structures. Cost leadership is a common and frequently necessary objective among process specialists that are assemblers of component and module subcontracted furniture, mainly in upholstery product segments.

Value-chain positioning (horizontal or vertical) is the third valid category dimension consisting of: (i) domestic production of SMEs with gradual capacity accumulation through original growth or through takeovers and fusions; (ii) partnerships in decentralized, local, national or international value networks; and (iii) subcontracting in a national or international business structure. All these major modes of value-chain positioning are visible in the European furniture industry. Large enterprises often grow through design or brand segmentation. Product or product-module outsourcing of individual enterprises has become typical in the second mode. The third mode, concerning open global competition of subcontracting participation among SMEs, is common in countries inside the EU as well as globally.

14.5 The New Innovation Approaches of the Furniture Industry Enterprises

14.5.1 Demand pull innovation view

Product orientation and supply push, which have long characterized traditional industry strategies, are rapidly changing. Furniture industry enterprises have started to adopt process, organizational and marketing innovations and invest in expensive high technology. There is ample empirical evidence from the intake of information and communication technology (ICT) through CAD/CAM solutions. They are important among process specialist enterprises (for optimizing production processes via CAM) but also among service-oriented enterprises (for improving customer flexibility via CAD). CAM has been stabilized as a common tool among furniture industry enterprises, with half of all furniture enterprises using CAM facilities (Gaboardi and Monti, 2008). The use of machine tools with computer numerical control (CNC) technology requires special training and can be useful if integrated with CAD activities (Elmhester, 2008). Only 15% of the furniture industry enterprises use ICT in their value-chain business activities. CAD processes have become an essential part of traditional industry product development. A current EU survey identified over 70% of furniture industry enterprises use CAD in their product design activities (Gaboardi and Monti, 2008). Research findings indicate that the integrated use of CAD constitutes an integrated planning process, i.e.

there are no CAD-specialized assistants in the design teams (Elmhester, 2008). The mode of CAD use is missing only from some micro-SMEs (Gaboardi and Monti, 2008). The majority of large companies use ICT-based management systems, whereas in small enterprises, this figure is 16% (Gaboardi and Monti, 2008). New e-business solutions provide organizational innovations and use Internet trading and marketing innovation opportunities, respectively. External communication with network partners, including transaction functions as well as ICT information transactions (called e-business), are common in large furniture industry enterprises; half of them use these technologies. E-business is also becoming common among SMEs because a third of all small enterprises use e-business and it is even more common among medium-sized enterprises. Process automation like ICT systems, e-business, as well as Internet marketing and trade are currently among the ordinary sources of CAs.

New process and planning tools support development towards enhanced customer orientation. There is R&D bias concerning the innovations and related capabilities among low technology industries and enterprises (Aslesen and Isaksen, 2007). The current systemic view challenges the distinction between product and process innovation. A new product is an outcome of: (i) an innovative material treatment or process; (ii) a modified design or application of new materials (new for the enterprise) directly necessitating changes in the production process; (iii) new machinery; or (iv) new competencies (Bender, 2006).

Innovative marketing and organizational solutions complement product innovation processes. Innovative organizational or marketing solutions can be identified within current products and production process technologies. The furniture industry shares wide production capacity by outsourcing experience to many European traditional industries (Cesaro *et al.*, 2006). Low unit costs together with short distances from the volume markets have supported outsourcing for simple, standardized products outside of Europe. The sources of cost advantages are based on standardized equipment and codified knowledge, and tend to be transferable and even adaptable to new work cultures. High-unit cost countries have counteracted these challenges by outsourcing capacity-building activities or developing new innovation policies.

Small, individual enterprises are reluctant to invest in technology development, which is easy and cheap for others to copy. Furthermore, incremental innovations frequently mean that new entrants to the industry 'explore' new trajectories in addition to those based on tacit knowledge. Fair research, development and innovation (RDI) budgets and capabilities among individual enterprises restrict the emergence of disruptive innovations and explain the rapid escalation of new knowledge among enterprises.

14.5.2 Configuration capabilities

Utilization of embedded knowledge (internal and external to the firm) characterizes innovation activities among traditional industry enterprises. It is an important part of the IECs, and in the literature is divided into two categories: transformative and configurative. Transformative capabilities concern the knowledge base to transform general knowledge (R&D outputs) into the modes usable in locally specific context. The configurative capabilities measure abilities to arrange new configurations from already established (mainly tacit) knowledge and skills (Bender, 2006). SMEs must manage knowledge bases from different disciplines (as an example, production machines and ICT tools) and synthesize that expertise, linking other actors and their relevant knowledge, technology and competence. The related complementary institutions and clustered services are also important. The configuration capabilities are related to: (i) synthesizing; (ii) organizing; and (iii) designing competencies (Bender, 2006).

First, synthesizing competencies comprise abilities to utilize knowledge from totally different areas and combine it. Available knowledge-intensive service activities (KISA) provide a wide set of opportunities to build up internal competence and knowledge. These activities extend over the services provided by KIBS enterprises (Aslesen and Isaksen, 2007). Secondly, organizational competencies cover abilities to use distributed sources of relevant knowledge and competence through linking actors together with relevant knowledge, technologies and competencies.

Thirdly, design competencies start from analysing customer needs and extending to processes with added customer value as an output. Design competencies provide an important source for strategic advances in the furniture industry. Design can be developed inside the enterprise or acquired outside, and it can be used to create new objects, artefacts (i.e. products), systems and services (Kristensen and Lojacono, 2002). Design consultants or separate design service providers use both tacit and codified knowledge that are difficult to separate.

14.6 Cluster Advances and Furniture Innovation Activities

14.6.1 Characteristics of furniture clusters in Europe

Regional cluster institutions and structures can support physical and mental investments when creating horizontal and vertical networks (as well as other coherent business related activities) (Alberti, 2002). A European cluster survey has identified that competition in a horizontal direction tends to speed up the emergence of horizontal value chains among SMEs (European Commission, 2002). Similarly, vertical cooperation is frequent when enterprises search for economies of scale in activities like trading or marketing. Cooperation among furniture SMEs typically relates to raw materials or other input-factor, but also to deliveries of special products. A cluster with a substantial number of specialized SMEs tends to benefit from joint service arrangements between companies. General learning ability is of crucial concern among traditional industry SMEs. Findings in Nordic SMEs (Asheim *et al.*, 2003) confirm the importance of a technological and organizational learning infrastructure generated from the supply of complementary skills and capabilities to ensure individual innovation processes. Organizational learning is the major challenge for the management of innovation processes among furniture industry SMEs. Innovation processes that rely on an analytical knowledge base ensure that entrepreneurs collaborate with R&D institutes, which can provide unique and specific knowledge. Frequently, the lack of a supportive system prevents small enterprises from adapting innovation R&D information.

The interplay between individual enterprises and innovation systems tends to be complicated in traditional industries. In this survey, regional innovation networks provide a basis for the evaluation of the innovation system. It is proposed that regional structures and policies are the major innovation systems that work for traditional industries.

The furniture industry clusters share many supporting characteristics that are common when developing a competitive structure and economic growth base. Enterprises in furniture clusters in both Europe and the USA face a rapidly changing and competitive global business environment. Furniture retailing has undergone a transformation, partly because of new consumer perceptions and the emergence of global production and distribution value chains. The furniture trade is increasingly based on fashion in upholstered products and cost efficiency in non-upholstered products. Specialized furniture exhibition halls and large interior stores, such as IKEA, have gradually substituted traditional marketing and distribution of furniture through local stores. Mass product retail has concentrated on national and international business networks that create their CAs through subcontracting and base their economies of scale in the up-flow of the value chain. The market share of Asian-based subcontractors has increased within non-upholstered products. The transfer of furniture assembling to the consumer has decreased vastly transportation and storage costs. European furniture enterprises have also outsourced much of their non-upholstery production offshore to low-cost countries. Easy and inexpensive long-distance ocean shipping has reduced the problems of location in relation to raw materials, such as wood, skilled craftspeople and deliveries.

This global development supports the development of businesses for European SMEs enabling them to concentrate on upholstered products that operate at short distances from the markets. Cutting, sewing and assembling for upholstered furniture have remained mainly domestic activities. These domestic, labour-intensive tasks are competitive alternatives in a fast-changing business mode (Acharya *et al.*, 2009). The availability of high-quality timber and other wood components and modules is a necessary condition to maintain CAs among furniture industry SMEs.

In addition, the enhanced purchasing power of young generations has enlarged the customer base but also created markets outside the traditional furniture shops. Young consumers are image conscious and their houses are getting larger (Acharya *et al.*, 2009).

14.6.2 Policy support on innovative milieus and clusters

Geographical space and embedded knowledge have been empirically identified as being behind the growth and profitability differences in traditional industries (Storper, 1997; Bender, 2006). The IECs discussed above have become important (European Commission, 2002; Malmberg and Maskell, 2005; Lundvall, 2007). IECs are analysed within the taxonomies, identifying high-, medium- and low-technology industries in the first dimension. Proximity and 'embeddedness' relate to the local milieu in the second dimension, respectively (Bender, 2006). Easy imitation and other sources behind low appropriability impede investments as activities of single enterprises. Research findings from traditional industries support the policy actions that strengthen the supply of complementary knowledge and capabilities through innovation policy actions (Asheim and Isaksen 1996; Isaksen, 1996; Asheim *et al.*, 2003; Bender, 2006).

Support for public innovation has diversified to cover systemic innovation processes (involving a network of stakeholders), and also to create knowledge and other resource supplies for innovating enterprises. The supply of supportive tools has promoted knowledge creation and competitiveness among the enterprises in the local clusters. Tacit knowledge bases with extended specification qualities encourage innovation only up to a point. The industrial business-district infrastructure can either support or prevent an individual innovation process (Boschma, 2005).

Active innovation policies can be directed to stimulate diffusion of interactive learning and provide a basis for radical innovation processes. Frequent modes of networking in the typical industrial district are based on interactive learning systems and developing relations between principal enterprises and their suppliers (Asheim and Isaksen, 1996).

Regional clusters in traditional industries perform socially and collectively through networking arrangements. The more actors who join, the more attractive and valuable the membership becomes. Social capital can be built from bottom-up activities, although top-down policy making has been more typical. The shared norms and cultural values among individual CEOs (and the community) support social capital creation. The recent triple helix actions and their progress have provided options for collaboration among SMEs, universities, research centres and other public sector stakeholders to contribute to the formation of regional strategies. They have also created concrete practical development projects between SMEs, R&D organizations and public bodies. The Finnish Centre of Expertise programmes (I 1999–2006 and II 2007–2013) provide an example of a feasible solution (Kavonius, 2008).

14.7 Case Studies – Croatia, Estonia, Finland, Poland and Sweden

14.7.1 Cluster infrastructures and innovation activities in the case-study enterprises

Table 14.1 summarizes the cluster background in the cases with the characterization of size, industry and enterprise structures. The cases concerning innovation activities are summarized in Table 14.2. They are discussed in their cluster context.

14.7.2 Findings from the case study clusters

Reactive innovation behaviour among furniture-producing SMEs is triggered by

Table 14.1. Cluster description in Croatia, Estonia, Finland, Poland and Sweden.

Country and cluster name	Cluster description
C1 Croatia: Wood cluster of north-west Croatia	The cluster was established in November 2005 by Varazdin County chamber. It consists of 15 enterprises and has 2600 employees. All 15 companies are in Croatian property. Firms in the cluster are specialized for sawmill production, carpentry products and furniture. The vision of the cluster is to increase competitiveness of the Croatian wood industry and to create a strong brand for the cluster, which will be recognized as a product with special characteristics and quality of distribution.
C2 Croatia: Cluster Slavonian Oak	The cluster was established in April 2004. It has 150 employees and 13 companies. Companies in the cluster are specialized for furniture, joinery (doors), parquet and sawn timber, and small hotels furnishing. The vision of all companies is to improve the competitiveness of the Croatian woodworking industry in a new form of organization between companies and by sharing operations both horizontally and vertically with the purpose of optimizing the production processes.
C3 Croatia: Tehnointerijeri	The cluster was established in June 2006. It has 966 employees and consists of 19 companies specialized for interior and hotels furnishing, furniture production and furniture and interior design. A main vision of the cluster is expanding export capacities and increasing added value of the existing export products of cluster members. The next vision is joint development of new producers as well as export to foreign markets and expanding export in the tourism sector via increased quality of family-owned hotels with the help of a cluster expert; also, joint promotion in the international markets.
C4 Estonia: south-east Estonia (Põlva, Valga, Võru Counties)	The cluster was established in 2005 by the initiative of regional public authorities and involves upstream production firms (sawn wood, panels, components), as well as forestry and transportation firms besides furniture producers. The cluster aims to eliminate different development barriers and encourage cooperation between local wood-related SMEs. From the start, the cluster activities were targeted to mainly facilitate knowledge and experience exchange between SMEs; however, in the furniture part this concerns traditional manufacturers and process specialists of home furniture.
C5 Finland: South Ostrobothnia	The cluster in dominated by SMEs (over 220 enterprises covering 15% of national aggregate and producing 11% of total annual national aggregate output, annual average turnover €630,000/enterprise, which is 26% below the country average). The majority of the furniture industry SMEs have limited capabilities to execute innovation processes, which explains the bottom up initials towards a regional I&R&D centre in 1997. Nikkarikeskus (Furniture Development Canter) is working in collaboration with School of Culture, University of Vaasa, and is currently connected to the Living Business action of the Centre of Expertise Program 2007–2013.

Continued

Table 14.1. Continued.

C6 Finland: Päijät – Häme	The cluster is the most influential in the counties in Finland, with furniture industry activities with 25% share aggregate national average. City of Lahti as an urban multi-industry centre of Päijät Häme hosts furniture cluster (see Chapter 7 in this book). Two large international furniture industry corporations with their Finnish subsidiary and a large national multi-product line corporation with extensive domestic furniture hall retail network exists in the region. The average turnover of the enterprises in the cluster (€1.9 million/enterprise) is 2.2 times the country average. The Lahti University Consortium, the Lahti City Science and Business Park Ltd are part of the cluster and the Lahti Development Agency Ltd (LAKES) acts as a regional facilitator for the regional cluster-based business and innovation policy (see Chapter 7 in this book).
C7 Poland: Furniture Cluster of Wielkopolska	The activity was withheld.
C8 Poland: West Pomeranian Wood-Furniture Cluster	Building a recognizable region brand in Poland and in northern Europe with production of high-quality furniture, competitive price and attractive design.
C9 Poland: Society Klaster Mebel Elbląg	Promotion of the companies related in Society and also the region of the Elbląg city, gaining funds in order to promote the cluster brand.
C10 Sweden	Component and flat-pack producing companies in a geographic region with a strong furniture tradition, and where many companies have supplied and grown with large retailing chain (IKEA). Competitive advantage from efficient production and logistics processes as competition has grown from low-cost countries.

competitors, demanding users and customers, or by a deteriorated competitive position in the market, rather than by formal R&D activities. Product innovation with integrated organizational and marketing features and improved reactivity to customer demand are among the major success-creating factors. These types of motivation and innovation outputs constitute a majority among the cluster cases in our study (Table 14.3).

Still, several cases bring forward CAs from process innovations. They relate to optimizing production processes, improving communication and knowledge exchange, but also aligning and improving configuration capabilities among innovating enterprises. Depending on the type of innovation sought (product, process or marketing), the cluster type solution follows similar patterns. Vertical clusters are commonly related to product innovations while horizontal clusters are connected to process innovations. The strengthened interconnections in the value chains, either in an up-flow (intermediary input producers) or a down-flow (customers) direction, provide relevant signals from major market changes and customer service needs (vertical types of clusters were mainly connected with product orientation in our case studies).

In our study, most of the enterprises were standard manufacturers or process specialists; one enterprise was classified as service oriented. However, better customer alignment through different services (logistics, brand promotion, complementary services) was a visible target in several clusters.

Public involvement activities tend to follow similar patterns in all case clusters studied: the public sector actively participated in cluster activities, mostly by funding them. Similarities can be explained by the global competition and outsourcing of production in all clusters. Interestingly, countries with previously strong domestic furniture industries have started to support the strengthening of global business clusters. Examples from other industries show that

Table 14.2. Description of case companies in clusters in Croatia, Estonia, Finland, Poland and Sweden.

Case company	Innovation mode product, process, business, network, marketing	The innovation	Source/drivers for innovation	Size (no. of employees)	Comments
C1 Croatia	Business network	Joint interest enterprises	Development of brand product and increasing competitiveness of Croatian wood industry	2600	http://www.drvni-cluster.com/
C2 Croatia	Business network/ innovation processes	Joint interest enterprises	New organization for between companies and optimizing production processes	150	http://www.zadruga-slavonski-hrast.hr/
C3 Croatia	Business network/ marketing	Joint interest companies	Joint promotion on international markets and increasing added value of existing products	966	http://www.tehnointerijeri.hr/
C4 Estonia	Business network/ process innovation	Technological process innovation	Demand (fashion) changes towards veneered products	90	http://www.puidklaster.ee/
C5 Finland: South Ostrobothnia (I)	Business network/ marketing	Joint interest enterprise for special furniture deliveries (able to jointly provide the whole hotel interior furniture assortment)	DIVAN furniture technology and innovation programme	8 (in owner enterprises, 100)	http://www.junet.fi/
C5 Finland: South Ostrobothnia (II)	Business network	Wholesale distribution of furniture equipment	Systematic R&D investments: new products, materials and business concepts, education and training to employees; R&D partners Seinäjoki Polytechnics (University of Helsinki section) and TEKES (Finnish Agency for Technology and Innovation)	70	http://www.ejh.fi/index.php?l=en

C6 Finland: Päijät – Häme (I)	Business network	Partner network creation with three retail dealers aimed towards a specific customer-oriented furniture line to be delivered through the retailer network	Joint interest network	30	http://www.pedro.fi/en/index.php
C6 Finland: Päijät – Häme (II)	Marketing	E-trade	TEKES (Finnish Agency for Technology and Innovation) project	25	http://www.htcollection.com/?page=&lang=en
C7 Poland: Furniture Cluster of Wielkopolska	The activity was withheld				
C8 Poland: West Pomeranian Wood-Furniture Cluster	Marketing	Product and process innovativeness, communication between partners, forest and wood product certification	Searching for suppliers of the materials and also receivers, searching for new technologies		Baumal Przedsiębiorstwo Wielobranżowe, Furniko Fabryka mebli Biurorowych, Makowski meble Sp zoo
C9 Poland: Society Klaster Mebel Elbląg	Business network	Partner network creation operating on a common market (cooperation of competitors)	European funds for development of clusters	2000	http://www.lupus.eu, http://www.meblekam.pl
C10 Sweden	Business network/process focus	Strategic network of firms in similar position and in relation to IKEA; Joint production efficiency project; CA in painting	Demands from customer (IKEA) to improve; Facilitated by companies and some support of regional officials	Cluster doubled its turnover in a 5-year period	

Table 14.3. Synthesis of cluster and firm level findings.

Case	Cluster structure	Cooperation orientation to innovation type	Case firm's type	Type/level of public involvement
C1 Croatia	Vertical	Product (brand) and distribution	Standard manufacturer	National – initiating and funding the activities
C2 Croatia	Network	Process (optimization)	Standard manufacturer	National – initiating and funding the activities
C3 Croatia	Horizontal	Product (export) and marketing (promotion)	Standard manufacturer	National – initiating and funding the activities
C4 Estonia	Network	Product and process (knowledge exchange)	Process specialist/ standard manufacturer	Regional – initiating role; national, EU – funding the activities, project-based
C5 (I) Finland	Horizontal	Organizational (joint marketing) marketing (complementary offerings)	Joint interest of standard manufacturers	National – initiating and funding for the activities, from DIVAN technology programme (TEKES)
C5(II) Finland	Vertical	Product and services (complementary accessories delivery)	Standard manufacturer/service oriented	National – funding the activities, project-based; TEKES and Polytechnical School
C6(I) Finland	Vertical	Product (design) expertise from the down-flow	Design manufacturer	None
C6(II) Finland	Network	Product assortment (intra-firm design products and partner delivery to std products)	Standard manufacturer	E-trade development aid TEKES
C7 Poland	n/a	n/a	n/a	n/a
C8 Poland	Network	Product and process (communication)	Standard manufacturer	National – funding the activities
C9 Poland	Horizontal	Product (brand)	Standard manufacturer	Local – funding the activities, project-based; EU – project-based funding
C10 Sweden	Horizontal	Process (optimization)	Process specialists	Regional – funding cluster leader

n/a, not available.

public institutions have taken a lead in crises by trying to retain industries in their own countries (Bender, 2006). This involvement may be a reaction to the role of the public sector in developing a systemic view of innovation. Under the Science, Technology and Innovation (STI) paradigm, the public sector had a clear role in directing funds into basic research (Asheim, 2009). The current paradigm, which assumes interactive learning and relations between actors, needs a more systematic policy view where the roles of different supporting policies are evaluated. By being more active in innovation, the public sector may not only fund but also initiate changes. In our study, it was only in the Swedish case that the public sector was not involved in initiating the cluster.

14.7.3 Discussion and findings

European furniture industry enterprises have pursued innovation strategies that generate creativity, quality and active differentiation of products. The major objective of improved customer service is accentuated in this agenda. Low in-house R&D, common among furniture industry SMEs, results from reactive innovation behaviour. A reactive approach is triggered by competitors, demanding users and customers, or a deteriorated competitive position in the market. Product innovations with integrated organizational and marketing features and improved reactivity to customer demand are among the major potential success factors. Strengthened value-chain interconnections to up-flow (intermediary input producers) as well as to down-flow (customers) in the value chain provide relevant signals from major market changes and customer service needs. Case enterprises in Finland and Sweden support the general findings from Europe (Elmhester, 2008; Pirc, 2009). Enterprises that have focused on product and customer issues related to their CAs have been able to stay in the market scene despite their high production costs. Innovative networking solutions as well as strong customer value identification are very visible in these companies. The communities hosting furniture clusters have invested in design and creation competencies, automation of production processes and other knowledge capacities. International commitments concerning issues like sustainability and carbon sequestration have gradually improved the CAs of wood-based solutions over those based on other materials.

14.8 Conclusions and Policy Recommendations

Qualitative flexibility in terms of quick responses to heterogeneous demand and continuous product innovations has generated competitiveness among local furniture industry clusters and SMEs in the surveyed clusters. Enterprises have identified constant reorientation to customers' needs in different markets with various style preferences. The prior Danish and Italian cluster feature of production flexibility has become visible in other European countries. The lessons learned from the development of the Finnish and Swedish furniture industry are that the more helpful an institutional endowment becomes for one type of activity, the less suitable it will be for another. Growth, based on global network subcontracting, has impeded the production flexibility that is necessary when subcontracting outside the local cluster. The significance of an appropriate fit between local furniture enterprises and the supporting institutions also suggests why the distribution of business activities is normally skewed in individual clusters. A cluster that produces fashion-adapted furniture tends to generate institutions that do not support enterprises utilizing positive scale economies, and vice versa. Business strategies are dependent on the value-chain commitments maintained by SMEs; the latter is even more true with business contracts. Research findings support positive relationships between strategic dependence and network dependence (Varamäki and Vesalainen, 2003). This positive dependence is, however, valid mainly in technology-related value-chain architectures.

The fast emergence in globalized value-chain solutions, characterized by component production outsourcing, has fundamentally changed the position of SMEs in local clusters. It has also introduced new solutions for joint operations. The Swedish example suggests that this networking is not related to marketing; on the contrary, it may be preferable to start for other reasons (Elmhester, 2008). The cluster cases from Croatia and Estonia support the view that high-level public sector involvement is important. Active public support can provide an additional initial step towards further bottom-up initiatives within enterprises. Local clusters and enterprise networks should be able to meet quickly new challenges and opportunities in the new global context. The fundamental revolution in business models, due to e-marketplaces, e-business and e-commerce in business-to-business trade, has created great competence requirements but also provided many opportunities for disruptive marketing and organizational innovations (Gaboardi and Monti, 2008). New CAD/CAM tools and related e-business activities are very important for design home furniture producers, bedroom-facility producers and office-furniture producers. Regional clusters are valuable because they can support joint innovation processes and innovation networks in the pursuit of 'knowledge platforms'. They can thus meet the development opportunities in cluster enterprises without challenging the open access to the independent intra-enterprise creation of innovative solutions. Policy actions that invest in knowledge generation are important in two subsystems: (i) the knowledge application and exploitation subsystem, principally occupied by enterprises in the cluster and knowledge generation sector; and (ii) the diffusion subsystem, consisting of polytechnics and research institutes that process knowledge platforms but also initiate, support and directly participate in enterprise-level innovation processes.

Innovation and learning infrastructures, characterized by locally embedded tacit knowledge, provide learning region boundaries through path-dependent craftsmanship that are characterized by traditional home, office and public furniture producers but also by interior wood furniture producers and joiners. The contribution to the innovation capability of an individual SME is determined by its ability to disseminate and adopt new knowledge in localized, interactive and socially embedded learning processes. The establishment and support of clusters and strategic networking are possible and often fruitful ways to obtain the required knowledge and competencies among local SMEs, as shown in this study.

The enterprise and cluster findings in this research clearly show the diversity of the European furniture industry. This diversity grows from the cluster-specific competition and CAs among enterprises and clusters. This diversity is valid inside individual countries but more so between countries. It will take time before the furniture industry enterprises in Europe will see themselves as colleagues and not as competitors, although the growth and development of some large multinational furniture retailers may help to solve these differences. There are some activities that can be considered necessary to support the continuation of a strong and healthy European furniture industry:

- Maintain and enlarge education within the furniture field especially for operators of advanced machinery. The future worker must combine computer knowledge with material sciences.
- Support local and regional cluster initiatives through rural and regional development programmes, such as Leader funding or similar. Professional external cluster managers or organizations could also be established.
- A first-generation cluster may require active involvement from the public sector, including funding as well as professional advice. Second-generation clusters should thus emerge from the established infrastructure as part of a bottom-up development.

Note

[1] The Skive carpenters guild in the Salling furniture cluster in Denmark is an example (Asheim *et al.*, 2003).

References

Acharya, S., Clayton, Z., Eriksson Giwa, S., Malinger, E. and Moura, A. (2009) The North Carolina furniture cluster – the microeconomics of competitiveness. Harvard University Press, Cambridge, Massachusetts. Available at: www.isc.hbs.edu/pdf/Student.../USA_Furniture_2009.pdf

Alberti, F. (2002) Industrial districts: inter-enterprise networks, entrepreneurial agency and institutions. Angelo Guerini e Associati Spa, Milan. Available at: www.insme.org/documenti/the_concept_of_industrial_district.pdf

Antonelli, C. (2005) Models of knowledge and systems of governance. *Journal of Institutional Economics* 1, 51–73.

Asheim, B. (2009) The Formation of Competitive Regional Innovation Systems. Presentation at COST E51 Meeting, Zagreb, Croatia, 15–16 January 2009.

Asheim, B. and Isaksen, A. (1996) Location, agglomeration and innovation: towards regional innovation systems in Norway? STEP report R – 13, Oslo. Available at: www.step.no/reports/Y1996/1396.pdf

Asheim, B., Coenen, L. and Svensson-Henning, M. (eds) (2003) Nordic SMEs and Regional Innovation Systems. Final Report. Department of Social and Economic Geography, Lund University, Lund. UEA 2003. European Furniture Manufacturers Federation. Available at: http://www.ueanet.com

Aslesen, H. and Isaksen, A. (2007) New perspectives on knowledge-intensive services and innovation. *Geografiska Annaler* 89B(S1), 45–58.

Bender, G. (2006) Policy and innovation in low-tech – knowledge formation, employment and growth contributions of the 'old economy' industries in Europe – PILOT. Final Report of the Project.

Boschma, R. (2005) Proximity and innovation: a critical assessment. *Regional Studies* 39, 61–74.

Brege, S., Milewski, J. and Berglund, M. (2001) Storskalighet och småföretagande – en studie av strategiska grupper inom svensk möbelindustri VINNOVA Rapport, VR. 2001:41. Available at: http://www.lime.se/Lime20Mobelstudie.pdf

Buehlmann, U. and Schuler, A. (2009) The U.S. household furniture industry: status and opportunities. *Forest Products Journal* 59, 20–28.

Buehlmann, U., Wiedenbeck, J. and Schuler, A. (2003) State of North American hardwood industry – how to compete globally. Presented to the National Hardwood Lumber Association (NHLA), Atlanta, Georgia, 18 September.

CEI-Bois Memorandum (2009) How wood products can help Europe achieve its political objectives while ensuring competitiveness and profitability. Brussels, September 2009.

Cesaro, L., Florian, D. and Padureanu, L. (2006) Delocalisation of wood working industries to Balkan countries: analysis of statistical data and case studies. Available at: http://joypub.joensuu.fi/publications/other_publications/niskanen_issues/niskanen.pdf

Drayse, M. (2008) Globalization and regional change in the U.S. furniture industry. *Growth and Change* 39, 252–282.

Drew, S. (1997) From knowledge to action: the impact of benchmarking on organizational performance. *Long Range Planning* 30, 427–441.

Elmhester, K. (2008) Små företag i strategiska nätverk – hur påverkas det enskilda företagets utveckling? Linköping Studies in Science and Technology Dissertations, No.1217 International Graduate School of Management and Engineering, IMIE Dissertation No. 118. Linköping.

European Commission (2002) Observatory of European SMEs, No. 5. Available at: http://ec.europa.eu/enterprise/policies/sme/facts-figures-analysis/sme-observatory/index_en.htm

EUROSTAT Business database (n.d.) Available at: http://epp.eurostat.ec.europa.eu/portal/page/portal/european_business/data/database

Gaboardi, E. and Monti, E. (2008) ICT and e-business impact in the furniture industry. Impact Study Report No. 03. European Commission, DG Enterprise and Industry. Available at: http://www.ebusiness-watch.org/studies/sectors/furniture/documents/Study_03 2008_Furniture.pdf

Grushecky, S., Buehlmann, U., Schuler, A., Luppold, B. and Cesa, E. (2006) Decline in the U.S. furniture industry: a case study of the impacts to the hardwood lumber supply chain. *Wood and Fiber Science* 38, 365–376.

Heanue, K. (2009) Aspects of the furniture industry in Ireland. PhD thesis, Dublin City University Business School, Dublin.

Hirsch-Kreinsen, H. (2006) Low technology – a forgotten sector in innovation policy. Paper presented at the International ProACT Conference, 15–17 March, Tampere.

Isaksen, A. (1996) Regional clusters and competitiveness: the Norwegian case. STEP report R 16. Oslo.

Kavonius, V. (2008) Finnish model for promoting cluster formation and regional-driven clusters: Strategic Centres for Science, Technology and Innovation, and Centre of Expertise Programme (OSKE). Paper presented in Lisbon, 2 December 2008.

Kristensen, T. and Lojacono, G. (2002) Commissioning design: evidence from the furniture industry. *Technology Analysis and Strategic Management* 14, 107–121.

Lorenzen, M. (ed.) (1998) *Specialization and Localized Learning. Six Studies on the European Furniture Industry*. Copenhagen Business School Press, Copenhagen.

Lundvall, B. (2007) National innovation system: analytical focusing device and policy learning tool. Working paper. ITPS, Swedish Institute for Growth Policy Studies. R 2007:004.
Malmberg, A. and Maskell, P. (2005) Localized learning revisited. DRUID Working Paper No. 05-19. Available at: http://www.druid.dk/wp/wp.html
Maskell, P. (1996) Localised low-tech learning in the furniture industry. DRUID Working Paper No. 96-11.
Maskell, P. and Lorenzen, M. (2003) The cluster as market organization. DRUID Working Paper No. 03-14.
Motik, D. and Pirc, A. (2007) Woorkbook – Tržište namještaja i ostalih drvnih proizvoda do 2006-te godine. Šumarski fakultet Sveučilišta u Zagrebu, Zagreb.
Motik, D. and Pirc, A. (2008) Woorkbook – Pokazetelji stanja na tržištu namještaja i ostalih drvnih proizvoda Republike Hrvatske do 2007-te godine. Šumarski fakultet Sveučilišta u Zagrebu, Zagreb.
Nybakk, E. (2009) Innovation and entrepreneurship in small firms: the influence of entrepreneurial attitudes, external relationships and learning orientation. PhD thesis, Norwegian University of Life Sciences.
Pirc, A. (2009) Report of a short term scientific mission at Finnish Forest Research Institute METLA. Finland.
Smith, K. (2002) What is the 'knowledge economy'? Knowledge-intensive Industries and Distributed Knowledge Bases. UNU INTEC Discussion 6/2002.
Storper, M. (1997) *Regional World: Territorial Development in a Global Economy.* Guilford Press, New York.
UEA (European Furniture Manufacturers Federation) (2006) Furniture in Europe.
UEA (European Furniture Manufacturers Federation) (2008) Furniture in Europe. Available at: http://www.ueanet.com/uea-extranet/IMG/pdf/FURNITURE_INDUSTRY_2008-2.pdf
Varamäki, E. and Vesalainen, J. (2003) Modeling different types of multilateral cooperation between SMEs. *Entrepreneurship and Regional Development Journal* 15, 27–47.

Appendix 14.1. Furniture business outlook in 2007–2008

	Croatia (2007)	Estonia (2008)	Finland (2007)	Poland (2008)	Sweden (2008)
Gross value of production	478	389	1,380	7,249	2,190
Gross added value	n/a	118	440	2,684	n/a
Export	279	543	240	5,718	1,650
Import	352	176	520	1,124	1,550
Apparent domestic consumption	551	22	1,650	2,655	2,090
Industry (production units)	506	516	1,570	25,000	890
Employment	11,904	8,781	10,570	170,300	17,800
Micro-enterprise	60%	11%	24%	14%	n/a
Small and medium enterprises	30%	75%	52%	42%	n/a
Large enterprises	10%	14%	34%	44%	n/a
Home furniture	208	260	1,070	4,144	1,070
Bedroom furniture	30%	n/a	23%	8%	23%
Home furniture	35%	72%	53%	n/a	54%
Kitchen, outdoor office and other home furniture	35%	28%	23%	n/a	23%
Share of *n* largest enterprises					
Kitchen furniture	(n = 4) 18%	n/a	(n = 4) 50%	n/a	(n = 4) 50%
Office furniture	(n = 3) 10%	(n = 2) 68%	(n = 3) 80%	n/a	(n = 3) 80%
Performance					(2002)
Gross margin in production	n/a	−1.7%	5.8%	n/a	(ROI) 4.4%
Gross margin in retail trade	n/a	1.3%	2.7%	n/a	n/a
Equity capital in production	n/a	30.7%	38.9%	n/a	38%
Equity capital in retail trade	n/a	27.6%	38.2%	n/a	n/a

Sources: Croatia: Financial Agency (www.fina.hr); Republic of Croatia – Central Bureau of Statistics (www.dzs.hr); Motik and Pirc (2007); Motik and Pirc (2008). Estonia: Electronic Database of Statistics Estonia (www.stat.ee); UN Commodity Trade Statistics Database (http://data.un.org/browse.aspx?d=ComTrade). Sweden: http://www.tmf.se/web/Mobler_1.aspx; http://www.finpro.fi/NR/rdonlyres/05979137-F079-4D87-BC02-867A90A53F0D/12665/Ruotsin_maarap-ortti221220091.pdf; Brege *et al.,* 2001.

(Companies rather than production units is the basis of the statistics. Companies without published balance sheets are omitted.)

n/a, not available.

15 Innovation in the Wood Bio-energy Sector in Europe

Erlend Nybakk,* Anssi Niskanen, Fahrudin Bajric, Gabriel Duduman, Diana Feliciano, Krzysztof Jablonski, Anders Lunnan, Liana Sadauskiene, Bill Slee and Meelis Teder

Abstract

This chapter aims to analyse sources of innovations for bio-energy at the firm level and to assess which policies have supported or impeded company innovations and business development. Our analyses are based on 14 firm-level cases from eight European countries. The case study data were collected in 2008 and 2009 using common interview guidelines. The cases represent companies engaged in activities that range from wood pellet and chip production to district heating, large-scale electricity production at a pulp mill site and other bio-energy business applications in Europe. As one might expect, the contents of the cases were diverse. Although the innovations analysed had different sources, in the majority of the cases the innovations were prompted either by emerging new market situations (e.g. rising fossil fuel prices) or new available woody biomass sources, or were driven by changes in company strategies or introduced and led by visionary managers. The case studies often stress the important roles of innovation champions, with their innovative charisma and excellent interpersonal skills. Also, policy and policy measures played an important role in many of the analysed cases. Based on the analysed results, we found that determined national policies and programmes supporting and providing favourable financial conditions for investments in the area of renewables boosted innovative activities in the use of woody biomass for energy purposes.

15.1 Introduction

There is often a gap between political ambitions regarding bio-energy use and actual bio-energy use in many European countries. The most obvious explanation is that bio-energy is not price-competitive because the costs of production are too high. Thus, the use of bio-energy has not grown as expected. There are two other potential explanations: first, the capital costs of wood-fuel boilers are high compared with those of alternative forms of heating or electricity generation; and second, the full environmental costs of use are not factored into the cost models for alternative heat and energy sources. Environmental factors, such as reducing greenhouse gas emissions, provide a justification for cost-share measures in bio-energy policy or carbon taxes. Some environmental benefits of bio-energy are internalized in energy prices, but policy support varies greatly from country to country.

* Corresponding author.

Many factors aside from price and environment also influence actual bio-energy use. In the popular literature, these other factors are referred to as *barriers* and *drivers* (W. White, 2007, 'Assessing drivers for bioenergy and biomass supply. The perspective of economics', unpublished manuscript). It is important to obtain more knowledge about these barriers and drivers because such knowledge can be used to design more efficient bio-energy policies. Many of the barriers and drivers are of the kind that economists call transaction costs. In mature, well-functioning markets, transaction costs are low. In relatively new and weakly developed markets, the transaction costs may be high. Most bio-energy projects involve companies and other actors who do not normally cooperate, and it takes time and effort to establish trust and a cooperative environment between them. The decision-making process used to establish any bio-energy system also takes time and involves several public actors. This process creates transaction costs associated with supply chain development (Roos *et al.*, 1999).

Another important factor influencing long-term bio-energy use is technological development and innovation (Audretsch and Feldman, 1996; Staffan and Bergek, 2004). All customers have alternatives to bio-energy, and the alternatives may become relatively less expensive over time because of innovation. Without innovation, bio-energy systems will not ultimately be competitive. A central concern is what may drive innovations in bio-energy systems faster or more efficiently compared with competing energy systems.

According to the Oxford English Dictionary, an invention involves the creation of something new, and innovation is the act of introducing something new. From an economic perspective, an invention must be advantageous, or it must minimally be thought to be advantageous, to be considered an innovation (Nybakk, 2009). It is important to distinguish between different types of innovation (Nybakk *et al.*, 2008). Jenssen and Nybakk (2009) found that product, process and market innovations are affected by different innovation drivers. Furthermore, it is common to distinguish between radical and incremental innovation (Damanpour, 1991). Radical innovation is innovation that is new and different from previous solutions, whereas incremental innovation implies smaller and more subjective new solutions adopted by a firm. Recent innovation research on the wood industry distinguishes between product, process and business system innovations (Hovgaard and Hansen, 2004; Hansen *et al.*, 2006). One might also add policy innovation to the list of systemic innovations. Product innovation is an act that leads to new products and services. Process innovation is an act that leads to new processes and refers to the process that is intended to produce something. Business system innovation is the act that leads to new business systems. A business system consists of everything in a company that is necessary to manage, structure, operate and administrate the business and its internal and external environment (Hovgaard and Hansen, 2004). Policy innovation is the introduction of new enabling policy measures (Rametsteiner *et al.*, 2010).

Drivers or antecedents of innovation are frequently debated in innovation literature. Largely, research concerning innovation and innovativeness can be classified in accordance with the level of the independent variable. Four levels can be distinguished: the individual, organizational, interorganizational and societal levels. At the individual level, the importance of individuals or 'innovation champions' is emphasized (Schön, 1963). Jenssen and Jørgensen (2004) define an innovation champion as a person willing to take risks by enthusiastically promoting the development and/or implementation of an innovation inside a corporation through a resource acquisition process without regard to the resources currently controlled.

The second level is organizational, which focuses on organizational culture and structure (Mintzberg, 1979). Factors at the organizational level that are assumed to influence innovation might be categorized as organizational structure and communication, organizational culture, strategy, incentives, finances and slack (e.g. Duduman and Bouriaud, 2007; Tidd and Bessant, 2009).

At the third level, the focus is on relationships between organizations, and it is assumed that relationships between individuals in different companies and networks of individuals at different organizations stimulate innovation within organizations (e.g. Burt, 1992, 1997; Granovetter, 1973). At the societal level, studies on the effect of regional clusters are an important example (Porter, 1990). Porter (1990) argues that the development of clusters is essential for national competiveness.

The innovation system approach is another perspective from which to study innovation. It was introduced by Freeman (1987) and is considered an approach rather than a theory. However, an institutional view of innovation is reflected in the literature on systems of innovation. Institutions shape and are shaped by the actions of organizations and the relationships among them (Edquist, 1997). The main components of a system of innovation are actors, institutions and their interactions. Actors are considered organizations, which are seen as formal structures with an explicit purpose that have been consciously created (Edquist and Johnson, 1997). Interaction among actors and institutional settings is important for innovation activities. The basic idea is that companies do not normally innovate in isolation. Instead, in the innovation system approach, innovations are seen as based on interactive learning between organizations or actors (Edquist, 1997). Several scholars have also applied an innovation system approach on the wood and forest sector (Rametsteiner *et al.*, 2005; Hansen *et al.*, 2006; Rametsteiner and Weiss, 2006a,b).

15.2 Research Question, Method and Introduction of Cases

Factors influencing the diffusion of certain energy technologies are the topic of several studies (Dieperink *et al.*, 2004; Madlener, 2007; Mahapatra, 2007). Unlike other studies, which focus on a particular country and on micro-data, we used case studies in different countries and compared the results from an international perspective. The aim of this study was to understand different types of innovations and their drivers in Europeanbio-energycompanies.Furthermore, we wanted to determine how external relations and institutional and policy factors affect the innovations and the innovation process. During the work of the European Science Foundation-funded COST Action E51, it became evident that innovations in the bio-energy sector are very different in various European countries. This chapter attempts to analyse sources of innovations for bio-energy at the firm level in different parts of Europe. Based on the similarities and differences found, the policy implications of the findings are discussed regarding how to support bio-energy business development.

This study was based on qualitative case studies that were built around the examination of different innovation activities in companies in various European countries. The case studies were used to investigate drivers of innovation, the degree and importance of networks, and policies that impede and foster development.

In the following chapter, we describe 14 firm-level cases from eight European countries: Finland, Norway, Lithuania, Estonia, Poland, Romania, France and Scotland. The research design, sampling, data and analyses are outlined. The results are then presented and discussed. Finally, some implications are considered, and suggestions are made for further research.

The case study data were collected in 2008 and 2009 using the same interview guidelines (originally developed by Thomas Rimmler) in all countries to increase reliability. Also, a similar semi-structured interview method was applied. The cases represented companies that engaged in wood pellet and chip production, district heating, large-scale electricity production at a pulp mill site and other bio-energy business applications in Europe. An intensity sampling strategy was used to select the cases. Themes touched upon in the interviews fell into the following four main categories:

1. What is the main innovation undertaken by the company?
2. What were the origins of the innovation?

Table 15.1. Case company descriptions.

Case no., country and firm name	Case description
C1: Norway: pellets case company	The company produces pellets from round logs; a garbage disposal company's surplus heat is used for the drying process
C2: Lithuania: pellets case company	The company produces pellets and briquettes
C3: Lithuania: wood chips case company	The company produces and delivers energy wood chips from logging residues
C4: France: electricity case company	Electricity production in pulp mills
C5: Poland: pellets case company	The company buys pallets, repairs them or chips them; the chips are used for the production of briquettes and pellets and sold to small industry and private users
C6: Poland: district heating case company	Town district heating, supplying heat to the town, installing biomass-fired boilers (to replace coal-fired ones) and partial use of own willow plantations for raw materials
C7: Estonia: wood chips and fuel wood case company	Company that sells different wood-made materials, recently has also expanded into combined heat and power (CHP) production from forest residues and fuel wood
C8: Estonia: district heating case company	Large-scale heat producer that began electricity production in 2009 (CHP)
C9: Estonia: pellets case company	Pellet producer that partly uses round wood with debarking (and sawdust from sawmills)
C10: Finland: pellets case company	Large-scale producer of pellets from sawdust
C11: Finland: district heating case company	District heating cooperative
C12: Scotland: wood chips case company	The enterprise produces wood chips both to be consumed by the estate mansion and to be sold in regional markets
C13: Scotland: pellets case company	The company produces wood pellets for fuel/energy and for horse bedding
C14: Romania: district heating case company	The company sells thermal energy for heating a town in north-eastern Romania; the thermal energy is produced from sawdust and small wood residue

3. What actors were involved, and how important were networks?
4. How did the respondent assess the potential for this innovation in the country (including the market and policy)?

A detailed overview of the studied companies is given in Table 15.1.

The data were collected by different researchers from the case study countries. The first portion of the data analysis was undertaken by the researcher in each country. This analysis resulted in one document from each case with 2–14 pages of text. In the second step of the analysis, these documents were compared, and the text was grouped into themes and subthemes. The results of the second step of the analysis are presented in the following sections and discussed in the last section.

15.3 Types of Innovations and their Drivers

15.3.1 Case studies descriptions

The innovations found in the 14 case studies are presented in Table 15.2. Even if we had selected more innovative companies for the case studies, the innovations found would probably have been predominantly incremental. The object of investigation was bio-energy companies (mostly pellet producers and district heating companies), and

Table 15.2. Innovations in the cases and their drivers.

Case company	The innovation	Source/drivers for innovation
C1: Norwegian pellets case company	Pioneers producing pellets from forest raw materials that were dried by surplus heat from a garbage disposal company	The CEO developed the idea behind the innovation and was the main driver during the start-up process; he was the internal/innovation champion. Synergy with a waste disposal partner company was key
C2: Lithuanian pellets case company	Briquettes and pellets from sawdust produced from round wood	Unstable sawdust deliveries from sawmills prompted the CEO to develop an independent sawdust production chain. Innovation was driven by good demand prospects in pellet importing countries
C3: Lithuanian wood chips case company	Fuel chips produced from logging residues	The idea born at seminars on bio-energy and exhibitions accompanying international projects on bio-energy. Environmental scanning: local supply of logging residues and local demand for fuel chips. Low transportation cost for fuel chips
C4: France electricity case company	Generation of electricity from wood biomass (pulping by-products, by-products of wood industries)	Governmental initiative (instrument: calls for tender, with 15–20-year fixed prices). Competitive pressure on production costs and on markets. Technical skills and characteristics of internal process of the mill
C5: Polish pellets case company	Pellets and briquettes are made from recycled pallets and sawmill waste wood. Pallets are either chipped or repaired. Chips are made into briquettes or pellets for heating purposes or sold to an OSB plant	There was an oversupply of wood pallets and wood waste, highly priced gas and coal, a market for pellets, CO_2 issues and rising eco-awareness
C6: Polish district heating case company	Heat energy supply from biomass-fuelled boilers that replaced old coal-fired boilers, a willow plantation project and a straw briquette line	Coal was replaced by wood because of rising coal prices and the need to replace coal boilers. Cheap land offered the option of establishing their own biomass plantations, and the chairman acted as an internal visionary
C7: Estonian wood chips and fuel wood case company	Started combined heat and power (CHP), both heat and the electricity production in Tallinn area from wood. Initiated the use of forest-felling residues and stumps	A visit to Nordic countries resulted in secured demand for heat from the district heating network 'Tallinna Küte', secured raw materials from Woodex, pre-negotiated sales of electricity
C8: Estonian district heating case company	Production of both heat and electricity (CHP) from wood and peat. First time the CHP was built on such a small territory. The total available area was 2.2 ha (site of the old boiler house); normally, the territory for CHP is *c.*4–5 times larger in size	Vertical monopoly: they owned the resource (peatland), production of heat, pipelines for delivery to customers and sales organization. Forest areas are nearby, but harbours where pulpwood is exported are distant, so occasionally pulpwood is sold for heat production. Increasing price of Russian gas. All of these factors led to the decision to build a CHP

C9: Estonian pellets case company	Chipping and grinding from round wood; stationary chipper; debarking of round wood; and flex heat brand	Lack of raw material (sawdust) and desired market attributes drove the debarking decision. High demand for premium quality (German standard). Cooperation with equipment manufacturers
C10: Finnish pellets case company	Technological development; investments in the efficiency of pellet production; and service innovations. The company increasingly focused on providing a full-service concept that included the delivery of pellets, heat production and the maintenance of power plants (households not included)	The company strategy shifted: more business from the processing of sawdust from the company's own mills and the dust bought from other sawmills turned into pellets. Market development was focused in Finland and abroad, because the company exports most of its production
C11: Finland district heating case company	The formation of cooperation: the owners of the cooperative increased the demand for otherwise non-marketable small-diameter wood; the community created new local business and jobs as well as a good image in replacing non-renewable oil-based energy production with bio-wood-based energy	The increasing price of oil improved the price competitiveness of wood-based energy. Thus, profitability increased for the energy producer. Income also increased for the landowners from otherwise low-value wood and jobs for the municipality (win–win)
C12: Scotland wood chips case company	Improved supply chain and process (chip boiler for heating), and business model	This activity began as a response to the rising cost of energy for heating a very large mansion house on the estate. The mansion is used for business purposes on a mixed highland estate that supports forestry, farming, quarrying, sport shooting and tourism enterprises. Wood procurement is characterized by opportunistic acquisition of low-grade fuel wood
C13: Scotland pellets case company	Wood pellet production with two markets. The wood fuel market was the intended market, and the horse bedding market was discovered by chance. Establishing a modern pellet production enterprise in high-quality buildings. 'Torrefaction' technique – it removes moisture from the woodchips in a vacuum. The brand name for the product: 'Stovies'. The raw material is sourced locally from the estate's own woods and from a nearby sawmill. The key innovation is to make the right choice of the raw material and to produce the right sort of fibre (without bark or twigs)	The CEO wanted to add value to his timber and picked up ideas for that while working for a landowners' association. He believed that in the future, wood fuel was going to be promoted by the Scottish government. They chose to produce pellets instead of chips, because pellets are more convenient and easier to handle. They started to produce wood pellets for horse bedding after noticing that another distributor was selling pellets to the horse trade. The plant was designed to use sawdust, but because it compromised the quality of the pellets, they stopped using it
C14: Romania district heating case company	Heat and hot water from wood residues	In a national setting, new logistical equipment for sawdust collection, new buffer stores for sawdust; new thermal power plant; new distribution network; the replacement of the old heating equipment in buildings' basements with new conduits; a new system for thermal energy production based on sawdust, chips or bark

this approach limited the possibility of finding significant product innovations.

As seen in Table 15.2, most of the innovations were basically incremental process innovations and often associated with supply chain innovations. Two of the companies, one Norwegian (C1) and one Estonian (C9), were both local pioneers, using round logs in pellet production. In these cases, the innovations were not related to the end-product but were process innovations enabling the utilization of round logs. In Norway, pellet production with roundwood as a raw material was normally seen as unprofitable because of the high energy cost related to drying. However, the Norwegian company was the first in Norway (and one of the first in the world) to produce pellets directly from raw material from the forest, ideally from round logs. The company obtained its lumber from the forest: tree trunks were ground into chips, and afterwards the raw chips were dried with warm air from the refuse disposal plant. After drying, these were manufactured into pellets. The company was also one of few pellet producers in the world to use low-temperature drying in pellet production. Several of the other case companies made larger investments in new equipment and improved processes as their main innovation (e.g. Lithuania and Scotland).

The Lithuanian case (C2) is an example of incremental innovation. The pellet and briquette producer exports 90% of its output. They were strongly dependent on sawdust deliveries from sawmills because the company was exposed to unstable round wood markets, resulting in varying quantities of available sawdust. Confronted with such a situation, the management decided to implement a process innovation by developing a sawdust delivery chain, with the sawdust produced directly from round wood. The company decided to make an investment. They bought a wood chipper and put it into operation, thus decreasing the company's dependence on uncertain sawdust deliveries from sawmills. The innovation was driven by a strongly rising wood pellet demand in Europe. The manager displayed innovative behaviour, if not features of an innovation champion, being able to analyse the present situation, as well as foresee the market developments and create a vision of the company's future.

The two Finnish companies developed a business system innovation. In the energy cooperative (C10), the most important innovation was the establishment of cooperation. The owners of the cooperative increased the demand for otherwise non-marketable small-diameter wood. Similarly, the community created new local business and employment and improved its image in replacing non-renewable oil-based energy production with bio-wood-based energy. This cooperation was possible because all key players (forest owner, community, inhabitants, forest transportation entrepreneurs and heat producers) had something to gain from the system created. The increasing price of oil improved the price competitiveness of wood-based energy (although fluctuating oil prices are able to alter market conditions markedly). In the other Finnish case (C11), the most important innovations were classified as both business system and process innovations. There were both technological developments, including investments in the efficiency of the pellet production, and service innovations: the company increasingly focused on providing a full-service concept that includes the delivery of pellets, heat production and the maintenance of power plants. These services did not emphasize households as main customers, because the company operated on a somewhat larger scale. The company also exported large amounts of pellets.

In one of the Scottish cases (C13), the most important innovation in recent years was a torrefaction technique (process innovation) that removes moisture from the woodchips in a vacuum. A further innovation was the spontaneous market innovation that emerged after the wood pellet producing process was initiated: the discovery of a major new market for the wood pellets in the 'horse bedding' market.

The Romanian case (C14) was also a combination of process and business system incremental (local) innovation. When it was initiated in 2004, the innovations introduced were of national significance because the technologies that were used, even if they were not the latest in this field, were the most modern in Romania. Examples of the innovations are as follows: (i) new logistical equipment for sawdust collection; (ii) new buffer stores for sawdust; (iii) a new thermal power plant; (iv) a new distribution network; (v) the replacement of the old heating equipment in the buildings' basements with new conduits; and (vi) a new system for thermal energy production based on sawdust, chips or bark. The main results and outcomes of implementing this innovation were: (i) the capacity to produce cheap energy from wood residue; (ii) a new price for the energy produced in this power plant, which was half the average price of energy in the country; (iii) the self-evident ecological benefits; and (iv) the improvement of the region's image, which became more attractive to tourists.

As stated above, most of the innovations in the Romanian case (C14) were process and business system innovations. In the Polish pellets case (C5), we also discovered product innovations that were new to the firm and a local market. The direct stimulus for establishing the firm was the possibility of utilizing wood waste in the form of pallets and sawdust. The oversupply of wood pallets and other waste wood was the basic reason for initiating the production of pellets. The waste wood was chipped and then pelletized, with a portion of the chips sold to a chipboard manufacturer. The CEO argued that the factors that were conducive to innovative activities in wood pellet production were ecological awareness in society and local demand. Local demand was enhanced because of the presence of areas without the natural gas infrastructure that would make it possible to use natural gas for heating purposes for inhabitants living there. The pellets also proved to be a cheaper fuel than heating oil and coal, and, because of their relatively low ash content, they were regarded as convenient to use.

The French case (C4) was one of product innovation at the organizational level. The case company operates three pulp manufacturing units, producing fluff pulp for absorbent products (nappies, etc.) and kraft pulp. The drivers of innovation are two constraints on the current production: increases in production costs and market pressures on the products. In 2007 and 2008, the company faced tighter fibre supply conditions, with higher costs for raw materials. An additional pressure came from a sharp increase of fossil fuels, leading to higher costs of chemicals and transport. The Law on Water, passed in 2006, raised water costs, which are important in pulp production. Conversely, the company's main market (kraft pulp) is a mature, worldwide market (dominated by multinational companies) that fixes the reference price leading the company to behave as a price-taker. The erosion of prices in 2009 has endangered the business. Moreover, the long-term trends of paper consumption (kg paper/inhabitant) are uncertain and may well decrease. The determinants of demand (the development of the information society, government policies related to the reduction of paper consumption, and competition from other materials) may also decline. These constraints drove the company towards new markets and oriented activities to innovation. In accordance with legal objectives, the public authorities issued three calls for tenders and brought an opportunity to develop the innovation in energy production.

Conversely, the other Polish case (C6) presented a typical process innovation. In the Polish district heating case, the main innovative activities begun in recent years were centred on three topics: new boilers fuelled with biomass that replaced old coal-fuelled ones; a willow plantation established in 2002 (with the initial idea that the firm could cover 60–70% of its heating needs with its own biomass sources); and a straw briquette-manufacturing line, with the straw briquettes sold to a large power station in the western part of the country. Each of the innovations proved successful, except the willow plantation, where major cultivation problems appeared. The replacement of old coal-fuelled boilers with new biomass-fired ones was the most important innovative activity and was directly tied with the willow biomass produced on land owned by the company. The main drivers behind the innovation were rising coal prices, cheap agricultural land with the potential to produce willow biomass and the need to renew the heat-producing infrastructure (boilers). What must be stressed here is the vital role of the company chairman, who acted as an engine and driving force and was able to stimulate the entire company management structure to formulate a new energy supply programme for the town.

15.3.2 Drivers for innovations

Previous research emphasizes that there are many different innovation drivers at different levels. Some of the literature concerns internal and external factors driving firms' innovative behaviour. Related to the case studies analysed in this work, one can distinguish four levels driving innovations: the individual, organizational, inter-organizational and societal levels. Here, the innovation drivers varied from country to country and among firms in each country. The Norwegian case (C1) clearly illustrated the importance of the 'innovation champion': the innovator and the engineer in a small business were the same person. Many actors were involved, but one person (or, at most, a few) turned the innovations into new business opportunities. Ultimately, it appears that several of the external actors were not very supportive. The champion theory is supported by several research findings (Jenssen and Jørgensen, 2004). Other cases (e.g. the district heating company in Poland) also emphasized the importance of the CEO and/or the owner. In at least one case in Scotland (C13), the existence of an 'innovation champion' was highlighted. This individual participated in as many activities as he could (e.g. shows, seminars and fairs around the country), but he recognized that much more could be done 'if there were six of him'.

Environmental scanning is often seen as an important innovation driver (Jenssen and Nybakk, 2009). It encourages CEOs' and other managers' awareness of emerging trends and can often lead to the recognition of innovation possibilities (Auster and Choo, 1994). First, the manager can learn about customers' preferences, demands and willingness to pay for different products and services. The Estonian pellets case company began by producing pellets from round wood with a new chipping and debarking machine. The innovation was mainly based upon the observation of market needs using scanning.

A similar situation was noticed in one of the Lithuanian cases (C3). A small family company doing forest contracting work in wood harvesting, forwarding and hauling saw a developmental chance in producing fuel chips from logging residues. The enterprise was aware of the local logging residue supply because it was a material that they used to leave behind in the clear cut areas where they performed wood harvesting operations. Also, a local heating plant was located in the area, securing a stable demand for woody fuels. Thus, environmental scanning enabled the enterprise to develop by diversifying the product it offered.

Secondly, CEOs and managers can learn about and be updated on technological changes and competitors. Companies in other regions or countries are more likely to have different information and ways of solving technical problems. In the Estonian wood chip and fuel wood case (C7), the ideas came from a visit to similar plants in Nordic countries. In one of the Scottish cases (C13), the CEO and a forester employee developed contacts with boiler builders, communicated with other people and conducted horizontal networking in a very practical way. In the other Scottish case (C12), the owner visited wood fuel plants in Germany, Northern Ireland and England before building his own plant. In procuring the plant, he made inquiries to several manufacturers. One particular engineering firm was especially useful because it immediately offered sensible advice and suggestions for the machinery that should be purchased.

Thirdly, the social and economic environment is changing, as are public regulations and support programmes (e.g. Auster and Choo, 1994). To gain a competitive advantage, it is therefore important to to systematically scan a company's surroundings for signs of possible opportunities and threats arising from new policy, which could reduce costs for new entrants or provide funding for new initiatives. Today, with the ongoing debate on how to promote bio-energy to reduce CO_2 emissions, this type of scanning is especially important. For example, in Scotland, the Climate Change (Scotland) Act 2009 created a statutory framework for greenhouse gas emissions reductions in Scotland by setting an interim 42% reduction target for 2020, with the power for this figure to be varied based on expert advice, and an 80% reduction target for 2050 (Part1). Similarly, in Lithuania, state promotion of wood heating plants created demand and local markets for woody fuels, indirectly helping woody fuel production chains to develop.

Organizational-level innovations (emerging trends, updates to technology and scanning the social and economic environment) were not the emphasis in the Finnish cases, though we directly asked about this element. In all likelihood, the managers and developers of the new bio-energy business did pay implicit attention to changes in the business environment (i.e. the managers and developers of the companies were so involved in developing bio-energy that they could not even recognize it).

15.4 External Relations in the Innovation Process

As described earlier, innovation is an interactive process within a company or among companies, organizations and other actors. These ties can be both weak (with occasional interactions, e.g. casual acquaintance) and strong (which implies more regular interactions within networks). The strength of a tie depends on factors such as trust and friendship. Stronger ties often go beyond short-term market transactions and include more long-term and stable partnerships. In Table 15.3, the importance of cooperation between different actors involved in the analysed innovation processes is emphasized.

15.4.1 Strong ties in the innovation process

These network ties, though often informal, are clearly visible in the Polish district heating case. The main actors involved in the principal innovative activity were higher educational institutions (two universities), businesses providing technical advice and sharing their experience, the town authorities, local banks and other partners with whom contact had been established while visiting technology fairs and exhibitions. The variety of organizations involved in the innovative activity resulted not only from the complex nature of the innovation but also from the devotion and charisma of the company chairman. The network that was built was a complex one, with the company's stakeholders and shareholders (the local municipality, the housing cooperative and a few industrial plants) standing at its centre. Vertically, the structure of the network mirrors the material flow. Wood biomass suppliers deliver chips from a number of wood industries to the heating plant where they are converted to heat and delivered to apartment blocks, public buildings and industrial plants. The lateral form of cooperation is much more complex, because it includes higher education institutions providing advice (e.g. on the willow plantation), banks and government organizations providing advice on the possibility of using public funds to pay for the innovation, and other (often small) businesses providing technical assistance, advice and support. As a result of this network, a relatively small district heating company was able to implement successfully a costly innovation and fund it largely using public funding. In the course of achieving innovative activities, a network of formal and informal links was established. The flow of information in the network happened in business and town council meetings, at exhibitions and fairs, and through personal, and often informal, contacts. Books and professional journals were named as important sources of information concerning innovative activities.

In the Norwegian pellet case (C1), the factory was built in connection with the refuse

Table 15.3. Descriptions of networking and the networks' importance for the innovation.

Case company	Network description	Importance for the innovation
C1: Norway: pellets case company	A separate company is buying all of their pellets, does the logistics further down the value chain, and assists with marketing and sales (domestic and exports). Local forest owners provide timber via a forest owners' organization (co-owner). A garbage disposal partner plant provides the heat for drying. Other energy companies are co-owners and, in some cases, customers	Medium/large degree importance: cooperation with the garbage disposal was of special importance
C2: Lithuania: pellets case company	Sawmills provide sawdust. Foreign wholesalers and end-users (90% of production). Association of businesspersons and financial institutions. A specific association for pellet and briquette manufacturers may be beneficial	Relatively low importance
C3: Lithuania: wood chips case company	Harvest residues from state forest company. All sales to local heating plant. Little contact with association. Owner does not believe in networking	Relatively low importance
C4: France: electricity case company	External collaboration with banks (leasing loans) and with the national professional organization of pulp and paper industries	Significant importance with regard to practical feasibility and realization of innovation
C5: Poland: pellets case company	No formal cooperation with any major institutions, authorities or businesses. Vertical chain providing feedback: wood pellet and waste suppliers (the pellet company), pellet buyers (private and small industry). Weak lateral network resulting from contacts with similar businesses (advisory role and exchange of experience). Important role of information flow from sources like wood machinery fairs, journals, the Internet and personal contacts	Low importance. The weak network was not vital for the realization of innovation. Important role of information flows
C6: Poland: district heating case company	Vertical network: local wood industries as biomass suppliers, local heat consumers (private residents, public institutions and businesses). Lateral network: local municipality, housing cooperative and industry, higher education and research institutions – advisory roles. Consultancies and banks provided economic analyses and prepared application documents for public funding. Smaller companies provided technical support. Vital role of personal contacts with science and business representatives. Important role of information sources: visits to fairs and exhibitions, access to trade journals	High importance. The innovation would not have been possible without the network. Numerous problems in the course of the innovative undertaking required cooperation on the part of very different agents
C7: Estonia: wood chips and fuel wood case company	Network for wood procurement: forestry firms, transport and logistics. In cooperation with the Estonian State Forest Management Centre control test regarding potential suitability of forest felling residues for chip production. Buyers of heat and electricity were mentioned as important partners, but not as much as network partners. Suppliers of technology were mentioned, but the contact was not frequent enough for them to be called an actor in a network (this contact was more a normal buyer/seller relationship)	Low importance

C8: Estonia: district heating case company	Network for wood procurement	Low importance
C9: Estonia: pellets case company	Research – University of Technology, equipment manufacturers, other companies, market representatives of end-users in other countries	Clients and marketing channels were important for changing technology
C10: Finland: pellet company	A network of company-owned and other sawmills that provided raw material (mainly sawdust) for pelleting. A network of different companies to increase the size of the plant. A company network to conduct research to improve the quality of pellets and extend its raw material base	Low importance. The main driver has been the company in working to develop its own business
C11: Finland: district heating case company	Network for wood chip procurement and heat service – local forest owners, wood procurement company, municipality and the cooperative, which charged the users of the heating services. The network was originally developed under a special project of the local Forestry Centre (District Government on Forestry), which invited different actors to discuss cooperation and later to agree to the establishment of the cooperative	Significant importance. Without the network, the cooperative would not have been established or been operational
C12: Scotland: wood chips case company	They rely mainly on an informal network of actors. Making contact with as many key players as possible (almost all of whom were private sector operators), visiting existing set-ups and learning from others' mistakes is considered more useful academic research on the subject. They never paid consultants, and their confidence on the expertise of institutions was almost zero	Strong importance at establishment of informal networks built to help scope project
C13: Scotland: pellets case company	The Forestry Commission and the Biomass Development Officer have been very helpful in promoting the enterprise outside and inside the municipal area. They are the core actors in the public sector – Aberdeenshire Council; Chamber of Commerce: Arbuthnott Wood Pellets applied for a business award that has some connections with the Chamber of Commerce	High importance for practical feasibility and realization of innovation
C14: Romania: district heating case company	Scouting and networking: making contact with as many key players as possible. The network is mainly composed of different power plants from the Sawdust 2000 project, environmental agencies, the town municipality, the Romanian government and the Danish Agency for Environment Protection. No collaboration with R&D or universities	Networks are very important. The existence of the analysed Thermal Power Plant is a consequence of network's existence. Networks are used to recruit new active markets

disposal plant. This co-location not only provided a good use for surplus heat but also allowed joint operation with the mutual utilization of the competence and manpower of the two companies (i.e. the waste disposal and pellet businesses). In addition, the forest owners' organization and several energy company co-owners were involved in pellet production.

The CEO in the Scottish pellets case (C13) also emphasized the importance of his network of contacts. The company belongs to the Aberdeen Renewable Energy Group, which has regular meetings. It also belongs to the biomass group that is part of the renewable energy group and is represented in shows, seminars and fairs around the country. The CEO also attempts to participate in as many networking events as he can. Other important actors who help the case company with marketing are the Forestry Commission and the regional Biomass Development Officer (a public sector official). They help to promote the company both outside and inside the municipal area. The raw material for wood pellets is sourced locally from the estate's own forest and a nearby sawmill, from which off-cuts are obtained. Additionally, there are several other important actors from the public sector. Maintaining a large network may be a key to success, but sustaining a large network of contacts can also be costly and very time-consuming. The CEO recognizes that he could do more if he had more time. This lack of time is a major challenge of setting up, running and building networks for an enterprise where one individual is pivotal.

In the French case company (C4), two main network partnerships were created. First, the company established three leases, one for each installation, with banks. The long-term contracts for electricity generation provide guarantees of revenues, reducing the risks associated and facilitating borrowing. The leasing form enables the company then to distribute installation costs over a period at a predetermined interest rate with an option on ownership, rather than a major one-off investment. The second partnership, with the national Paper and Board organization, facilitated resolution of difficulties in collaboration with public authorities, enabling practical implementation.

At the inception of the Norwegian company, different private companies, associations and public organizations had an important role, but the entire process was driven by the inventor and CEO. The Estonian power and electricity production company (C8) generally did not consider suppliers of technology to be network partners, because (for them) it was more a weak, short-term buyer/seller relationship. The forestry companies, suppliers of round wood or chips, were mentioned as network partners, but they were assigned low importance. There was competition in the timber market, and if one supplier was having problems, then the others would be available.

At the inception of the Romanian company (C14), the process was driven more by a system of actors. The project or case company is based upon cooperation between the Romanian government and the Danish Agency for Environment Protection, as well as the Environmental Protection Inspectorates in the different counties involved. The Danish Agency for Environment Protection has been involved in developing and co-financing a large number of projects in Romania, with Neamt County as the specific target area. Sawdust, bark and wood waste had been stockpiled in many places, causing a considerable impact on the natural environment. The new plant reduced local pollution from the wood industry and changed the heating system from a reliance on fossil fuels to the use of bio-energy. In 2000, the project was developed by the Romanian Agency for Energy Conservation, the Danish Agency for Environment Protection and local Romanian agencies for environment protection from different counties, together with a consulting firm with engineers. There were five locations selected. With respect to lateral cooperation, the Danish Agency for Environment Protection and the Romanian government were, collectively, both the initiator and the finance institution on the basis of the Kyoto Protocol. At the same level, other actors were involved: the European Union, which funded 'Sawdust 2000' through a Phare Project; the municipality (as the proprietor and finance institution); and two companies that offered consultancy and coordinated the activities (because they were permanently in contact with the municipality and the Danish Agency for Environment Protection). Vertical cooperation existed both before and after implementation between the municipality and the staff of the power plant to assure the proper functioning of the plant. The Sawdust 2000 project allowed the establishment of five power plants in Romania. Thus, it created an interesting network for horizontal cooperation: the five plants remain connected through their managerial staff. In addition, as part of horizontal cooperation, the plants communicate with environmental agencies and customers that provide the feedback needed for permanent adaptation to new innovations.

15.4.2 Weak ties in innovation process

In contrast to the networks presented above, both Lithuanian cases showed relatively insignificant or underdeveloped networking. In the case of the smaller family enterprise doing forest contracting work and producing fuel chips, this is understandable. However, in the other case (the pellet manufacturer) the network observed was insufficient, and the company's management was aware of it. Although the pellet manufacturing company (C2) was a member of a local business association and established contacts with banks and leasing companies, no consulting or advisory bodies were present in the network. This absence led to a situation in which the management had to gather almost all the necessary information by itself. The CEO expressed a strong need to establish an association of woody-fuel-producing enterprises.

The Polish case (C5) may be regarded as nearly the opposite of the network in the previous case (C2). The pellet manufacturing company, employing only six people, did not build a strong formal network. However, in the course of the delivery of wood pallets and other waste wood to the company, before the innovation to the pelletizing line was initiated, certain vertical ties with suppliers were established that facilitated the information flow. Ties with pellet buyers, however, appeared later when the pelletizing line began operations. A weak lateral network was also noted; the company management, being active in waste wood processing, had contacts with other similar businesses that served as sources of information and advice. Thus, the information flow proved decisive in the process leading to the innovation. Here, different wood-processing machinery fairs, exhibitions and trade journals played an important role in delivering information on technical solutions. Finally, the company was aware of the potential local market for wood pellets before the idea of the innovation was even developed. This awareness indicates that information on the local market demand must have been available to the management, and it was sourced informally through unofficial contacts. Indeed, external relationships are often not fully perceived because they have the character of relationships based on acquaintance, friendship or accidental meetings.

Also, the low importance of the network was highlighted in both Estonian case studies (C7) and (C8) and in the case of the Finnish pellet company (C10), where the companies developed the business and innovations without the support of other actors. Here, the networks are important only in terms of the companies' efficiency regarding the provision of raw material and the maintenance on the market.

15.5 Institutional and Policy Factors Affecting Innovations in Bio-energy

The supporting and impeding actors related to the different cases are presented in Table 15.4. The institutional settings in the different countries differ. The CEOs' views of the policy instruments also vary between companies in the same country.

In the Estonian electricity production cases (C8), political 'lobbying' was needed to change the local legislation Energy Act, wherein electricity production from local

Table 15.4. Positive and negative impact(s) of policy.

Case no.	Supporting factors	Opposing factors
C1: Norway: pellets case company	Subsidy at inception. The grant was an important factor in the endeavour to undertake the project. Subsidies were provided to private households for the purchase of pellet stoves. Subsidies were also provided to trade and industry for investment in pellet stoves	Support to households was not sufficient. Subsidized electric prices in many municipalities. Subsidized addition to the electric grid but no corresponding goals for bio-energy
C2: Lithuania: pellets case company	Open markets enabled the company to develop by exporting the majority of production. Financial support through EU programmes	No direct subsidies were provided, and the policy has not been the main supporting factor. Underdeveloped local market for pellets. No promotion of woody fuels use in households
C3: Lithuania: wood chips case company	The innovation within the company was not supported directly by any political means. However, indirectly, the state promotion of wood heating plants created demand for wood chips. Additionally, the EU programmes supported the purchase of wood chip production equipment	No direct support of investment into wood chip production from logging residues
C4: France: electricity case company	Calls for tender instruments (not direct subsidies like subsidies as part of the investments) but fixed prices guaranteed for electricity produced	Visibility of investment is reduced because guaranteed prices between calls for tenders are different. Uncertainty of investment in R&D activities in future
C5: Poland: pellets case company	Price relations are favourable for woody fuels; prices are a basic tool for supporting RES. Rising eco-awareness in society	Insufficient support to small companies innovating through eco-friendly projects. Still insufficient advertising of wood as a source of energy
C6: Poland: district heating case company	Government subsidies for investments in eco-friendly projects (95% of the investment was in various ways supported by the government). Plantations with energy plants are financially supported by the government. Obligatory renewable energy quotas for companies dealing in energy	Application procedures for public funding too complicated
C7: Estonia: wood chips and fuel wood case company	EU legislation and energy policy: the use and production of renewable energy should be increased; the opening of the electricity market in Estonia (which occurred more slowly than in the rest of the European Union). Estonian energy policy: by 2010, the share of renewable electricity would be 5.1% (in 2007, it was 1.75%). Support from Tallinn City government (is it a political issue?)	In the planning stage, the legislation did not support the idea; another factor was the proximity of the Iru Power Plant, a subsidiary company of the state-owned monopolistic electricity and power production company Eesti Energia
C8: Estonia: district heating case company	Their own management team made a significant effort and took the initiative to change the Energy Act so that local fuels could be used for electricity production. When the law was changed, the opportunity emerged to start planning the CHP. In the current 'Development Plan of the Estonian Electricity Sector until 2018', more attention is paid to energy production from renewable energy sources (including bio-fuels). The Sustainable Development Act in 2005 set some limits on monopolistic activities by Eesti Energia (at that time an electricity producer using oil shale; since 2009, this company also uses wood for electricity production)	The old version of the energy act, which regulated the principles of energy price and return on investment (investment payback) in electricity production. The Earth's Crust Act (earlier, it was important for peat production, with permits from local communities), not important from a forestry point of view. Waste Act: currently the ash from CHPs cannot be used for further production (e.g. construction, road construction)

C9: Estonia: pellets case company	Availability of different support programmes and grants offered by Enterprise Estonia that are co-financed from the EU structural funds. The company studied different programmes. However, during that time, it did not find one that could meet its needs and, thus, it did not use any of them	In end-user markets (in different countries), CO_2 quotas and support programmes for obtaining kettles/boilers that use pellets will create additional demands for pellets
C10: Finland: pellets case company	Public support for R&D: mainly knowledge on pelleting technology. In the future, the extension of the raw material base for pellets will be emphasized in R&D. Investment support for households if they replace oil boilers with pellet boilers	No impeding policies identified, though the slow growth in the number of households using pellets for heating has caused debates regarding whether the policy measures have been sufficient and efficient
C11: Finland: district heating case company	State subsidies for harvesting young stands (the purpose being to support forest growth) and for chipping. Without these subsidies, the chip-based energy would not be price competitive. Public support for R&D: knowledge of small-scale heating ability and technology. A person at the Forestry Centre (district government Office on Forestry) made the network work. He ran the plan and enabled different partners to cooperate. His salary was partly paid by regional development funding for a special project coordinated by the Centre	During the municipality planning stage, the use of pipelines for hot water heating was not obligatory or even an option to consider. Even now, this use is not obligatory, but since the municipalities learned that district heating using wood chips is an option, they have invested in pipelines
C12: Scotland: wood chips case company	Grant aid for biomass boilers was available	Weak support from the state authorities that comprised the official support system. Regulatory and planning environment limiting the renewable energy options of the estate
C13: Scotland: pellets case company	The funding for the plant was 40% grants from the Scottish Biomass Support Scheme and 60% from the company's own capital. A policy 'window' was open for just a few months, but AWP managed to capitalize on it to provide support for setting up the plant	Needed licence to handle the waste (sawdust) produced
C14: Romania: district heating case company	Danish Agency for Environment Protection, the initiator and finance institution (28% of the total cost). EU support through the Phare Project (36% of the total cost). Romanian state institutions: the necessary legal framework was created for the implementation of the Sawdust 2000 project. The Romanian government, through the Romanian Agency for Energy Conservation and the town Municipality, supported 36% of the total cost of the project	No opposing factors identified

(bio)fuels (including wood and peat) was not allowed. In this way, the state protected its own electricity production monopoly. After changes to the Energy Act, the final planning of combined heat and power plants by private businesses could be initiated. When in the Estonian case enterprises began electricity production from wood, the state-owned company also started to use wood in addition to oil shale (in 2009). After that change in the local timber market, the biggest forest management organization, the State Forest Management Centre, created a new subdepartment: the Wood Energy Division.

In the French case (C4), the public authorities issued three calls for tenders for electricity generation from biomass in respect to the national objectives for increases in renewable energy sources (RESs). The national objectives are reinforced by the latest European and national legislation. The revised directive for the promotion of energy from RESs establishes an objective for France of 23% of energy being produced from RESs in 2020, strengthened by the new national legislation increasing the objective of electricity production from biomass (MEEDDAT, 2009). The calls for tenders created an opportunity for business, ensuring the sale of electricity produced at a predetermined price for a certain period (15–20 years). It provides guarantees on future revenues and increases the security of the investments. In electricity generation, the case company possesses both the technical skills necessary to run the process and opportunities in internal processes with chemical pulping. In this context, the calls for tender brought opportunities to realize the constraints.

The stimulating role of policy in supporting innovative activities that led to the implementation of clean technologies (e.g. heat production from biomass) can be clearly noted in the Polish district heating case. New biomass-fuelled boilers that replaced the old coal-fired ones were funded up to 95% with public money set aside by the government in the form of different eco-funds for eco-investments. Various policies that are often very general and only set directions for future development usually result in more specific documents specifying rules and financial means and measures that support the activities described in those policies. Such a situation enabled the district heating company to make use of financial means of eco-investment, thus contributing to the realization of the eco-policy. Using these methods of funding is often perceived as difficult, because the application procedures are complicated and require experience and knowledge. With this in mind, the importance of building networks, including financial consultancies, cannot be overestimated. Another stimulating instrument that had its origins in the policy and was specified in greater detail in resulting legal documents was the financial support for the willow plantation. Growing willow, which is considered an energy plant, entitles the company or farmer who wants to establish an energy plantation to certain financial assistance from the government. This measure encourages firms and formers to set up energy plantations, thus contributing to innovative activities. Yet another factor that contributed to the successful performance of the district heating company was the obligatory renewable energy quota that has to be met by all firms that deal in energy. This quota results indirectly from the energy policy. On one hand, this obligation requires solutions and innovations related to the use of renewables among firms dealing in energy. However, it also provides them the opportunity to sell their bio-energy production to energy giants, for example, who simply have to meet their obligatory quotas, leading to better financial results for the innovator. This dynamic can also be perceived as indicating the stimulating role of centrally adopted measures and policies.

The Polish pellets company case (C5) was not able to use public funding, because its application was not strong enough to secure such funding. When asked about political factors supporting innovative activities in the bio-energy sector, the CEO mentioned the favourable price relationship between woody and fossil fuels and the rising eco-awareness of the public at large. This eco-awareness could still be increased if proper advertising activities promoting woody fuels were used. Another opposing factor that was indicated was the insufficient financial support to small firms innovating in the bio-energy area.

15.6 Discussion, Implications and Future Research Needs

A summary overview of the findings from the case examples is presented in Table 15.5. The case companies can be classified into pellet or chip producers and electricity and/or heating companies.

Most bio-energy companies in Europe, and most cases in this study, are SMEs. Several studies show that SMEs, in general, are less innovative than larger companies, but there is considerable variation among them (Kaufmann and Tödtling, 1999; Asheim *et al.*, 2003). However, there are also studies that indicate the opposite (Acs and Audretsch, 1988) and studies that find no differences in innovation related to company size. Some of the differences can be explained by varying definitions of innovation. Indeed, certain scholars have a more radical view of innovation and others a more incremental view.

Another complex relationship is the effect of innovativeness on performance. Many studies indicate a positive effect (e.g. Damanpour *et al.*, 1989; Deshpandé and Farley, 2004). Nybakk *et al.* (2008, 2009) studied forest owners and nature-based companies and found a significant relationship between innovation/innovativeness and performance. Thus, even if there is less innovation among SMEs and one-man firms, the importance of that innovation for economic performance can be just as significant. A last point is that there are considerable differences between industries. This result must

Table 15.5. Summary of the cases and findings.

Aspect	Pellets, chips	Electricity, heating
Innovative activities	Incremental improvements of autonomous processes Widening applications Business model	Systemic technological innovations
Size of firms	Smaller in comparison	Larger in comparison
Dominating drivers	Broader drivers are similar: environmental concerns, relative cost advantage, etc.	
	Internal (personal) initiative and knowledge	Combined initiative: internal (personal), network, government
Supporting networks	Mainly direct business networks: • Supplier involvement (equipment, other inputs) • User involvement (new applications)	Business networks: • Supplier involvement (cooperative procurement channels) • User involvement (local acceptance) Broader cooperation with (municipal) government offices, professional associations, banks and higher education institutions
	More informal networks	More formal networks
	National differences: network failures	National differences: network failures
Public policy	Subsidies for technology and R&D	Support more through regulations of price, quantities, etc.
	Effects of existing policy measures	Effects of both existing and anticipatory measures/legislation
	National differences	National differences

be taken into consideration in the presentation of the different innovations in Table 15.2. Most of the innovations described in this study can be considered incremental. These innovations are new to the country (or perhaps only new to the firm) but do not mean that that they are of lesser importance for economic performance and growth.

Our study demonstrates that innovation champions are often very important for the realization of projects. The innovation champions may be employed at either private or public organizations or could be self-employed. Most projects are complex and require the power and initiative of a champion. From a policy perspective, an important issue is how the institutional framework for the establishment and operation of bio-energy plants can be made as simple as possible. This framework would create space for champions to operate more effectively. It should be stressed here that champions, often possessing not only innovative charisma but also good interpersonal skills, are usually very effective, even under adverse conditions. A general economic and legal framework can be more or less favourable to innovativeness, but even the best possible conditions can prove unhelpful when there is no personal leader who is keen on driving innovation forward.

Regardless, such an innovation champion does not exist in a vacuum. Several of the case studies indicate a positive relationship between external relationships and innovation. This finding is consistent with different traditions in the management and economics literature (e.g. Granovetter, 1973; Burt, 1992; Jenssen, 1999; Weiss and Rametsteiner, 2005; Nybakk *et al.*, 2008, 2009). Many of the bio-energy companies are small and do not normally innovate alone but rather do so in cooperation with others. Networking can contribute to innovative capacity and innovativeness among small firms by giving companies novel ideas and access to resources, as well as by transferring knowledge (Nybakk *et al.*, 2009). Accordingly, small firms that invest in networking with local actors will obtain an advantage via exposure to new ideas, by concentrating on core expertise and by finding new and better ways to run their businesses. External relationships and a larger social network are also linked to learning. A larger network improves the firm's understanding of its environment, resources, markets, customers and suppliers. This network fosters the implementation of new ideas, new products and new ways of running the business, and is consistent with both organizational management literature (e.g. Calantone *et al.*, 2002) and systems of innovation literature (e.g. Lundvall, 1992; Isaksen, 1999). Inter-organizational factors and networks are also of great importance among small wood bio-energy companies. An interesting research topic would be to examine the role of consultants, industry organizations (e.g. bio-energy associations), environmental organizations and business networks in creating framework conditions. By developing new policy instruments to promote networking and clustering in rural regions in Europe, policy makers can help to develop innovativeness among bio-energy companies.

Consistent with the industry structure of bio-energy companies in Europe, most of the case companies are small-scale. For these small companies, few significant organizational factors influencing innovation were identified. For example, most of the firms would not have their own research and development (R&D) departments or an employee working full-time on R&D. However, the organization of a small family business does present different challenges. Nevertheless, none of the drivers of innovation was linked to the cooperation within the owner's/managers' family in the present study.

Policy and policy measures played an important role in all cases. Energy policy is one of the most important policy fields for all governments, and there are numerous policy measures and public programmes concerning all energy systems. One observation is that policy measures are said to be decisive in many of the cases. To study the role of policy measures in more detail, one would have to consider each case in greater depth, but it seems that countries with an explicit policy regarding alternative energy have had greater success with bio-energy innovations. Furthermore, a tax on CO_2 emissions would make substitutes less competitive and would be important to increase the use of bio-energy. Environmental policy drivers are obviously important, with policy regarding the reduction of GHG emissions (replacing coal firing or oil prices) especially essential. This policy is part of the rhetoric in many cases, but for there to be a real impact on decision making, the rhetorical arguments must also translate into policies, such as taxes on fossil fuels or capital and ongoing support for bio-energy (e.g. subsidies and calls for tenders). Another important factor is the supply of cheap raw materials. In some cases like Romania (C14), the abundance of raw material constituted an environmental problem and was a very strong driver of bio-energy innovation.

A range of other factors emerged as important in different cases. Price development and expected price development of substitutes obviously play an important role because energy markets are volatile and future price development paths are difficult to foresee. Technological development was important in some cases, but until recently, more research money has been used for conventional energy sources than for bio-energy. Thus, it is difficult to foresee if there will be any major innovations in the near future concerning bio-energy systems. In many of the cases, the bio-energy solutions were quite new to the market, markets were poorly developed and many actors that did not usually cooperate had to cooperate to arrive at a solution. All of this indicates high transaction costs. Another indicator for high transaction costs is relatively long planning times for the projects. Comparing transaction costs for bio-energy investments in different countries would be an interesting future research topic.

Much rhetoric is deployed at the European Union level about a 'level playing field' for European producers, but, in the case of European wood energy production, this does not exist. Some countries have provided substantial policy support for wood energy over a long period. In other cases, positive policy developments have been more recent. This study did not explore the policy means in great detail, but it can be concluded that the conditions for innovation created by policy vary widely.

The wood energy developments explored are predominantly, but not exclusively, small-scale. Some larger-scale operations were examined in Romania and France based on the exploitation of waste or by-products. It seems probable that wood energy production will continue to embrace both small firms operating in predominantly local markets and much larger undertakings. This creates space for different types of innovation, but in most cases, the innovation will be incremental rather than profound and based on process and supply chain development. Innovation in the policy environment can also be an important supporting factor. Indeed, it is the unwillingness of those in the policy community to put a price (or the right price) on carbon emissions that is probably the biggest single obstacle to the development of bio-energy from wood. However, the widespread availability of low-grade wood products, wood waste or waste heat from paper-making creates an opportunity for bio-energy production that has been exploited, more or less successfully, by the operators of our case studies.

In summary, although supply availability was occasionally instrumental in stimulating innovation, the decisive factor in wood heat developments is more often changes in the market or policy environment and a champion who is prepared to make the effort (either as a private or a social entrepreneur) to develop local projects. In general, the innovations observed were incremental rather than fundamental but, in sum, represent a significant development in wood energy systems.

References

Acs, Z. and Audretsch, D. (1988) Innovation and firm size in manufacturing. *Technovation 7*, 197–210.

Asheim, B., Isaksen, A., Nauwelaers, C. and Tötdling, F. (2003) *Regional Innovation Policy for Small–Medium Enterprises*. Edward Elgar, Cheltenham, UK.

Audretsch, D.B. and Feldman, M.P. (1996) R&D spillovers and the geography of innovation and production. *The American Economic Review* 86, 630–640.
Auster, E. and Choo, C.W. (1994) How senior managers acquire and use information in environmental scanning. *Information Processing & Management* 30, 607–618.
Burt, R. (1992) *Structural Holes. The Social Structure of Competition.* Harvard University Press, Cambridge, Massachusetts.
Burt, R. (1997) The contingent value of social capital. *Administrative Science Quarterly* 42, 339–365.
Calantone, R.J., Cavusgil, S.T. and Yushan, Z. (2002) Learning orientation, firm innovation capability, and firm performance. *Industrial Marketing Management* 31, 515–524.
Damanpour, F. (1991) Organizational innovation: a meta-analysis of effects of determinants and moderators. *The Academy of Management Journal* 34, 555–590.
Damanpour, F., Szabat, K.A. and Evan, W.M. (1989) The relationship between types of innovation and organizational performance. *Journal of Management Studies* 26, 587–601.
Deshpandé, R. and Farley, J.U. (2004) Organizational culture, market orientation, innovativeness, and firm performance: an international research odyssey. *International Journal of Research in Marketing* 21, 3.
Dieperink, C., Brand, I. and Vermeulen, W.J.V. (2004) Diffusion of energy-saving innovations in industry and the built environment: Dutch studies as inputs for a more integrated analytical framework. *Energy Policy* 32, 773–784.
Duduman, G. and Bouriaud, L. (2007) Managing innovations in wood harvesting and primary processing firms – case study of Suceava. *Forestry Studies* 48, 89–101.
Edquist, C. (ed.) (1997) *Systems of Innovation: Technologies, Institutions and Organizations.* Pinter/Cassell Academic, London/Washington, DC.
Edquist, C. and Johnson, B. (1997) Institutions and organizations in systems of innovation. In: Edquist, C. (ed.) *Systems of Innovation: Technologies, Institutions and Organizations.* Pinter/Cassell Academic, London/Washington, DC, pp. 41–63.
Freeman, C. (1987) *Technology Policy and Economic Performance: Lessons from Japan.* Pinter, London.
Granovetter, M. (1973) The strength of weak ties. *American Journal of Sociology* 78, 1360–1380.
Hansen, E., Korhonen, S., Rametsteiner, E. and Shook, S. (2006) Current state-of-knowledge: innovation research in the global forest sector. *Journal of Forest Products Business Research* 3, 27.
Hovgaard, A. and Hansen, E. (2004) Innovativeness in the forest products industry. *Forest Products Journal* 54, 26–33.
Isaksen, A. (1999) *Regionale innovasjonssystemer [Regional Innovation Systems].* Oslo.
Jenssen, J.I. (1999) *Entrepreneurial Networks. A Study of the Impact of Network and Resource Access on the Start-up of New Organizations.* Norwegian School of Economics and Business Administration, Bergen.
Jenssen, J.I. and Jørgensen, G. (2004) How do corporate champions promote innovations? A literature review. *International Journal of Innovation Management* 8, 63–86.
Jenssen, J.I. and Nybakk, E. (2009) Inter-organizational innovation promoters in small, knowledge-intensive firms. *International Journal of Innovation Management* 13, 441–466.
Kaufmann, A. and Tödtling, F. (1999) *SME Innovation and Support in Upper Austria.* Department of City and Regional Development, Vienna University of Economics and Business Administration, Vienna.
Lundvall, B.-Å. (ed.) (1992) *National Systems of Innovation: Towards a Theory of Innovation and Interactive Learning.* Pinter Publishers, London.
Madlener, R. (2007) Innovation diffusion, public policy, and local initiative: the case of wood-fuelled district heating systems in Austria. *Energy Policy* 35, 1992–2008.
Mahapatra, K. (2007) *Diffusion of Innovative Domestic Heating Systems and Multi-story Wood-framed Buildings in Sweden.* Mid-Sweden University, Östersund.
MEEDDAT (2009) Grenelle Environnement: 50 mesures pour un développement des énergies renouvelables à haute qualité environnementale (accessed 01 February 2001).
Mintzberg, H. (1979) *The Structuring of Organizations.* Prentice-Hall, Englewood Cliffs, New Jersey.
Nybakk, E. (2009) *Innovation and Entrepreneurship in Small Firms: The Influence of Entrepreneurial Attitudes, External Relationships and Learning Orientation.* Norwegian University of Life Sciences, Ås.
Nybakk, E., Vennesland, B., Hansen, E. and Lunnan, A. (2008) Networking, innovation, and performance in Norwegian nature-based tourism. *Journal of Forest Products Business Research* 5, 1–26.
Nybakk, E., Crespell, P., Hansen, E. and Lunnan, A. (2009) Antecedents to forest owner innovativeness: an investigation of the non-timber forest products and services sector. *Forest Ecology and Management* 257, 608–618.

Porter, M. (1990) *Competitive Advantage of Nations*. MacMillan, London.

Rametsteiner, E. and Weiss, G. (2006a) Assessing policies from a systems perspective – experiences with applied innovation systems analysis and implications for policy evaluation. *Forest Policy and Economics*, 8, 564–576.

Rametsteiner, E. and Weiss, G. (2006b) Innovation and innovation policy in forestry: linking innovation process with systems models. *Forest Policy and Economics* 8, 691–703.

Rametsteiner, E., Weiss, G. and Kubeczko, K. (2005) *Innovation and Entrepreneurship in Forestry in Central Europe*. European Forest Institute, Joensuu.

Rametsteiner, E., Weiss, G., Ollonqvist, P. and Slee, B. (2010) *Policy Integration and Coordination: the Case of Innovation and the Forest Sector in Europe*. OPOCE, Brussels.

Roos, A., Graham, R.L., Hektor, B. and Rakos, C. (1999) Critical factors to bioenergy implementation. *Biomass and Bioenergy* 17(2), 113–126.

Schön, D.A. (1963) Champions of radical new inventions. *Harvard Business Review* 41, 77–86.

Staffan, J. and Bergek, A. (2004) Transforming the energy sector, the evolution of technological systems in renewable energy technology. *Industrial and Corporate Change* 13, 815–849.

Tidd, J. and Bessant, J. (2009) *Managing Innovation: Integrating Technological, Market and Organizational Change*, 4th edn. John Wiley, Chichester, UK.

Weiss, G. and Rametsteiner, E. (2005) The role of innovation system in non-timber forest products and services development in Central Europe. *Economic Studies* XIV, 23–36.

16 Policy and Market-related Factors for Innovation in Forest Operation Enterprises

Laura Bouriaud,* Edgar Kastenholz, Lukáš Fodrek, Zbigniew Karaszewski, Piotr Mederski, Thomas Rimmler, Arto Rummukainen, Liana Sadauskiene, Jaroslav Salka and Meelis Teder

Abstract

Micro-, small and medium-sized enterprises (SMEs) carry out most of the forest harvesting activities all over Europe. They often act as service providers only and do not own land or wood-processing capacities. They are under pressure from the bargaining power of the wood suppliers (the forest owners, regulating the access to the resource) and the bargaining power of the customers (timber-based industries, regulating the access to the global markets). Innovation in technological development or in business practices is required to improve the situation of forest service enterprises in a difficult market. This chapter studies eight cases of innovation, which have been identified in micro-, small and medium-sized firms in the forest harvesting sector in central and eastern Europe. The aim is to identify the market and policy-related conditions of innovation. It was asked whether there is a relationship between the innovation-related policies and the market in which the firms operate and innovate. It was also of interest to explore if any market-based mechanisms have helped the firms to implement innovation, and if any policies created enabling conditions for innovation. The results suggest that most innovation policies have not yet led to a considerable uptake of innovation among forestry contractors. The innovations that could be identified are mainly incremental applications of already existing technical solutions, and in some cases departing from and/or embarking on additional tasks and new product markets. These innovative activities can be characterized as reactions to market developments and customer demands. There is, moreover, a need for supporting investment in technical innovation in forest operations; for reacting to the increasing demand for harvesting and transport capacities; and for curbing the forecasted lack of workers in many countries. At this stage, a higher degree of mechanization cannot be achieved without public support of investments. An even more important political goal should be to build up or maintain the capacities for support, training and commercial advice of SMEs to develop further the human resources and entrepreneurial capacities needed.

16.1 Introduction

Most forest operations in Europe are carried out by micro-, small and medium-sized enterprises (SMEs), which are firms with up to ten workers or, respectively, 25–250 workers according to EU Recommendation (2003). They typically offer services in harvesting or in other forest operations such as silviculture, forest improvement, biomass harvesting, road

* Corresponding author.

transportation and ancillary forestry-related services. Statistical information or scientific studies about this sector are rare. A rough estimation shows that some 50,000 mainly small enterprises employ a workforce of about 250,000 people, which today forms the bulk of the professional forest workers in European forestry (ENFE, 2008). A huge difference can be observed in the degree of mechanization of forest harvesting. It greatly varies from 97.5% in northern Europe, to 72% in western central Europe, down to 3% in eastern central Europe. In areas where mechanization is used, the average age of the machines rises from the north to the east: from an approximate average of 4 years of age in Finland to some 15 years of age in eastern Europe, where even timber lorries that are older than 40 years can be found (COMFOR, 2008).

However, SMEs in forest operations all over Europe have a number of issues in common. Contrary to the ambitious goals for the development of rural areas and the sustainable use of forest resources, policy makers do not seem to pay much attention to the precarious situation of most of the forest operation enterprises in Europe. These businesses struggle, their margins being very low. Because of their size, their low capacities for investing in technologies, and so their limited or weak participation in technology transfer, SMEs are the weakest and most vulnerable link in the global and internationalized forest-based value chains. Furthermore, they are more and more under economical constraints with the increasing requirements to undertake environmentally friendly forestry work (Brogt *et al.*, 2007). In addition, small sole traders may raise a serious problem in terms of competition to larger enterprises because they can dramatically lower the price level of services (Austroprojekt, 2008).

The regulatory framework of forest management and of forest harvesting activities entails high transaction costs (e.g. the fragmentation of forest operations hinders the economies of scale, yearly contracts increase uncertainties, natural hazards, including weather conditions, increase the risks, etc.). Neither forest owners nor industrial processing companies are willing to deal with such costs and they convey them down the chain towards the forest contractors. The sector is usually subject to high competition and to pressure on service costs. Timber procurement markets are often based on yearly contracts and the policies for timber procurement may change from one season to another, hampering investment prospects.

Moreover, looking at the forest-based value chain, contractors are tied in between the forest owners, the timber trade and wood processing industries (Westermayer, 2006). They are continuously under pressure from the bargaining power of the suppliers of the raw material (the forest owners, regulating the access to the resource) and the bargaining power of the customers (timber-based industries, regulating the access to the global markets). With increased business competition, innovating in technological systems or in business practices could help SMEs to improve their situation and the forest-based value chain as a whole. While the business of micro-, small or medium-sized forest contractors is precariously balanced, gaining competitive advantages through new technologies or services is not only a matter of profit or gaining markets; it is a matter of survival.

Innovation, in general, has a particular importance because it represents a major response to intensifying competition by enhancing the learning abilities of firms and workers (Lundvall and Borrás, 1997). Innovation is more challenging in SMEs than in larger firms. SMEs generally have fewer resources at hand and spend less on research and development (R&D), they are also faced with more uncertainties and barriers to attain innovation. At the same time, relations to science and technology transfer are rare and SMEs make only a limited use of the full potential of their respective regional innovation systems (Tödtling and Kaufmann, 2001). SMEs rarely interact with universities, contracting research organizations or technology centres and training institutions (Kaufmann and Tödtling, 2002). To innovate, SMEs rather use tacit knowledge, get support from local networks and common learning, local institutions and resources (Malmberg and Maskell, 1999; Lewark *et al.*, 2010), or they rely on customers in directing their innovations (Tödtling and Kaufmann, 2001). Some studies also reveal the importance of learning by doing while innovating among service performing firms, which differs from innovating in manufacturing firms (Pires *et al.*, 2008). Thus, SMEs can be innovative without having a specialized R&D department, and, more generally, without strong connection with R&D infrastructures.

This chapter focuses on the development of innovation in forest-related SMEs. The factors, or mechanisms, that facilitate innovation within the firm are market-related and policy-related, according to a classification inspired from the Porter's five-force model (1979). A change in any of these forces (threat of substitute products, threat of established rivals, threat of new entrants, bargaining power of suppliers and bargaining power of customers) normally requires a business unit to re-assess the marketplace. This represents a key moment for seizing innovation opportunities. Two main questions guide the study of innovations in the sector:

- Is there a relationship between the market in which the firm operates and the innovation? Which market-based mechanisms have helped or pushed the firm to implement the innovation?
- Is there a relationship between the policies fostering innovation and the innovation in the firm? To what extent have policies helped or created enabling conditions for innovation?

In the attempt to explain how and why innovation appears, this chapter presents eight cases of innovation in micro-, small and medium-sized firms performing forest operation services in Germany, Lithuania, Poland, Slovakia and Estonia.

16.2 Analytical Frame

According to Tödtling and Kaufmann (2001), innovation implies 'modifications and improvements of products or services, changes in a firm's range of products or services, as well as adopting or developing new technologies'. Innovation is also identified as the implementation of a new or significantly improved physical or service product, a technological process, a new marketing method, or a new organizational method in business practices, workplace organization or external relations (OECD, 2005). Because of the particularities of the SMEs in forest contractors' work, innovations in this sector are seldom new to the market as a whole. In this study, innovation is defined as any change in the firm's range of products or services or any technological development that will have a positive impact on the firm's daily business. Such a wider interpretation of the concept of innovation, which still stands in the frame of Tödtling and Kaufmann (2001) and OECD (2005) definitions, draws the attention to examples of innovation that may be considered as well as merely strategic changes when firms are affected by market conditions. But in the strong competition and regulatory frame of the sector, strategic orientation may ensure competitive advantage and market survival.

Innovative activities of SMEs can be related to products, processes, research activities and resources, especially human resources (Tödtling and Kaufmann, 2001). To ensure the successful implementation of a new innovative idea, the concept should, first of all, be feasible, operational and possible to implement in practice. Secondly, the concept should be viable from an economic point of view. Finally, a successful innovation requires personal involvement, effort and trust in its successful outcome (Drucker, 1998).

The driving forces behind innovation can be the rent-seeking behaviour of the owners, who invest in innovation with the aim of making a profit, but also the policies that create innovative opportunities and that foster the capability of firms to exploit new business opportunities and enhance competitive advantages. Very often, developments depend on the financial contributions of the owners, but also on policy (Kaufmann and Tödtling, 2002; Lundvall and Borrás, 1997). As Drucker (1985) states, sources to innovate arise from different challenges and opportunities both within and outside the firm. What should be retained is that high-impact innovations are the result of a 'discipline' to innovate – a continuous effort inside the firm to look for innovation, yet sometimes the innovation requires years to yield results (Drucker, 1998).

An explanation of how innovation occurs in an SME is offered by the interactive model of innovation. Innovation is conceived as a process of *interactive learning* in which a wide array of institutional mechanisms can play a role (Lundvall, 1992). Learning occurs in the relationships between the firms and the basic science infrastructure, between different functions within the firm, between producers and users at an inter-firm level and between firms and the wider institutional environment. In this context, the market position of a firm, new technology, new entrants and industry competition, learning and adapting inside the firm all lead to innovations (Porter, 1979). Tödtling and Kaufmann (2001) describe four main characteristics of the interactive model of innovation:

- Innovation is considered an interactive process: inside firms and with other firms and organizations.
- Knowledge is the most important resource for innovation, brought forward by R&D, marketing, distribution, production and other activities.
- There are various kinds of actors involved in the innovation process that interact in innovation systems, mainly: customers, suppliers, competitors, service firms, universities and research organizations, technology centres and transfer organizations, finance and training organizations.
- The exchange of tacit knowledge is particularly favoured by face-to-face contacts and by spatial proximity.

Porter's five-force model (1979) has been used to figure out the relation between the firm, policies and markets. Among the many potential factors for innovation, the market position of the firm, the business and cooperation network and industrial relationship, on the one hand, and policies, on the other hand, have been analysed.

The market position of the firm is defined by customers, competition and the position in negotiations for prices, participation in decisions and participation in the wood supply chain.

Policies are considered under the aspect of general existing policies relevant for enterprise development, and policies that have specifically helped or hindered the implementation of the innovation in the firm.

16.3 Methods

To identify the driving forces of innovation, this chapter uses a case study approach. The description of the innovation itself and the process of the innovation are put in the context of the characteristics of the firm in terms of market position, human resources, technical equipment, social relations in the enterprise and policies affecting the development of the firm. Timber procurement services for industrial users are the most important single markets for forestry contractors, whose activities are led by a business-to-business logic frame. Therefore, enterprise characteristics, technology development, market position, barriers to market entry and the role of institutions have been examined especially from the point of view of timber procurement markets.

An exploratory case study approach (Yin, 1994) was considered suitable to understand how innovation is implemented and how it appears in the firm. Different geographical and policy contexts are described in the case studies. The data collection protocol was established in 2008 and implemented during 2009. The protocol contains procedures and general rules such as:

1. Overview of the case study investigation (objectives, issues, topics being investigated).
2. Field procedures (rules of the interviews, sources of information).
3. Case study questions (specific questions to be asked).
4. A guide for case study report (Yin, 1994).
The unity of the analysis is centred upon innovation in the firm. The innovation needs to be new to the firm and to comply

with the definitions provided by Tödtling and Kaufmann (2001) and OECD (2005). The guidelines for the case studies take into account the selection of the firms (and, if possible, the innovations) that should be as typical as possible for the forest sector in the country. The data collection has been structured into four main parts:

1. Description of the enterprises, their business environment, changes ongoing therein and the resulting pressure for the firms to innovate.
2. Information about the situation of typical forest operation enterprises. The description covers the role and position of typical enterprises in terms of activities, resources, social relations and sources of competitive advantage, their position in the market and their role in the wood supply chain.
3. Policies fostering innovation, as mentioned by the representatives of the firms in the data collected.
4. Description and evaluation of an innovation process in terms of fostering and impeding factors and policy support.

The information about the innovation case was collected by semi-structured interviews with owners of the firms or other representative and, in one case (Germany), as an expert assessment based on a number of interviews and enterprise observations. The interview addressed standard questions related to the given protocol and data collection. All information provided during the course of the interviews was, however, recorded. Thus, as the study of the innovation among forest contractors is only in an incipient phase, it was essential to collect as much information as possible about past or ongoing innovation processes, keeping in mind that the case studies should offer the opportunity to maximize what could be learned in a limited time period (Stake, 1995).

16.4 Results

16.4.1 Overview of the case studies

The firms that were analysed are different in size, yet all except one are still within the range of micro- and small-sized enterprises (Table 16.1).

Apart from two exceptions, the firms operate on local or regional markets. Access to the market is mostly tender based, with 1-year-maximum contracts.

Policies fostering innovation in the case studies were based on the entrepreneur's

Table 16.1. Key aspects of the studied innovation cases.

Case study	Size of the enterprise	Access to the market	Type of market
Entering bio-fuel harvesting (GE)	4–5 employees, and the owner (family)	More and more tender based	Local, but expand because competition
Purchasing a forwarder (PL)	5 workers	Tender, state forests	Local
Purchasing a harvester (LT)	24 and the owner	Tender, state forests	Local, around 50 km
Custom-made logging tractor (SK1)	Division of bigger firm, 20 employees	Tender, negotiations, salvage timber market	Regional, all forests, and abroad
Packing timber for conservation (SK2)	Division of bigger firm, 36 employees	No tender for packing timber	National, military forests, abroad
Purchasing a logging tractor using subvention system (SK3)	22 employees, plus 15 sole traders	Tender, negotiations	Local and regional, occasionally abroad
New uses of the harvester (EE1)	10–12 employees plus 10–12 sole traders	Tenders in state forest, negotiation	Local, 30–50 km
Use of forwarders out of season (EE2)	4–5 permanent and sole traders as needed	Tenders, negotiations	Regional (Estonia mainland)

perception of the issue. In general, the firms' representatives were not aware of policies fostering innovation except when they applied for financial support in the frame of specific policies. However, at the EU level and then at the national level, some policies may stimulate innovation in the forestry contractors sector, e.g. the European Regional Development Fund – Regional Operational Programme, or EU Rural Development Policy 2007–2013, energy efficiency policy for increasing the supply of energy from bioenergy sources by 20% until 2020, or forest/nature conservation-specific policies to enhance biodiversity protection.

16.4.2 Description of the case studies

Case study 1 – Germany: Entering bio-fuel harvesting (GE)

The German case describes a typical forest harvesting enterprise for the region. A number of interviews were carried out in previous studies by researchers at the University of Freiburg. The main activity is timber harvesting and transport (skidding) to the roadside. Mechanized systems have become the leading system. Silvicultural works (planting, clearing, pre-commercial felling) are carried out but represent a small share of the overall turnover. The enterprise employs four or five workers. Occasional higher workload is handled by using subcontractors or partner enterprises. All members of the enterprise are able to work both on machines and motor-manually. However, the harvester operators and forwarder operators do not rotate. Management is the task of the owner. Office work is mainly carried out by the owner's spouse. The main equipment is a harvester–forwarder system. The enterprise has two harvesters (one new, one used) and one forwarder. When one of the machines is written off and when it is replaced, the most recent technical generation will be bought. If in the past machinery was mainly one's own capital, more and more machines are financed by bank loans. New investments will be facilitated by increasing offers for leasing machines. There exists a traditional relationship (both private and public) with forest owners in the region where the enterprise is based. Competition in the region, but also from outside the region, encourages the contractor to take on contracts outside the region to maintain a decent utilization rate of men and machines.

INNOVATION. An important innovation for the firm was the entry into bio-fuel harvesting. The market for forest harvesting services is becoming saturated, prices are falling and competition is increasing. This has resulted in enterprises searching for additional sources of revenue. Bio-fuel is seen as profitable. Innovation is implemented by purchasing a chipper and by making efforts to find and sign contracts with larger customers, such as pellet producers and/or local heating plants. Also, contacts with local forest owners and communities are made in order to negotiate the demand and supply. The owner considers shifting more capacities into this sector in the future.

MARKET. In the wood supply chain, the general procedure is that the contractor gets a contract for harvesting the timber, which is marked by forest owners or foresters. The timber is forwarded to the roadside. A different enterprise carries out the haulage to the customer. There is generally no direct business relationship between the harvesting enterprise and the hauler. The enterprise is paid by the forest owner and the hauler by the timber buyer. The main customers are the forest owners in the region, approximately within a radius of 100 km around the base of the enterprise. Occasionally some contracts are signed beyond the region. Buying and harvesting standing timber is a rare business. It can be done but it is considered occasional. The market in this case is local or consists of suppliers of large industries with a local base. Competition has been increasing particularly after the recent storms. Now the timber demand has dropped. Contractors offer services all over Germany. New entrants, who dramatically underbid cost-covering price levels, have penetrated local market relationships. More and more contracts are based on tenders. Yet, some contracts are still negotiated directly by the contractor with the forest owners. In these cases, the negotiation power is weak or non-existent. This is further enforced by an increasing pressure from competitors. The participation in the decision regarding timber procurement policies does not exist. The enterprise is hired to do the harvesting operations and its job is limited to that.

POLICIES. Within the institutional framework, there are no evident factors to foster development but also no real barriers. The hindering factors are related mainly to market situations. Industrial relations are poor, because there is no direct supportive structure in the customer relations (e.g. contractor empowerment by industries). In the case studied, the policies did not support the innovation.

Case study 2 – Poland: Purchasing a forwarder (PL)

This enterprise is a typical enterprise for Poland, undertaking harvesting, forwarding, silviculture and forest protection operations. It is a micro-enterprise with five employees, specialized in forest and transport operations, engaged as machine operators (of chainsaw, forwarder and tractors). The equipment consists of five chainsaws, two used agricultural tractors, one used forwarder and one plough. The company intends to and has even started to embark on new technologies by exploring available products, such as harvesters. In general, forest operation companies work rather separately, with little business partnership. Since 1992, an Association of Forest Entrepreneurs (SPL) has provided mutual assistance to contractors in the country.

INNOVATION. For years, harvesting and logging were performed using chainsaws and agricultural tractors. Recently, the problem of finding new forest workers has arisen because of difficult working conditions. Thus, the owner of the Polish enterprise made the decision to invest in a forwarder. After getting an interesting offer from a machinery supplier, a used forwarder (3 years old) was rented for 1 year, and a leasing contract has been signed for the next 4 years. The seller of the machinery also provided information, technical support and training of the staff. The cooperation with the machine supplier has been a fostering factor of innovation. In fact, the entrepreneur built up a good releationship with the machine supplier. The owner of the firm expected more support from the state forests to implement the necessary organizational changes in harvesting and forwarding.

MARKET. The biggest customer is the state forests, with early-based tenders. The local market is usually in the area of one or two forest districts. Usually, the contractors who work in one forest district do not compete with each other, keeping a local *status quo*. When one firm wants to develop, it must compete fiercely with small companies. The winner often hires the losing tenders as subcontractors. Business cooperation and participation in decision making regarding the timber supply/timber procurement market is rather weak. Meetings with the contractors are organized only by forest districts. These are mostly carried out to exchange information, but they do not to give a real opportunity to participate in decision making processes concerning further harvesting operations.

POLICIES. Occasionally, a company can get support for the training of new employees, e.g. for chainsaw operator courses. After that, the contractor has to guarantee a workplace to the qualified entrants for some period. However, there was general support from the main partner, the state forests national company, for introducing new methods of harvesting, e.g. switching to cut-to-length systems. For example, in the period 1994–2003, the state forests national company helped enterprises financially to purchase special machinery for forest operations, though they no longer do so at present. Nevertheless, owning new machinery offers tenders advantages in the tenders in forest harvesting services. The impacts from policies regulating timber harvesting and yearly based tenders were further mentioned as negative factors.

Case study 3 – Lithuania: Purchasing a harvester (LT)

In the case of Lithuania, the activities of the firm cover the whole wood procurement chain, mechanized and manual (with chainsaws) wood harvesting, mechanized forwarding, timber trade and road transport. The income from wood production amounts to about 55% of the total turnover. The firm has one (new) harvester–excavator, two self-loading trailers for wood harvesting, and one truck. The owner of this enterprise and his wife (as bookkeeper) do the main part of management work. Harvester and forwarder operators have a tractor driving licence but did not attend any special courses for harvester or forwarder operators.

INNOVATION. Three years ago a lack of forestry workers (especially chainsaw operators) encouraged the owner to buy a harvester as he had a harvesting contract with the state forest enterprise. At that time, there were only a few harvesters in Lithuania. The use of harvesters was an innovation to the firm and a new development in Lithuania. The enterprise doubled the harvesting volume after purchasing the harvester. To implement the innovation, the owner of the enterprise consulted with the supplier of machinery. The investment was completely private.

MARKET. The enterprise participates in tenders for timber harvesting services mostly in state forests. The winner of the competitions is the enterprise that offers the lowest price for timber harvesting. Additionally, this enterprise buys timber from the state forest enterprises as well as from private forest owners and sells it to customers. The enterprise has a long-standing relationship with one state forest enterprise, but the enterprise must participate in the bidding for tenders each year. The enterprise works on the local market, about 50 km around the base of operations. Having enough volume of work in a very compact area provides some advantages in relation to competitors. The owner states that there is a high level of competition in the field of timber harvesting because of an oversupply of forest operation services. The enterprise does not participate in the decision-making processes related to forest harvesting services. In some cases, the enterprise participated in seminars where the requirements of legal acts were discussed. The owner of the enterprise has a negative opinion of cooperation with or participation in any business network.

POLICIES. There are some political programmes dealing with innovation but they do not have any impact on forest enterprises. The answer of the interviewee was 'we would like politicians not to disturb us'. No state policy or support measures for cooperation of contractors were reported.

Case study 4 – Slovakia: Custom-made logging tractor (SK1)

The Slovakian enterprise was established in 2001 (and the forest-timber division was founded in 2005). It is one of three enterprises in the country that processes more than 100,000 m^3 of timber annually. The enterprise is divided into three departments: (i) forest-timber; (ii) transport; and (iii) pellet production. The forest-timber division has its own technician and production manager. The forest-timber division represents 14% of the total turnover of the enterprise. The division performs various forest services: timber felling by an assortment of methods using harvesters and a processor cableway; picking, transport and processing of biomass for energy use; and public services on water courses. Timber is mostly sold at the skid way. One of the main trends in this business area is the merging of firms that provide forest services with timber-processing firms. In its cooperation and business network, the enterprise concentrates on processing salvage timber felling. In buying and improving the forest machinery, there is a close cooperation with a machinery producer. Long-time businesses are established with some wood processing firms. The firm also cooperates well with municipal forest associations of the Slovak Republic.

INNOVATION. The enterprise bought an innovative, custom-tailored tractor with a 50% subsidy of the tractor's price based on an assessment measure from National Strategic Plan for Rural Development 2004–2006 (axis 2.1 – improvement of the economic value of forests). The idea of innovation appeared when the owner was faced with the lack of qualified workers for animal skidding. The tractor was ordered on the basis of the enterprise's concrete requirements resulting from practical experience. The innovation is based on foreign R&D done by the machine supplier. This tractor is new, not only for the enterprise, but also for the entire Slovak market. The first orders of this tractor indicate its potential for a wider application in Slovak forests. It could provide advantages against competitors as it can be perceived as an environmentally friendly skidding tractor as far as it gets close to horse skidding. The main actor of this innovation was the director of the forest-timber division. After consultation with the workers, he concluded that small tractors with a long cable can effectively skid timber in Slovakian terrain conditions. The director addressed the machine supplier with the question whether it would be possible to conceive a tractor with a 200-m cable. Practical demand and experience were the sources of this innovation, rather than research. The fostering factor of the innovation was at first the subsidy provided via the Agriculture Paying Agency, covering 50% of the price of the tractor. The next positive factor was the long-term and good cooperation between the firm and the machine supplier, e.g. at exhibitions the firm presented the machine supplier products for free. One of the impeding factors was a lobbying action of the environmental NGOs, who favoured animal skidding against machines. The other impending factor for further investment is that the subsidy from structural funds is limited to SK 200,000 and can be provided only once in 3 years.

MARKET. Generally, the activities of the firm are located in forest districts with occurrence of calamities in recent years. Competition among over-regional tenders is relatively high, thus the forest division has built its competitive advantages by investing in equipment, qualified operators, experience in working abroad and quality of services – clean, undamaged, correctly measured assortments delivered to the skid way. The biggest customers are state forests.

The competitors are a branch of one forest machinery supplier and foreign competitors coming from the Czech Republic, Austria and Germany. The supply of forest services is sufficient, but most firms or sole traders use old, ineffective and not very ecological machines. In fact, the quality of forestry work is not very high, and herein lies the competitive advantage of the firm. The firm plays no part in making decisions, because decision-making processes about procurement policies can be significantly influenced by state forests only, and the situation is not likely to change.

POLICIES. Some policies could help the enterprise if the firm was able to apply for financial support, although the firm representatives consider the whole process as requiring too complicated procedures. Thus, the firm has directly benefited from provisions of the Council Regulation (EC) No. 1698/2005 (National strategic plan for rural development SR 2004–2006, axis 2.1) and from the provision of the Act of the National Council of the Slovak Republic No. 239/2001 regarding 'red diesel' that is subject to lower taxation (11.5% cheaper than normal diesel). Red diesel can be used by a registered subject only and only for machines working in agriculture and forestry. Procurement policies are perceived as barriers because they are inflexible and slowly implemented.

Case study 5 – Slovakia: Packing timber for conservation (SK2)

The firm was founded in 2003 as a branch of a bigger firm. It provides various services: soil preparation; afforestation; slash cleaning;

plant protection against game, insects, weeds, rodents; fungi control; forest tending; harvesting operations; advisory service, including forest management expertise; biomass collection, including the purchase of standing timber, etc. The technical equipment is modern and relatively new. The firm faced problems with finding qualified staff for operating a cableway. As much as possible, the firm is trying to integrate the whole timber value chain, from stump to mill, even if sometimes the firm has to contact other sawmills in order to process all the timber bought or cut. The enterprise further wants to expand timber transport and timber trading. Besides, the firm has rented some forest land. Although the enterprise is cooperating with few sole traders, it does not force its own employees to turn into sole traders, like many other enterprises in Slovakia do.

INNOVATION. The innovation is based on foreign R&D. The idea of innovation comes from the situation of over-supply of unmarketable timber after a calamity. One of the solutions is to conserve the timber following a development by German foresters after the Lothar storm in 1999. The logs are packed airtight in a customary silage film and can be conserved up to 5 years. The method is brand new in Slovakia, and implementation seems to be difficult. The innovation was achieved with private investment sources.

MARKET. The enterprise is already relatively well known on the market and active in the whole of Slovakia on both local and regional markets. The biggest advantage against its competitors is the fact that the firm holds several sawmills and contracts with some other sawmills, which enables them to guarantee the customer that the harvested timber will be bought.

POLICIES. The innovation case does not depend on any direct influence from policies.

Case study 6 – Slovakia: Purchasing a logging tractor through a subsidy system (SK3)

The firm was founded in 2005. The forest services provided are: timber felling and skidding; planting new forest stands; and services connected with treatment of forest stands. The firm has some experience of working in Poland, but not on a large scale. Long cooperation exists with one timber processing firm. The technical equipment is relatively new, but no innovative technologies are implemented. The main trends in this business area are the lack of effective personnel able to work with forest cableways and the rising demand of fuel-wood in the future. The enterprise plans to enlarge its machinery endowment with a new logging tractor.

INNOVATION. The enterprise is waiting for its application for a subsidy to be granted in order to buy a new logging tractor. The main innovative ideas were about light forest cableways. The director tried to obtain more information about them but he did not find what he was looking for. Therefore, the enterprise started with applying for a subsidy to buy a new tractor, also with the idea of further increasing the production capacities of the firm. It was the first time that the firm tried to obtain a subsidy; all the other machines had been bought with their own resources. What was new for the firm was not the fact of buying one tractor, but the fact that they were granted external financial support and the fact that they outsourced the application work. The application had been elaborated by a consultancy firm and sent to the Agricultural Paying Agency. The owner of the firm appreciated as a fostering factor the professional approach of the consultancy firm and its troubleshooting ability to write the application. The main impeding factor was the lack of properly provided information at the level of Agricultural Paying Agency.

MARKET. The firm works on local and regional markets. In the region, the firm is one of the biggest in its business area. Competitive advantages come from the good level of mechanization on the local scale, and a high productivity of skilled employees. The main competitors are small sole traders, who own old, un-ecological machines. They are able to push the prices for their services dramatically down. Actual market trends are indicating that customers will be looking for more effective (e.g. smaller damage on standing forest stands) and more ecologically sound services (smaller contamination of forest soil by oil products). The enterprise has no opportunity to participate in decision-making processes about procurement policies.

POLICIES. Some policies, which are generally relevant for the firm's development, are mentioned in the case study. In the present case the enterprise attempted to get relevant financial support within the frame of Council Regulation (EC) No. 1698/2005 (National strategic plan for rural development SR 2004–2006, axis 2). The main institutional barrier was the fact that the APA (Agricultural Paying Agency) was not able to provide the relevant information properly or in a timely manner. One other barrier was the fact that the enterprise had to own, or had to rent, at least 10 ha of forest to get the subsidy.

Case study 7 – Estonia: New uses of the harvester (EE1)

The firm established in 2003 is situated in the south-west of Estonia. Until 2003, they dealt with forest operations only, mostly felling and skidding trees to the roadside in state forests (State Forest Management Centre – RMK). In 2003, they diversified their activities to include maintenance services and infrastructure work using their bulldozer equipment, as well as dump truck services since 2005. At present, forest services represent 40% of the turnover. The firm offers mainly felling and skidding services to the roadside, with occasional roundwood transportation by subcontractors, afforestation services or digging forest ditches. Cooperation and business networks are not strongly developed. In forest harvesting operations, the firm has only two main partners: RMK and StoraEnso. Because of strict requirements in RMK tenders, the firm has no direct access to the RMK forests, except via subcontracts with other companies, who have won RMK tenders for forest harvesting operations.

INNOVATION. In 2006–2007, the firm found new uses for the forwarder in the low season, e.g. recovering a former wasteland/garbage area, destroying beaver dams on forest ditches, and forest ditch soil/earth digging for reconstruction or cleaning old ditches. Everything was done at their own initiative, based on intensive exploration of opportunities – researching on the Internet, reading different procurements of tenders, making projects and trying to implement them.

MARKET. In forest harvesting operations, the firm operates only in Estonian markets, within a radius of 30–50 km from its home location. High competition exists, especially after 2005, when annual Estonian allowable cut decreased by nearly half. Also, the tax system for forest owners makes them harvest smaller volumes of timber now. In the state forest tenders, the prices for forest operations have sometimes been extremely low, to the point where outsourcing was carried out to avoid too high costs linked to the use of their own machines. In this way, the firms have been working in the state forests as subcontractors for other companies. Generally, the owners cannot participate in decision making about timber procurement, which implies that the situation is 'take the job or leave it'. At the same time, one of the company owners is very active in different lobbying activities; he is a member of the board of various organizations, like local forest owners' association, so that he can influence some decisions.

POLICIES. Specific policies directly related to the firm innovation have not been recorded in this case study. Indirectly, the firm is influenced by transportation restrictions. Another limit is related to the use of specially marked (blue colour) agricultural diesel fuel that is cheaper, but its use is not allowed in forests and forest machinery, e.g. harvesters in forests and fork-lifts in sawmills.

Case study 8 – Estonia: Use of forwarders out of season (EE2)

The firm is situated in the south-east of Estonia. Until 2002, all the turnover came from forestry operations (felling and skidding), but the firm has evolved, and in recent years (2008–2009) only 20% of the turnover has come from forestry operations, the remainder resulting from participation in landscaping and infrastructure works. In the forest market, the entrepreneur works with the Estonian private forests only. Lately, the forest market has developed towards the use of the felling residues (mostly from mixed forest) to supply combined heat and power production units. If the use of wood for electricity production is an innovation in Estonia, it is the same for the collection of branches and felling residues. In forestry-related areas, the firm is not a member of any network, but it is a commercial partner working for larger forestry companies (such as StoraEnso) and small private forest owners. In landscaping and infrastructure activities, the firm is a member of some business networks. The owner is also a member of the local private forest owners' association.

INNOVATION. Innovation appeared accidentally, when once in the low season the firm was asked to do roadside fortification. In the beginning they used a lorry and manual work and they soon realized that the efficiency was not very high. Within 1 week, they solved the problem of low efficiency by starting to use forest forwarders. Later, harvesters were also used for different types of work. Since then, in low season the forestry machinery (forwarders and harvesters) is used for roadside slope fortification and other landscaping activities, which might take place during road reconstruction or the building of new roads.

The regulation on the use of blue, cheaper diesel fuel now appears as an impeding factor. This fuel can be used in agriculture on the condition the machines stay on properties and are not used on public roads. In forestry (especially concerning forwarders), the use of cheaper, agricultural-intended diesel has always been a problem. On top of that, the recent changes in labour-related legislation do not allow working in 24-h shifts. Working contracts with employees have to be changed to adapt to this regulation, yet the practice up to now was that in wintertime employees worked long shifts, compensated by longer summer vacations.

MARKET. The firm is operating mostly locally. In the forest sector, there is a very high competition. In road construction and landscaping-related activities, competition exists, but it is not as high as in forestry. In infrastructure works, the partners of the firm are big road construction companies, winners of public procurements competitions in the area. The firm has worked with them all over Estonia.

POLICIES. There are no specific policies for innovation in this case. Generally, the forestry contractors and machine owners are faced with financial difficulties when they must stop work because they have to pay their loans for leased machinery. Surviving in summertime pushes for inventive solutions as the firm developed in entering the infrastructure market activities.

16.5 Discussion

16.5.1 Categories of forest service enterprises

The central element of analysis in this chapter relates to innovation among enterprises performing forest operations. Each enterprise was selected to ensure that it was representative of the sector in its region. No attempts were, however, made to determine the representativeness of the firm for the whole domestic sector. The results of this suggest that there is a further demand for a typology of forestry contractors. It may be suggested that further analysis should consider the following strategic categories of forest contractors in Europe:

1. Large highly mechanized firms operating nationally. Tender-based contracting, eventually subcontracting in the case of high work volume.
2. Medium-sized firms with niche market focus. Direct contact with customers and

tender-based contracting, subcontracting if needed.
3. Small firms with high specialization. Local market contract, local tenders, participate as subcontractors for larger firms.
4. Medium or small-sized firms with a diversified market focus.

16.5.2 Type of innovation

Most of the innovations described in the cases are mainly linked to the investment in and the utilization of new machines or techniques. This can be interpreted as symptomatic for the sector, where the day-to-day challenges are characterized by practical operational tasks, and where the search for an optimized technical solution is a priority in decision making processes. This shows, among other things, where the focus and main interest of forestry contractors lies. In general, these innovations are incremental rather than representing new ideas. It may be considered that they represent a business development diversifying the normal service into a new field according to the market opportunities or changes, or a way of improving the competitive advantage and gaining economies of scale rather than an innovation. However, for the individual firm the named technological or product innovations bring a radical change in their management. In the cases of bio-fuel harvesting, purchasing a forwarder or purchasing a harvester, and using subsidy systems to buy a logging tractor, the innovations are radical and new to the firm. In the case of the custom-made tractor, and packaging timber for conservation, the innovations are new to the market as well. Estonian cases with innovative use of forest machines are new to the firm and incremental in the sense that the existing machines are improved to make a better use of them.

In all the cases where new machines were bought, mechanization was an innovation because it meant a new way of working for the firm. Usually, purchasing machines represents a normal upgrading of the enterprise technical equipment that cannot be considered innovation because of its permanence. However, the first purchase of a forwarder or harvester completely changes the production system of the firm and should thus be considered an innovation new to the firm only. Purchasing a forwarder or harvester for the first time often means an innovative approach to acquiring funding (e.g. because of lacking private financing or the difficulties associated with obtaining a loan). It additionally means having to contact machine suppliers and learn about performance and using and maintaining the machinery, which in eastern countries may be challenging considering the lack of qualified agents to maintain the machinery or the recruitment of qualified workers, given that few professional schools in eastern Europe can train workers for these types of machines.

16.5.3 The driving forces of innovation

The driving forces of innovation are the profit-seeking opportunities offered by the implementation of the innovation, market-related opportunities in most cases, and policy-related opportunities in the two cases dealing with the use of the subsidy scheme for purchasing new harvesting equipments. In both cases, the innovations were triggered by the need to find a solution to the lack of manpower (forest workers). According to Drucker (1998), the opportunities for innovation are limited: unexpected events, discrepancies, specific needs of some activities and changes in the sector and in the market. External to the company, there are demographic changes, perception changes and new knowledge.

Unexpected events came in the Estonian case from the policy side when it was decided that for a sustainable management of forest resources, the annual allowable cut should be reduced. The sharp drop of the timber supply tightened the competition, and therefore the firms have been looking for new ways of using the expensive machines they bought. Also, an unexpected event was the storms in 2001 and 2004 that favoured firms with pro-active management to forecast the scarcity of timber, and to adapt by using machines in alternative ways, or by developing parallel activities such as infrastructure works or machinery repair. Changes in the sector and in the market inspired the innovation described in case one – entering the bio-fuel market. Changes in the policies of the timber provider enhanced adaptation and some innovation in the forest contractors firms, but mostly they are perceived by the firms as having a higher potential risk. The demographic changes (lack of labour because of immigration or to the changes in the profile of the young generation) are bases of innovation in the purchase of a harvester in the Lithuanian case, and the purchase of a forwarder in the Polish case.

New knowledge-based innovations are the most spectacular innovations, yet they require long-term development, finalization and the synthesis of different kinds of knowledge, called the discipline of innovation (Drucker, 1998). This is the case with the conservation of timber and with the custom-made tractor; both innovations require time, and they are not really implemented in the market yet. Drucker (1998) argued that innovations are not just some events, but the result of a systematic search for innovation. It may be seen in the second Estonian case that the innovation comes from a firm actively searching for market opportunities in the domestic market or abroad; it may be seen as well in the custom-made tractor that the product came under the frame of a long and stable cooperation between the machine supplier and the firm. It should, however, be stressed that cooperation requires good personal relationships (e.g. between the owner of the firm and the manager of the machine supplier) and not merely an institutional, de-personalized cooperation between the firm and the machine supplier.

As stressed in many case studies, competition in forest contractor works is high. The intensity of competitive rivalry leads the firm to find sustainable competitive advantage through innovation (Porter, 1979). Competitive advantages were found in:

- The integration of the wood value chain through the diversification of services, such as timber transportation or saw milling. Innovation is related to the change in the market segment.
- Providing quality services, e.g. via modern technologies, as in the Slovakian and Polish cases, and increasing the labour productivity – innovation is related to investing in new machinery as a choice criterion to access the timber procurement market.
- Specializing in niche markets, such as salvage (SK1) or storm cuttings.
- Finding new markets for services by using machinery for other purposes than forest-related activities in the low harvesting season – innovation here related with entering a new market (the infrastructure market) changed the usual business environment of the firm.

The firms' capacities in negotiating prices and the influence on public timber procurement policies and decisions are very weak compared with the bargaining power of suppliers. Through the regulation of timber procurement, through spot transactions, short-term delivery contracts, relatively small values/quantities and few if any substitutes, the suppliers of forest services tend to have a significant power over the firm. Therefore, when the supplier requires machinery to be less than 10 years old, it compels the firm to innovate. The requirement for quality services is a driving force of innovation in several cases.

On a more general level, the policy-related milieu of the forest sector can facilitate the horizontal and vertical integration of forest activities. The strong regulatory climate, as in the Polish case, would not allow the same horizontal development of activities as in Lithuania, Estonia or

Slovakia. In these three latter cases, the portfolio of activities is very large, from harvesting, forest protection to forest management services, which allows the firm to be flexible and to face the uncertainties of the timber markets. When diversification of activities is allowed, there is a better climate for product or services innovation. When the firm undertakes only forest harvesting services, as in Poland or Romania (Austroprojekt, 2008), not only is the innovation in new products hindered, but the firm is also affected by the drop of the production flow in the low seasons. Some forest contractors considered the integration of the wood chain as a solution. They bought the timber they have been contracted for harvesting, they process it in their own mill (or in a partner's mill), and they provide transportation and trading services.

16.6 Conclusion

16.6.1 The role of markets and cooperation

Innovation was financed by subsidies in the case of SK1 and SK2; all other innovative activities were financed by private investment. Yet, the cases show a trend towards loan-based investment. In all cases, business cooperation and networks between contractors are rarely mentioned or they are regarded as having little importance. This is in fact not typical of the contracting sector, since there are frequently contractor/subcontractor business relationships already in place. Furthermore, most interviewees did not regard the market-related commercial partnership as a business network. They have a negative perception of professional associating bodies and related lobbying activities. Therefore, one should classify cooperation (as recorded in the case studies) as being market-relationship-based and professional-associations-based.

The existing market-relationship-based cooperation shows that the firms have their own strategic or operational relationship, such as:

- *Long-term relationship with clients.* Certain forest districts or private owners are the 'usual' customers of forest services and this partnership with the customers is a smooth solution to the uncertainties of a yearly based market and/or tender-based systems (in public forests). However, the case studies show a trend aiming at replacing the long-term relationships by tendered contracts for services, which fairly often do not respect or favour trustful past cooperation, but just the actual price offer. This is a huge threat for locally existing business networks.
- *Stable partnership with suppliers.* The cooperation with the suppliers of machinery leads in one case to the development of innovation (custom-tailored tractor; SK1).
- *Subcontracting as an operational cooperation strategy to face extra workload.* When work volume or need for specialized equipment is required, the contractor resorts to partner enterprises which work as subcontractors, or he gets subcontracted by them if required.

Subcontracting is used either to share advantages when faced with larger demand (GE), or to push some costs to smaller firms or sole traders, as mentioned in general in the Slovakian case (the studied firm cooperates with sole traders, but it does not force its employees to turn into sole traders, contrary to the trend). None the less, small sole traders might be a problem since they are able to push down the price level for services (SK3, PL). In none of the studied cases was subcontracting used as a vector to implement the innovation; the possibility should, however, not be excluded. A second form of cooperation is the membership in professional bodies and organizations, such as Associations of Forest Entrepreneurs, whose aim it is to lobby for the members or for the sector. The cooperation in professional organizations and bodies represents a 'power sharing' approach of the firm development, based on non-binding membership and loose commitments. In the cases studied there were, however, negative perceptions of the benefits of membership in professional bodies and of the real impact of lobbying activities. One explanation can be that most of the cases were observed in eastern European countries where, after the shift to market economy, a strong resistance to social and economic control in associations persists.

16.6.2 The role of policy milieu characteristics

Results from the case studies suggest that policies do not have an effect, or (if any) the impact on the activities of the firm is negative. However, the financial support that the firms received from public policies helped them to invest in new machinery via the rural development programme (SK1, SK3), to employ young or long-time-unemployed people (SK3) or to pay for training qualified people (PL). Even though the employment and education policies were not directly aimed at innovation, they helped the firm to develop, particularly in the context of increasing demand from the timber supplier (state forests in Poland, Estonia and Slovakia) for high skill and modern environmentally sound machines. These policies can consequently play an important role in fostering innovation, as training and qualifications are amongst the crucial prerequisites for innovation.

The negative effects of policies were identified within the regulatory barriers for forest harvesting operations in legislation or within the internal regulation of timber providers. Examples in the case studies were: shortening the harvesting period; requirements for the quality of the services; age of the forestry machinery to be used (e.g. Estonian RMK requires less than 10 years old forest machinery); the regulatory provisions concerning transportation limits (truckload length and gross weight); and the rules regarding the use of cheaper agricultural diesel fuel (Slovakia, Estonia). These regulatory provisions have had an impact on the general performance of the firms, or can be regarded as general frame conditions or technical specifications of the contracted operations. Therefore, they may be considered only as indirectly hindering factors of innovation. It is none the less worth noting that the contractors themselves see these requirements as 'politically' imposed restrictions to their daily business. They furthermore blame 'policies' for the higher cost and lower profit, which result from environmental or procurement regulations.

The policies of the main timber providers, such as state forests in Poland aiming to change from tree length to cut-to-length systems or RMK in Estonia requiring high standards of mechanized work and quality services, have a high direct impact on the harvesting technologies used. The policies of the main timber suppliers can provide strong incentives for innovation in some firms, *provided* they are able to adapt in a short time. This conclusion fits with the recorded case studies, where the state forests still have an important role to play on the market. Technological innovations are stimulated, but also organizational ones, when the state forest radically changes the timber procurement procedures.

16.6.3 Lessons learned and recommendations for policies

It can be assumed that a majority of innovation policies have not yet led to a considerable uptake of innovation among forestry contractors. The innovations that could be identified are mainly incremental applications of already existing technical solutions, and in some cases departure from/embarkation on additional tasks and new product markets. These innovative activities can be characterized as reactions to market developments and customer demands. The question remains whether the urgently needed development of contractors could be considerably fostered by national or European policies. Advice and support by public or semi-public bodies, which are based on public enterprise development policies, can be found in most countries. However, it seems that forestry contractors are addressed by such organizations in rare cases only, or the offers are rather directed to craft or trade businesses. Herein lies a considerable potential for identifying the needs for development and innovation among forestry contractors, and for identifying supportive measures, which would reach those enterprises that often work in remote rural settings.

At this stage, one of the major requests is quite obvious, particularly in eastern and central European countries: there is a crucial need to support investment in technical innovation in forest operations, to react to the increasing demand for harvesting and transport capacities, and to curb the forecasted lack of workers in many countries. Reaching a higher mechanization level cannot at this stage be achieved without public support investments. It is worth discussing whether a direct subsidizing system would be politically desirable, since direct subsidies are not in any case the best means to encourage entrepreneurship and innovation. There are still challenges and opportunities to be found in the EU rural development funds to direct the grant systems in such a way that supported forestry measures can be connected to the use of innovative technique. To develop market-opportune measures aiming at a desired impact on innovation in forest operations is a huge challenge for political decision makers in the near future.

An even more important political goal should be to build up or maintain the capacities for support, training and commercial advice of SMEs to develop further the human resources and entrepreneurial capacities that will be crucially needed in the future in order to enable contractors to fulfil their important role in the forest-based value chains.

Acknowledgements

The work was partly supported by the Slovak Research and Development Agency under the contract No. APVV-0692-07 (2007), by Estonian Environmental Investment Centre project 'Innovation in Forest Sector' (2008) and by COST Action E51: Integrating Innovation and Development Policies via the Short time scientific mission (2009). Edgar Kastenholz and Piotr S. Mederski would like to thank Prof. Siegfried Lewark for his valuable help and comments necessary to make the analysis of the German and Polish cases.

References

Austroprojekt (2008) Forestry and forest industry in Romania. Wood sector study. Final Report April 2008 elaborated by AUSTROPROJEKT Agency for Technical Cooperation Ltd, Vienna.

Brogt, T., Lewark, S. and Kastenholz, E. (2007) Learning cultures in small enterprises in forest operations. *Proceedings of the 3rd Forest Engineering Conference*, 1–4 October, Mont-Tremblant, Quebec, Session 9 – Who will log in the future? Available at: http://www.feric.ca/FEC2007

COMFOR (2008) SME and micro businesses in forest operations – Research Demands. Delivery No. D 1 in the project COMFOR: Collective work science approach to solving common problems of occupational health and performance in European forest operations SMEs. Available at: http://www.enfe.net/comforopen/pdfopen/D1-SME-Research-Demands.pdf (accessed February 2010).

Drucker, P. (1985) The discipline of innovation. *Harvard Business Review*, 95–102.

Drucker, P. (1998) *Peter Drucker on the Profession of Management.* Harvard Business School Press (translated from Romanian (*Despre profesia de Manager*)) by Dan Criste). Meteor Press, Bucharest.

ENFE Press Release (2008) Sustainable forestry needs sustainable contractors. European Networks of Forest Entrepreneurs Press Release, 15 December. Available at: http://www.enfe.net/archiv/sberichte/press_release_dec08.pdf (accessed February 2010).

EU Recommendation (2003) nr 2003/361/EC. European Commission, Brussels.

Kaufmann, A. and Tödtling, F. (2002) How effective is innovation support for SMEs? An analysis of the region of Upper Austria. *Technovation* 22, 147–159.

Lewark, S., Brogt, T., v. Kutzschenbach, M., Steinert, S. and Westermayer, T. (2010) Learning, knowledge transfer and competence development in forestry operations small and medium sized enterprises (SMEs). *Arbeitswissenschaftlicher Forschungsbericht* Nr. 10. Albert-Ludwigs-Universität Freiburg im Breisgau, April 2010.

Lundvall, B.-A. (ed.) (1992) *National Systems of Innovation: Towards a Theory of Innovation and Interactive Learning*. Pinter, London.

Lundvall, B. and Borrás, S. (1997) The globalising learning economy: Implications for innovation policy. DG XII, Commission of the European Union, Report. Available at: http://cordis.europa.eu/tser/src/globec.htm (accessed February 2010).

Malmberg, A. and Maskell, P. (1999) Guest editorial: Localized learning and regional economic development. *European Urban and Regional Studies* 6, 5–8.

OECD (2005) *Oslo Manual: Guidelines for Collecting and Interpreting Innovation Data*, 3rd edn. Available at: http://213.253.134.43/oecd/pdfs/browseit/9205111E.PDF

Pires, C., Sarkar, S. and Carvalho, L. (2008) Innovation in services – how different from manufacturing? *Service Industries Journal* 28, 1339–1356.

Porter, M.E. (1979) How competitive forces shape strategy. *Harvard Business Review*, March/April.

Stake, R. (1995) *The Art of Case Research*. Sage, Thousand Oaks, California.

Tödtling, F. and Kaufmann, A. (2001) The role of the region for innovation activities of SMEs. *European Urban and Regional Studies* 8, 203–215.

Westermayer, T. (2006) Outsourcing of work in Germany's forestry. Rural social structure and identity in transformation. *Arbeitswissenschaftlicher Forschungsbericht* Nr. 3, Institut für Forstbenutzung und Forstliche Arbeitswissenschaft Albert-Ludwigs-Universität Freiburg im Breisgau, Germany. Available at: http://www.freidok.uni-freiburg.de/volltexte/2924/pdf/AwiFb003.pdf

Yin, R. (1994) *Case Study Research: Design and Methods*, 2nd edn. Sage, Beverly Hills, California.

17 Innovation in EU Forestries: a Science–Policy Dialogue

Filip Aggestam* and Gerhard Weiss

Abstract

This chapter analyses a cross-sectoral discussion forum – a science–policy dialogue – held at the final conference of COST Action E51. The design of the forum was as a moderated fishbowl discussion with invited discussants and the audience. The aim was to discuss the problem situation and possible strategies for policy improvements among researchers and policy practice. The discussants included policy makers and stakeholders from research policy, forestry, forest industry and regional development. The discussion forum was video recorded. This enabled the extraction and review of all major issues raised throughout the discussion, and the transcription of parts of particular importance. The main questions were directed at the success and gaps of current policies and which policy improvements are needed in order to support innovations in the sector and for rural development. These questions were discussed for territorial goods and services, and for the wood value chains. Results suggest a need for improving the interface between research and practice. Support for innovation should move away from project-based support to long-term cooperation and collaborative learning, and from a focus on research to the support of the whole innovation process and better interaction of all parts of the innovation system. Policy measures need to define characteristics and address different types of micro-, small and medium-sized enterprises (SMEs) such as innovative front-runners and/or traditional firms. For the forest sector, it is also relevant that rural development policy considers all sectors that interact in the landscape, is multi-layered and stimulates connectivity and integration at and between the appropriate levels.

17.1 Introduction

This chapter presents a summary of discussions held at the Final Conference of COST Action E51 – The role of policy in European forest-based innovation: bridging sectoral and territorial approaches – in Brussels, Belgium, on 1 June 2010. The objectives of the conference were to present results from the Action E51, and derive conclusions and policy recommendations for integrated innovation support in forestry and the forest-based sector.[1] The conference also aimed at initiating a cross-sectoral discussion and a science–policy dialogue on how better to coordinate policies for an innovative forest sector, concerning territorial goods and services of forests as well as the wood value chains.

17.1.1 A forum for a science–policy dialogue

As the conference had the ambition to not only present the research results but also to stimulate a lively exchange between

* Corresponding author.

research and practice, it gave ample room for inputs from policy makers and stakeholders. In the first session on the problem situation and knowledge demands, keynote speakers presented the perspectives from research policy, integrated rural development and the forest industry. Then the main results from the COST Action were presented related to the field of territorial goods and services and to the wood value chains. This concluded with a classical plenary discussion.

The afternoon was then dedicated to the moderated science–policy dialogue. The design of the forum was a fishbowl discussion with invited discussants and the audience. Discussants and the audience from policy practice and research were allowed to take a seat in the discussion circle at any point. The invited discussants included policy makers and stakeholders from different sectors. Policy makers in the conference represented the European Commission (DG RTD, DG ENT and DG AGRI) and the Austrian Ministry of Agriculture, Forestry, Environment and Water Management. Stakeholder organizations included the Forest-Based Sector Technology Platform (FTP), Forest Association of Portugal, Confederation of European Forest Owners (CEPF), Union of Foresters of Southern Europe (USSE) and European Network Rural Development (ENRD).

The discussion forum was divided into two successive rounds of 1 h each, dedicated to the two major themes of the Action: wood value chains and territorial goods and services. The first round asked what policies are required for an integrated wood value chain management. The second round focused on integrated rural development. Both rounds concerned policies aimed at supporting innovation in the many fields related to forestry and the forest sector. It was of interest to explore practical experiences and discuss gaps in current policy measures; how current policy instruments could be improved; and at what level (e.g. EU or national level) and/or who (public–private) should become active for a better support of innovation in these sectors. The cross-sectoral discussion forum is available to view online (http://www.boku.ac.at/coste51).

17.1.2 Aim and method of this chapter

The aim of this chapter is to analyse views presented by the participants of the discussion forum at the Final Conference of COST Action E51; the invited discussants are given at the end of the chapter.

Since the discussion forum was video recorded, it was possible to: (i) extract and review all the major points brought up during the discussion; and (ii) transcribe parts of particular importance. The qualitative method aimed to distil the most important issues that are related to the topic of innovation in the forest sector and rural development, and to compare the views of the represented groups.

The results are presented under a number of headings that appeared to be central to the participants in regard to innovation support in the sector.

17.2 Results

17.2.1 Policies: how effective are they for SMEs?

The COST Action E51 cross-sectoral discussion forum opened up with a discourse related to policy measures affecting innovation among micro-, small and medium-sized enterprises (SMEs) in the forest sector. Below follows a brief summary of some factors that were considered to be restricting the impact of innovation policy.

The first problem that was raised concerned the understanding of what types of SMEs should be targeted by the policy measures. Current policies apply a definition of SMEs that is too general. In the discussion, an important distinction was made between traditional SMEs and innovative front-runners, the argument being that while policy measures often cater to innovative front-runners, having an impact on (or gaining access to) traditional SMEs can be much more difficult. Policy measures for innovation in the forest sector thus need to address and define characteristics inherent in the type of SME that they aim to influence or support.

The effectiveness of policy measures that are aimed at stimulating innovation often suffers from having only a short-term impact. There was a general consensus that, while direct engagement of SMEs in research is (and can be) successful, the impact of measures is most often not long-term and only effective while funds are available. One obstacle is that innovation support measures are often project based and do not focus so much on the innovation process itself. Innovation policy may accordingly gain more by orienting its support to the whole innovation process rather than just to research. In addition to engaging SMEs through projects, innovation policy instruments should go beyond short-term collaboration with a view to establishing a systematic and continuous collaboration.

A factor that restricts the impact of innovation policy is linked to access to SMEs (e.g. at a local and/or regional level). It was stated that it may be more resource-effective to simply focus on SMEs that themselves would be interested in enhancing their innovative performance, such as front-runners. Alternatively, policy measures need to involve existing or creating new intermediary agents that have the capacity to connect external organizations with local and/or regional SMEs. One example was the Forest Technology Platforms (FTP)[2] Innovation Task Force direct engagement of wood clusters, meaning that they proactively engaged and involved forest enterprises at a local and regional level through cluster organizations. This is particularly important for SMEs that would otherwise not have the capacities to engage in innovative activities by themselves (e.g. traditional SMEs in the forest sector).

In summary, the key issues raised concern the necessity to define the type of SME and instruments a policy measure should utilize. Linked to this need, policy measures should ensure that the effects are not only episodic but rather continuous and long term. This could be accomplished by providing more process support rather than having project-based innovation support. Depending on the aim of the measure, SMEs need to be engaged and integrated at a local and regional level. Effectively, regional and national administrations need to adopt innovation practices that aim to get as close as possible to the SMEs' base (e.g. local level) to gain access and have an impact. In this process, intermediary organizations play an important role in facilitating communication and cooperation among the various actors in the innovation systems.

17.2.2 Context specific: are there regional differences?

'There is not one solution or one shot for all. That is very important! It needs to be customized to the groups or even for the regions or local areas where the SMEs are located and operating.' This statement from the discussion highlights the importance of tailoring innovation policy to the context in which it will be applied. Different types of ownership structures, fragmentation and varying interests in the forest define needs (e.g. the need for knowledge and/or access to a wood market), which in turn influence the degree of participation in the wood market. Policy measures and instruments should reflect these different local and regional contexts and problems across Europe.

It was further emphasized that SMEs do not necessarily need to engage in more research, but rather need more local and/or regional cooperation. The point highlighted was that policy measures need to focus more on the regions, not only at the EU and national level (both in terms of innovation support and cooperation). Coupled to this call for integrating SMEs at all levels, the need to extract learning from other regions was also stressed. While it was repeatedly noted that it is context-specific factors that determine the type of policy measures that should be applied, these factors cannot be defined without the exchange of ideas and 'good practices' across Europe. There is the need for a framework in which we can define these context-specific factors and that fosters a collaborative learning approach.

17.2.3 Research–practice relations

Research versus practice

There is a gap in terms of the questions asked (or priorities) and the flow of dynamics in practice versus research. The differing interests and modes of operation were brought up as a major barrier to innovation and cooperation. For instance, SMEs most often do not see the relevance of questions pursued in research ('ivory tower of research'), or the long timescale and limited reward for public research projects make them lose interest. In contrast to large enterprises that have more resources (e.g. central research and development labs), these problems are particularly linked to SMEs in which innovation and research are usually less structured and *ad hoc*. The 'reality' that confronts SMEs in terms of priorities (e.g. growth) and available resources (e.g. time and finances) is simply too different from that of larger enterprises and the research community.

At the EU level, it was additionally noted that the administrative burden associated with participating in public projects prevents SMEs from participating. The procedures and practices used in many tenders still have the effect of disadvantaging SMEs over larger enterprises. As argued by one discussant, 'reducing the red tape' and simplifying the procedures could speed up participation by industry and SMEs in particular. Stimulating SME-friendly practices should be a priority and, as stated previously, one way of achieving this may be through policy measures that provide more process support through national and regional administrative bodies. It was also noted that intermediary agencies (e.g. clusters) may provide a wider, more varied and holistic approach to supporting the innovation process. In policy terms, intermediaries may improve connectivity within a system (e.g. connecting actors and sectors) as well as create new possibilities and dynamics within a system. This may contribute to establishing more long-term collaboration between SMEs and research organizations.

Perhaps most importantly, the access to knowledge was vocalized by several discussants as a barrier to innovation. At one level, this relates to the translation of research results into a common and easily understood language: the 'message needs to be simple, clear and focused'. At the other level, it relates to the facilitation of knowledge. It is not enough only to produce research results, and not enough to provide access – research findings also need to be brought to the market. Knowledge needs to be wrapped in an attractive package, as stated, 'we need to make our results sexy'. SMEs (or the wider audience outside a given project consortium) need to be reached. It was, for instance, suggested that participatory learning processes could be utilized to disseminate knowledge and to generate an interest in collaborative research projects.

Communication and interaction

Failure to communicate and interact across the wood value chain can cause a range of problems. This was related to mistrust and missing confidence in the value chain. There is, in some cases, simply no trust between producers and industry, particularly the confidence between fragmented forest owners and industry was considered important. Mistrust similarly constitutes an obstacle for collaboration between research and practice. It is plausible that institutions (or intermediary agencies) could address this failure in communication, both between participants of the value chains and between industry and research. This could, for instance, be achieved by inviting local organizations to engage SMEs (also cross-sectoral), to allow producers and industry to 'get to know each other'. This may not only result in improved communication and interaction, but if successful, the increased utilization of local resources.

Linked to the difficulties surrounding communication and interaction, it was argued that the forest sector often represents a closed and traditional sector. It is a sector that is not prone to innovation and is not likely to engage in cross-sectoral cooperation, but to foster an innovative sector it is essential that it reaches out and interacts with other disciplines and sectors at all levels (EU, regional, national and local). The forest sector should look across its boundaries to account for the full value chain (e.g. from forests to recreation to public health policy), as well as to evaluate and compare the impact of other policy fields across Europe (e.g. national differences in energy policy). The role of research in this should be to collect and review relevant cases and to disseminate any findings to forest owners. Furthermore, as stated by one discussant, to 'see if [the forest owners] have any ideas, or to compare their business with businesses in other regions', stressing the need for research to proactively pursue participatory and collaborative learning and exchange with practice. This need to communicate and engage other sectors was also linked to the forest sectors' lobbying activities. Arguments were made that the forest sector needs to make its lobbying more appealing.

17.2.4 What policies are necessary for an innovative forest sector?

The discussants were asked what they consider necessary for an innovative forest sector. Three topics were raised: the first was financing, the second concerned public valuation and the third related to the establishment of a framework for a landscape policy.

Regarding monetary policy, besides the general lack of funds available to support innovation in the forest sector, the issue was how funds could be allocated more effectively. It was previously mentioned that policy measures should focus more on providing process support. At the institutional level, this means that support measures for innovation need to move away from the project level and aim to provide more long-term support. This does not mean that traditional subsidies should be abandoned but that additional institutional support measures for small owners and companies be added to the repertoire of tools. One example that was brought up was the Leader initiative,[3] designed to help rural actors consider the long-term potential of their region. It represents an institutional structure that provides a platform for participation and exchange among rural actors and that could also facilitate access and translation of new research findings. Accordingly, it is not sufficient only to provide more funding, policy needs to look beyond the provision of subsidies and episodic project support. As stated, 'to stimulate the atmosphere for innovation, and also mobilize the emotional part and conviction of innovative leaders [...] you need to stimulate that they become passionate about something. Otherwise it will not work.' This could, for example, be achieved through outreach and training programmes aimed at providing new tools, techniques and methods that boost local capacities and confidence and that reflect the needs of the innovation challenges in the given context.

Indirectly linked to the financial viability of the forest sector is the need to make society value ecosystem functions and territorial services, and to find measures and mechanisms that would enable this valuation. It was noted that public valuations mostly envelop services (e.g. biodiversity) and not goods (e.g. firewood), regardless of whether people are from an urban or rural area. This is combined with an unwillingness to pay for services (considered a public good) and associated with the public having a limited perspective on how the forest functions. As expressed, 'Society is not ready to value territorial services. We talk a lot about them, but society is used to public goods and is not willing to do more than provide subsidies.' There is essentially a conflict between what the public value and what they are willing to pay for. This calls for a clear policy framework and government initiative to assist in the application of monetary value to ecosystem functions and territorial services.

The need for a policy framework concerned with public valuation was discussed in terms of the range of sectors that have an interest in, and an effect on, European forests. It was argued that several sectors in effect meet in the landscape (e.g. rural development, agriculture, forestry and energy), which in turn causes conflicts over the limited resources available in the landscape (one example being the conflict between the bio-energy sector and the wood industry over the use of forest resources). This suggests the need for a 'landscape policy' that can be utilized as an instrument to streamline other sectors (Fig. 17.1). Developing a policy framework for the landscape could furthermore be useful in the process of making the public value territorial services.

Communication and interaction

The process of streamlining policies that compete in the landscape is inherently linked to communication and interaction between individual actors, industries and/or sectors. It was argued that cooperation is key within this context, especially when considering innovation in the rural environment, particularly as it is often the local and human potential that often drive innovation in the forest sector. It was suggested that this should be addressed via rural development policy and small-scale efforts targeting the human potential present at the local and regional level.

The need to motivate cooperation is also reflected in the need for networking to enable public policy. However, it was stressed that networking is not sufficient if it only occurs in a top-down fashion but requires a significant bottom-up element to be successful.

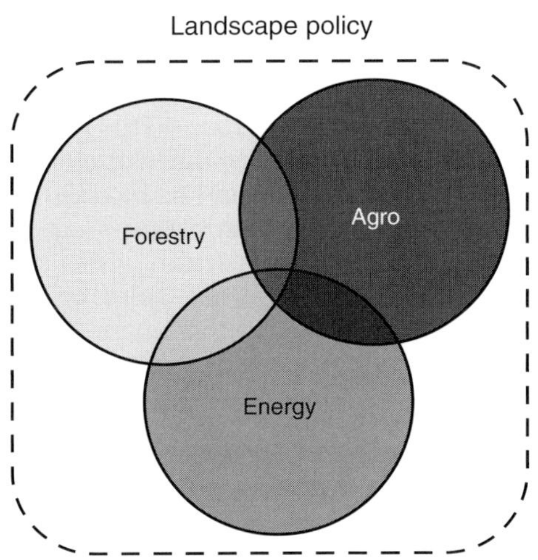

Fig. 17.1. Defining a landscape policy.

Enabling creativity – enabling innovation

Despite efforts characterized by public policy, innovation in the forest sector often occurs sporadically and in instances where we find a local concentration of innately creative and engaged people that in turn stimulate and feed into rural development. This questions the importance of an institutional framework that encourages innovative activities. While initiatives such as Leader were highlighted as important in this context, the discussants stressed that the forest sector needs to involve itself more. Involvement in this case is defined as the creation of an enabling environment that stimulates bottom-up efforts.

Enabling conditions were not specified further as they were presumed to be context specific and, more importantly, there is a need for more research to define what constitutes enabling conditions. They do none the less relate to conditions that help to enable people, both inside and outside the forest sector, to be creative and take initiative. Further, policy measures need to make sense (e.g. accounting for regional specific conditions), be 'user friendly' and aim to integrate other sectors.

Rural development

Central to policies for an innovative forest sector is rural development, but despite its importance, it was consistently argued that we do not have a truly rural policy yet, principally as rural development is still very agro-centric. The forest sector is, as stated by one participant, 'lost in the middle of agricultural policy'. It was emphasized that we need to move beyond this point to embrace forestry in rural development. It was even argued that we need to 'embrace the whole rural economy and the value chains therein'. Rural development policy should in essence help to create jobs in the landscape and generate income by taking the local context and profitability into account.

Aside from the need to incorporate more forest-related issues in rural development, it was indicated that multiple goals and services should be a central element when considering the rural economy. There is currently an absence of conditions that would allow alternative sources of revenue to come through. It is in this case not a matter of expanding subsidies, but it is important to recognize that other goods and services can generate an income.

Connected to the debate on multiple goals and services, it was further discussed whether there is room for legitimizing the payment of environmental services within the context of the EU Common Agricultural Policy (CAP). Given the ongoing discourse as regards the revision of the CAP, it was argued there might be some room to incorporate multiple goals and services. One stated: 'it seems to me much easier to legitimate payments for environmental services than it is to legitimate single farm payments under Pillar 1'. Within the scope of this discussion, however, it was stressed that the forest sector, as a whole, needs to be more vocal and raise these issues at the EU level to have an impact.

17.2.5 Who should become active?

When asked to define who should become active, the first issues that were brought up concern the fact that public policy is not enough to create an innovative forest sector; rather, identifying the motivation that drives innovation is key. This brings us back to the previous issue of generating enabling conditions (at the local and/or regional level). At one level, it is thus relevant that the research community takes an active role and focuses more attention on characterizing and defining these enabling conditions in more detail and to facilitate the access to this knowledge.

At the level of the state, to enable the development of a system for the payment of environmental goods and services requires initiative by Member States or the EU to develop a clear policy framework. It is, for instance, relatively easy to charge for goods (e.g. picking mushrooms), but there are few cases of payments made for services (e.g. biodiversity). This needs to be coupled with initiatives by the state as well as the market to develop frameworks (e.g. commercial networks) that enable forest owners to access these future markets.

On the whole, it is consequently not enough to only have a top-down approach, and neither is it sufficient to only have a bottom-up approach; both are in fact needed. From the top down, there is the need for leadership and policies that enable an innovation friendly environment, create markets and provide services. However, this would be for naught if there were no innovative people that would take up these opportunities and push new ideas forward from the bottom up.

17.3 Conclusion

17.3.1 Defining a landscape approach and system integration

The call for a 'landscape policy' can essentially be equated to the need for a balanced and authentic approach to integrated rural development. However, this cannot be accomplished unless:

- we develop a policy framework that truly envelops the complete landscape as well as streamline and integrate all the sectors therein;
- we confront the current view on the role of the forest – this relates to the multiple goals and services that should be part of the rural economy (e.g. environmental services should be incorporated more strongly into the Common Agricultural Policy);
- we enforce a more level playing field between the sectors that interact in the landscape – this requires that we move away from the currently agro-centric focus in rural development policy; and
- all the actors and sectors that compose the rural economy need to communicate, network and cooperate more actively at all levels, from the local, regional, national and EU level (Fig. 17.2).

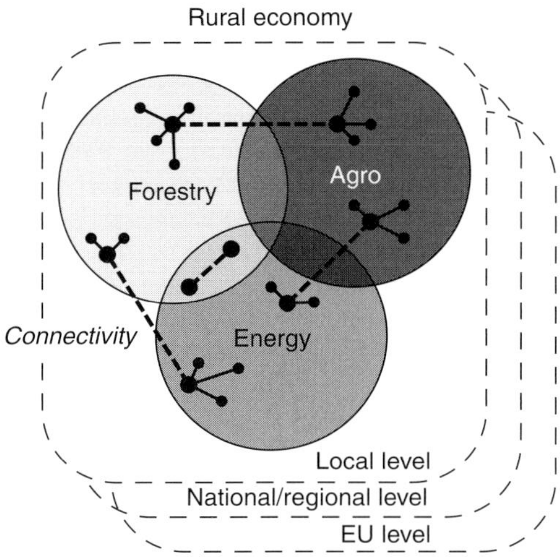

Fig. 17.2. A 'landscape approach' to integrated rural development policy.

Connectivity in Fig. 17.2 envelops such factors as cooperation, networking and the flow of knowledge between actors within the various sectors. Figure 17.2 aims to illustrate that rural development policy needs to consider the complete landscape and all relevant sectors (there are certainly more than the three given as an illustration), be multi-layered, and stimulate connectivity and integration at (and/or between) the appropriate levels. In broad terms, this may help to encourage more involvement, build confidence and improve research, education and training.

17.3.2 Research–practice

When considering policies and instruments for innovation in the forest sector, it becomes clear that there is a need to improve the interface between research and practice. The setting of priorities and establishment of collaborative efforts should be more strategic and tactical in terms of catering to the needs of practice. It was argued that increased communication, cooperation and networking is vital in this context, particularly when considering innovation in the rural environment. This is coupled with a need to reach a wider audience with recent scientific findings (e.g. outside the immediate project consortium). It was suggested that this could be accomplished through instruments, such as participatory learning processes that facilitate access to knowledge. There was also a consistent demand for better project management practices (e.g. at the EU level) for projects to reduce the administrative burden for SMEs. Stimulating SME-friendly practices should be a priority.

17.3.3 Policy measures

Inherent to the need for considering the complete landscape, policy measures need to be layered and tailored to target specific types of forest owners or businesses. Thus, policy measures need to address and define characteristics for the type of landowners and SMEs that they aim to influence. Support for innovation should further move away from project-based support to the support of the whole innovation process with an emphasis on long-term cooperation. This would also require that issues related to mistrust (e.g. between managers and owners) are addressed.

In addition, there is the need to develop a 'cost–benefit' framework that can help policy makers decide when and where it is worthwhile to invest in measures such as cooperation or wood mobilization. This relates to, for instance, the motivation underlying forest ownership (e.g. non-profit oriented forest owners) and the potential impact associated with different policy measures for the context in which they would be applied.

Finally, it was repeatedly argued that what is currently lacking in innovation practice are policies that create enabling structures that allow people to pursue innovative ideas (allowing for a bottom-up approach) and that truly stimulate creativity. In this context, it was also stressed that the forest sector should not be afraid of pursuing bold new ideas.

Acknowledgements

We want to give thanks to all who participated at the final conference discussions of the COST ACTION E51, and particularly to the invited discussants: Andreas Kleinschmit (FTP), Attila Lengyel (CEPF), Gert-Jan Nabuurs (EFI), Inazio de Arano (USSE), Rosario Alvez (Forest Association of Portugal), Tamas Szedlak (EC DG AGRI), Martin Greimel (AT Ministry of Agriculture), Fabio Cossu (ENRD: European Network Rural Development) and Zuzana Sarvasova (National Forest Centre in Zvolen).

Notes

[1] For more information on COST Action E51 – Integrating Innovation and Development Policies for the Forest Sector – see http://www.boku.ac.at/coste51.

[2] See http://www.forestplatform.org/.

[3] See http://ec.europa.eu/agriculture/rur/leaderplus/index_en.htm.

18 How to Support Innovation in the Forest Sector: Summary and Conclusions

Gerhard Weiss,* Pekka Ollonqvist and Bill Slee

18.1 A Differentiated View on Entrepreneurship and Innovation in the Forest Sector

The Schumpeterian entrepreneur is still often seen as the driving force behind innovation and economic growth. Entrepreneurs are often innovators, and innovators are often entrepreneurial persons. A differentiated view that assesses the potential and the limitations of entrepreneurship for the sustainable development of the sector, however, is needed in order to: (i) effectively support entrepreneurship and innovation activities in the forest sector; and (ii) utilize the existing creativity for both market and non-market goods and services of forestry.

18.1.1 Entrepreneurship – a personal trait or depending on a supportive environment, or both?

Entrepreneurship is often seen as a personal trait but can also be developed for an organization (for example, an 'enterprise'), a sector or a region. For entrepreneurs or innovators to develop their potential, they need an innovation-friendly environment within their organization, sector and region. Firms need to be managed to be or become innovative ('innovation leadership'). At a policy level, the task is to build a creative, innovative milieu that invites learning. Entrepreneurship can be observed in enterprises – small or big – but also in public or semi-public organizations, such as research institutes, training centres, interest groups, local governments and within networks of differentiated actors.

Innovative projects are often driven by extraordinarily entrepreneurial persons called 'innovation champions', who take the promotion and implementation of a novelty on to their personal agenda (Jenssen and Jørgensen, 2004). They enthusiastically strive to bring the new idea to fruition by acquiring the necessary resources and, if necessary, embarking on changing unfavourable organizational or policy rules. It is important to note that they are not only willing to take risks but also responsibility (Morris, 1998). They also need enthusiasm to calculate their risk effectively and they are persistent in following up their ideas because the successful development and implementation of innovations often takes years and often needs adaptations of legal rules. The answer to the above question, therefore, is: entrepreneurship needs to be seen as a personal trait that requires assistance from a positive and supportive environment.

* Corresponding author.

18.1.2 What is the role of entrepreneurship for the sustainable development of the sector?

As we know, there are multiple demands on forests and many of these demands have public good characteristics. While some of the forest goods are easily marketable (provisioning services such as timber or fuel wood), others are not, or are only to a limited degree (regulating or cultural ecosystem services such as biodiversity conservation or recreation). Thus, there is a risk that innovation in the delivery or internalization of public goods could be subverted by a more individualistic entrepreneurial profit-seeking vision of forestry.

For non-marketable services, Slee describes in his conceptual chapter that it may be that effective institutional design and the design of new policy instruments in turn form a potential arena for innovation that is pushed aside if a more market-centred entrepreneurial approach is taken. In practice, social entrepreneurship may be important in an institutional design, but this potentially differs from more red-blooded market-oriented entrepreneurship.

For the marketable goods and services of forestry – and this includes recreational and other ecosystem services to some extent – the need for entrepreneurship lies primarily with the owners (although here, too, the innovation-friendly environment is necessary). In the case of public goods, the crucial activities are at the institutional level – creating markets for carbon sequestration, developing payment schemes for biodiversity conservation, etc.

Many forest-related services are found in between, and are not necessarily pure private or pure public goods (Glück, 2000; Mantau *et al.*, 2001). Biodiversity conservation may lie far on the public good side, but eco-sponsoring by companies is a mechanism through which private engagement can bring income to forest owners and that improves ecosystem quality. Recreational demand can be tapped into in various ways by land owners, as described by Maso *et al.* (Chapter 10) and Matilainen *et al.* (Chapter 11) in this book. Even if the forest laws allow everybody to enter the forests for recreational purposes (especially under the so-called ‘everyman's right’, the Swedish *allemansrätt*), income may be generated though add-on services such as guided tours, adventure, sports or gastronomy services. We have collected many examples that show that there is much more room for private business than is often thought and, indeed, than is actually utilized by land owners.

At the same time, the field of environmental or recreational services often require innovations on a higher level, e.g. by new institutional rules (carbon markets, Ciccarese *et al.*, Chapter 9) or simply the engagement of institutional actors in order to enable certain activity: making a region, for instance, attractive for tourism often needs enhanced services and infrastructure at the landscape level. Examples of such are mountain bike routes, long-distance hiking trails, nature conservation in biosphere reserves, etc. In these cases numerous land owners are concerned, and numerous tourists benefit. How to channel the resources from the tourists to the land owners who provide the services such as trails or adapted forest management is a challenge that almost always requires public or semi-public actors to step in. Land owners, nature conservation or tourism interest groups have special roles in coming to solutions in these fields. Regional/rural development agencies may also engage in promoting new innovative activities that give rural regions new attractiveness. Entrepreneurship is required here in public or semi-public agencies. In sum, there is room and need for entrepreneurship in a purely market-oriented mode, but also in a more collective engagement by multiple actors and public or semi-public organizations for a common benefit, connected to the sector and/or the region.

18.2 Innovation in Multi-functional Forestry

18.2.1 Changing demands for forest goods and services

In research, policy and practice, there is an overwhelming recognition of the multifunctionality of forestry. However, the balance of multifunctional outputs varies significantly from country to country and the territorial component is rather weakly developed in some countries' forestry discourses. Whether this reflects the relative importance of the provisioning services or established power structures is less certain.

European visions for forest sector research represented by the European Commission or the Forest-Based Sector Technology Platform (FTP) are, by and large, wood-product-driven visions. As a consequence, the more diffuse and complex roles of forestry in rural development are in danger of being ignored, both as a research topic and as a development strategy. There is a danger that the issues relating to innovation in relation to territorial goods and services are marginalized or ignored. There is a need for strengthening policy in forestry to support public goods, and to strengthen this issue in European and national research agendas.

The forest sector focus on the provisioning of wood is the dominant way of thinking about forestry in many countries. This does not mean that forests are not multifunctional, but it does suggest a degree of capture of forestry interests by the commercial timber business and it may have implications about the discourses of forestry that shape research agendas and innovation possibilities. This takes us back to earlier COST actions (E3, E30) and to MULTIFOR.RD, which was an EU-funded project that showed that different discourses of forestry prevail in different parts of Europe (Elands and Wiersum, 2001; Kennedy *et al.*, 2001; Niskanen *et al.*, 2007). In other countries, especially the more densely populated countries, other ecosystem services (supporting, regulating and cultural) tend to be given greater prominence in shaping forestry discourses. Therefore, as a general observation, we need to understand the forest policy community and the forestry and forest products lobbies.

The general contention that the mainstream provisioning services from the land-based industries – wood and food – have been in sectoral decline, as components of rural economies, is incontrovertible. However, the extent to which emergent shortages of raw materials could have an effect on the primary sector is unknown. Many argue that the rise in demand for bio-fuel, Asian economic growth and climate change will fuel a raw materials crisis that will reinvigorate the primary sector. The timing may be uncertain, but there is a widespread perception of a pending shortage of primary commodities. In other words, it is not impossible that a resurgent (neo-)productivism could be an arena of innovation in forestry in countries that have had, until recently, a weakly developed productivist forestry discourse.

With the Leader instrument, now firmly established within Pillar 2 of the Common Agricultural Policy (CAP), there is a well developed model for delivering multi-sectoral territorial development in rural Europe (Feliciano *et al.*, Chapter 6). However, in terms of CAP Pillar 2 support, its budget is relatively small. Furthermore, although some Leader groups are well connected to the forest sector, there is also evidence that forestry was 'off the radar' of many Leader groups.

18.2.2 Changing role of policy in innovation support

There is a need to get out of a 'we-must-support-forestry' way of looking at the forest sector to a vision based much more on delivering value and service to the population. Like other primary industries, forestry still suffers from a supply-based productivist vision. A greater sensitivity to consumer demand clearly lies behind some of the impressive examples of the territorial goods and services developed by the more entrepreneurial characters in the forest sector. That this should be deepened and extend across both territorial and wood-based products is seen as highly desirable.

Where there are public good needs and where there are clear needs to get a large number of land owners to manage their land similarly to deliver the ecosystem service, there is a need to consider how to bring recalcitrant participants into the fold. A steeply tapered system of support that rewards the collective provision of club goods or common pool resources might be one way of achieving multi-actor engagement, where a limited cooperative action is weakly rewarded but a stronger alliance of partners would receive a higher reward per unit area under the scheme. The public good character of many territorial services creates complexity and a challenge for individualistic entrepreneurial innovation. However, that same characteristic creates scope for innovation with respect to policy and institutional design.

Even within the constraints of quasi-public good characteristics, it is not impossible to conceive innovation associated with the capture of some of the public good values – a type of internalization of the externality. Forest owners are not advised to wait expectantly for the payment for ecosystem services from cash-strapped state budgets, but a better understanding of the full value of the suite of ecosystem services delivered by forestry may provide a source for innovation at the firm or policy level.

We should not neglect the scope for forestry to deliver to sustainability agendas more prominently than it does at present. Given the salience of climate change, the recognition of new forms of intervention from Reducing Emissions due to Deforestation and Degradation (REDD) to natural flood management, it seems that forestry is well placed to address the issues of water quality, biodiversity and climate change. These issues were all picked up in the recent CAP Healthcheck. This may require the addressing and overturning of a range of path dependencies in the energy sector and in the policy arena.

In addition, forestry as an industry needs to assert its multiple values to a greater degree and to do so in ways that help shape CAP Pillar 2 policies and helps create policy instruments to deliver to these important contemporary agendas. Whether we are dealing with sustainable building materials, bioenergy, flood management or habitat corridor expansion, the case for new policies and institutions to support enhanced delivery is compelling.

It is almost impossible to conceive of significant innovation taking place without a supportive milieu or cluster. Individual entrepreneurs may develop isolated examples of commercially successful business ventures and there are some outstanding examples of such successes. However, behind most apparently entrepreneurial individualistic innovation is a network or milieus of multiple actors and interests, comprising state bodies, local authorities, non-governmental organizations and private sector actors. We remain unclear about the transferability of good practice from areas that have made enormous progress in the multifunctional development of their forest sectors to areas where little has changed. Issues of context specificity dog the transfer of best practice, and recipe-based approaches are likely to confront unforeseen local obstacles.

The most compelling argument for the stimulation of innovation in forest-related territorial services is the future prospect of market and policy driven demand for sustainable materials, sustainable recreational space and sustainable management of environmental quality from biodiversity to landscape to flood management. These are not optional extras from the forest system but rather core supporting and regulating services that forestry can provide, probably rather cost-effectively. The challenge is to enhance the policy environment and institutional conditions in which these core human and planetary needs can be better met. Such innovation is not only desirable, it is also indispensable for the future sustainability of the planet.

18.3 Mode of Innovation in the Forest Sector and Implications

18.3.1 A traditional sector with new challenges

Traditional industries are often described as non-innovative. As an indication of this, they are often classified as low-tech, the indicator being the rate of investments into R&D activities. This view should be changed as research shows that traditional sector enterprises are very significant in economic growth and productivity, provide the greater share in employment and do in fact innovate, however in different ways (Tunzelmann and Acha, 2005; Hirsch-Kreinsen *et al.*, 2008). They often take in novelties from other sectors and they innovate much more through interpersonal communication. Tacit knowledge has a greater role than formal R&D. Research shows that these other forms of innovation are generally as important (or even more important) than R&D activities, and even more so in traditional sectors and small businesses. All that is also proven for the forest sector in Europe (Rametsteiner *et al.*, 2005; Nybakk, 2009).

Most innovation efforts in the forest sector have been directed towards increasing the productivity throughout the production chain. The new requirements seem to be much more in qualitative improvements of the products within a global competition situation. The increasing global interconnection and the need to orient the direction of innovations more and more to customer needs and fashions at the same time alters the innovation processes and requires new skills within the firms and networks. Interdisciplinary skills and cross-sectoral competence shall help wood-based industries and value chain enterprises to cooperate with other sectors. Not all small businesses have the capacities to acquire those networking skills, which is why institutional actors increasingly have to provide those services as intermediaries that may act as the link with new research and business partners.

A growing challenge among the enterprises lower in the value chain is to get hold of the relevant market information that is more in the hands of enterprises with stronger customer interaction. That information is becoming inevitable to make adjustments in products and services based on the knowledge about changing tastes and fashions. The potentials achieved from creativity are important in traditional industries and related value chains when adopted into mass customized products and services, in analogy to the textile and cloth industry (Böheim, 2008).

18.3.2 High-tech in wood: the example of ICT

Enterprises in wood-related value chains, although characterized as low and medium tech when measured with direct inputs to R&D, apply high-tech production process components developed outside the value chain. The latter has become even more true with the intake of information and communication technology (ICT), as can be seen in a recent survey of the furniture industry (Gaboardi and Monti, 2008). The requirements concerning basic ICT infrastructure, broadband Internet connections and wireless LAN technology vary between sectors. They are required in timber transportation, important in furniture and bio-energy production and emerging in the timber construction component industries. CAD and 3D tools are playing an increasingly important role in furniture companies of all sizes and product lines in order to keep their efficiency and customer-driven production and innovation activities. The same is true with business-to-customer timber construction and is fast becoming true also in business-to-business construction value chains. Design tools can be used to enhance the customer experience at the point of sale, providing virtual visualization of product options and, ultimately, leading to customer-driven production and innovation. A further goal would be to increase the efficiency of the design process and foster overall integration with manufacturing, marketing and sales activities.

18.3.3 Systemic policy intervention on regional level

Sectoral cluster initiatives are an example for sector-specific support of innovation processes, mostly on the regional level. As analysed by Rimmler *et al.* (Chapter 7) and Ollonqvist *et al.* (Chapter 14) in this book, they take advantage of regionally clustered firms, make use of the regional proximity of actors and recognize the importance of tacit knowledge in the development of novelties.

They aim at intensifying the cooperation between the firms and try to improve the institutional knowledge infrastructure around the industry by initiating or strengthening research, education, training, etc. Through their capacities, they can act as intermediary organizations to connect firms, develop strategic projects and to bring in research organizations. In this way, they provide innovation system functions to the companies that would not have enough capacities to engage in international knowledge transfer or research projects.

Regional development agencies may exert similar functions, in a more cross-sectoral perspective. The Leader instrument would provide systemic support through various ways, including training, financial investment, connection to consultancy and advice, etc. The forest sector, however, is thus far underrepresented in the use of those services.

18.3.4 A role for institutional actors in the connection of the local and the global

For innovation support and regional development, the better connection of the local and international level becomes a crucial factor. This strongly relates to the role of external knowledge sources in innovation processes that are often underutilized (Ukrainski and Kajanus, Chapter 4). Institutional level actors have special roles as they have to provide such connections to international and/or scientific knowledge sources for small businesses that do not have enough skills and capacities to screen for such cooperation themselves.

18.4 Interactive Learning and the Local–global Challenge

18.4.1 Need for both: local sticky knowledge ...

Innovation research and innovation policy have discovered that innovation needs an exchange of knowledge and experiences among various actors: the studies of industrial districts and clusters describe company networks and complex interrelations of various parts of the production system. The innovation system approach emphasizes the importance of various actors from public and private spheres in national, sectoral and regional innovation systems – besides firms, also including public administrations, interest groups, education and research organizations, etc. The concept of the triple helix particularly mentions three actor groups: industry, research and public administration. The stakeholder approach adds all kinds of stakeholders, including customers, neighbours or civil society initiatives. This is also the notion within the learning region and creative milieu approaches. Clusters, networks or creative milieus play important roles in many of the empirical chapters in this book (Maso *et al.* (Chapter 10) on the role of networks in non-wood forest products and services; Rimmler *et al.* (Chapter 7) on cluster initiatives in the wood sector; Ollonqvist *et al.* (Chapter 14) on the networks and local milieus related to the furniture industry). The importance of an active management of stakeholder relations is described by Matilainen *et al.* (Chapter 11).

Altogether, these analysis approaches emphasize the importance of knowledge and learning for the innovative processes. Both codified knowledge (e.g. study book or scientific knowledge) and tacit knowledge (knowledge from experience that is communicated through personal, face-to-face training or interaction within or between firms/organizations) are important. Particularly the role of tacit knowledge explains why the local proximity of actors is often important for innovations.

The support of clusters became an important policy tool that was used in numerous industries. Clustering as an empirical phenomenon was described as a regional agglomeration of firms with a supporting institutional environment such as education and research institutes (Porter, 1990). The phenomenon of clustered innovations or innovative clusters was explained by the exchange of tacit sticky knowledge, the use of collective resources such as public education and research, knowledge spillovers, and unintended knowledge and competence transfers.

Small and micro-sized enterprises heavily depend on their socially embedded economic relations. Spatially focused explanations on competitive advantages name mutual trust and the expectation that parties will work for the mutual gain and refrain from opportunistic behaviour (Heanue, 2009). The institutional thickness associated with local agglomerations of entrepreneurship is frequently connected to family, ethnic or other characteristics among the exchange partners (Lorenzen, 1998). However, the emergence of global business relations has challenged these locally based tangible and intangible benefits of associations. Entrepreneurs must increasingly build trust on relationships among geographically dispersed enterprises.

The theories surrounding embeddedness and an innovative milieu imply that proximity and trust among actors matter and can be crucial in both innovation and the diffusion of innovation. In a world with much-enhanced communications at a distance, innovation and collaborative action can proceed at a distance as well as through local collaboration. If contiguity and proximity are now potentially less important for some types of innovation, they are certainly important where forestry is being used to provide catchment-scale flood management or habitat network creation.

18.4.2 ...and global relations

The emergence of international business networks triggered the concept of functional clusters, which then means the proximity in an organizational context, i.e. along the value chain. Such functional clusters also have strong connections across distances.

Knowledge exchange across regions becomes increasingly important in a globalized economy (compare Ukrainski and Kajanus, Chapter 4). Parallel innovation models with regional linkages and global and alliances among the suppliers and customers ascribe new roles for various actors with regard to innovation activities (European Commission, 2004). Interactive and parallel models indicate the importance of innovation capabilities and information and knowledge absorbing competencies. The interaction between the research community, manufacturers, suppliers, delivery systems and customers on local and global levels becomes a platform where innovations are generated (Ollonqvist, Chapter 12). With increased globalization, knowledge-intensive business services (KIBS) have emerged as a traded source for learning and innovation activities. Expert knowledge from KIBS enterprises are on the market and a wide set of players offer knowledge-intensive services providing complementary knowledge and capabilities for innovation processes (Aslesen and Isaksen, 2007). International hierarchies where local knowledge and competences are only a part in the complex pattern of learning, knowledge and competence networks have rapidly emerged in low- and medium-tech (LMT) industries (see the Norwegian shipbuilding analysis) (Bjarnar, 2008). Organizational proximity presupposes the existence of shared knowledge and assembling from fragmented information. Technology-related international links of knowledge and competencies has partly substituted the position of local or tacit knowledge and its diffusion. The management of a variety of knowledge and competence channels from local to global sources have proven to be a successful development in LMT industries (Heanue, 2009). Within LMT industries, research describes the interactive processes of knowledge generation, diffusion and implementation where local, regional, national and international knowledge interactions for innovations are utilized (Bjarnar, 2008).

18.4.3 Complementary knowledge and capabilities: the example of the wood-based industry

The emergence of global business networks has challenged the knowledge base among the SMEs and their local business infrastructures. Knowledge flows and competence building are often provided by KIBS within the regional or sectoral innovation system. Commercial KIBS take place on markets that require sufficient and stable demand. The requirements of new knowledge in the traditional sectors (environmental sustainability and network connection infrastructures as examples) have widely enlarged the complementary knowledge and competence demand among the enterprises. Public–private partnerships are common solutions for the adequate supply of the knowledge-intensive service activities in the regional or sectoral innovation systems. Organizing knowledge-intensive service activities through networks allows sharing risks that are connected with radical innovations.

The specific kinds of knowledge interaction in the wood-related examples of this book (Bouriaud *et al.* (Chapter 16) on timber harvesting, Nybakk *et al.* (Chapter 15) on bio-energy generation, Ollonqvist *et al.* (Chapter 14) on furniture industries and Nord *et al.* (Chapter 13) on timber construction) are analysed in the following scheme, which is based on different types of knowledge transaction and different types of knowledge creation and transfer (Tödtling *et al.*, 2009). Knowledge may be transferred on the basis of contracts or through informal solutions. Knowledge processes may be static (transfer after the creation) or dynamic (joint development).

The demand for complementary knowledge and competencies for the innovation activities among the enterprises in the four forest-related businesses are shown in Fig. 18.1. Entrepreneurs in forest operations apply high-tech machinery and information technologies, making them dependent on individually tailored KIBS activities. Local knowledge spillovers provide an innovation relevant information source among traditional furniture clusters. Joint interest activities to create the innovative milieu have also been important for them. Wood-based energy production involves new technologies and the entrepreneurs depend on KIBS services and joint efforts in knowledge-intensive service activities.

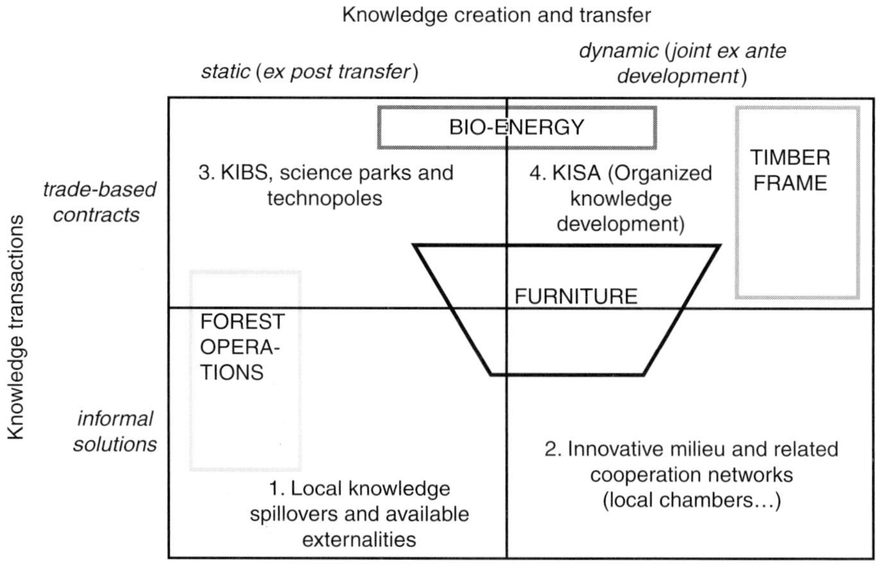

Fig. 18.1. Knowledge transfer among enterprises in forest operations, bio-energy production, furniture industries and timber construction industries (based on Tödtling *et al.*, 2009).

The knowledge on construction systems in timber construction industries is delivered in formal networks among architects and construction engineers. They rely on joint building information modelling, covering architecture, engineering and construction, exchanged in digital form among the partners. It facilitates information exchange and interoperability in digital format with 2D or 3D representations.

The needs among the enterprises in the four business sectors for the future knowledge and competence supply were identified in the respective book chapters and are shown in Fig. 18.2. Entrepreneurs in forest operations would benefit from the KIBS through supporting tacit knowledge communication and transfer. In the wood-based energy sector, the knowledge of the 'innovation champions' should be transferred among enterprises that could be organized on the basis of joint knowledge-intensive service activities or through local KIBS enterprises. Local furniture clusters are challenged by the knowledge and competence requirements in order to be able to interconnect into global business (design, ICT and e-business through international networks). Enterprises would benefit from knowledge-intensive service activities to jointly build up international competitiveness. The requirements are much the same irrespective of the mode of network connections: partner, subcontractor of the global locomotive enterprise, or member in a mutual interest network. Timber construction module industries for business-to-business market segments face strong competition from the value networks based on other materials – concrete and steel. However, the recent development in the field of construction systems is even more challenging for those competing sectors. The supply of complementary knowledge and competence services related to the adaptation of lean construction principles and methods is the most acute need to be supported through knowledge-intensive service activities.

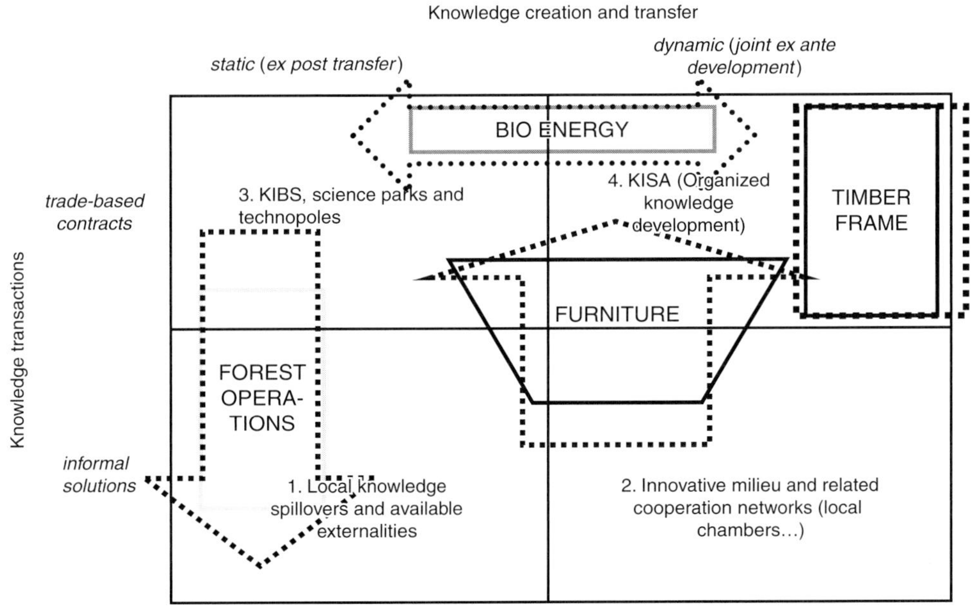

Fig. 18.2. Needs among enterprises for future knowledge and competence supply among the enterprises in forest operations, bio-energy production, furniture industries and timber construction industries (based on Tödtling *et al.*, 2009).

18.5 Systemic Innovation Support for Multi-sectoral Regional/Rural Development

18.5.1 Do policies create an innovation-enabling environment?

In this book we have asked the question of how far do relevant policies support innovation in the sector and how far do they create an innovation-enabling environment? The case studies presented in the empirical chapters did include these questions in various ways and have come to different answers, most of them resulting in yes and no. They often remark that direct influences of policies in innovation processes were hardly visible. Methodologically, there were two approaches in the thematic case studies: (i) studies of certain policy instruments such as Leader or cluster initiatives. These studies reveal the positive impacts of the policies but also deficiencies. For both instruments it is true that the conceptual set-up of the instruments allows comprehensive and systemic innovation support but the practical realization of this potential depends on the specific institutional framework and operational design for the implementation.

Naturally, it is relatively easy to discover policy influences when the method selects examples where the instruments had been applied. With such a method, the quality of the influence can be studied but not the overall significance. Another methodological approach (ii) started at innovative projects, for which the innovation processes were analysed, including the question of whether a policy influence was visible. In those examples, a direct policy influence was often not visible, a result that has to be carefully evaluated. First, it says clearly that many innovations happen without direct influence of policies or an 'innovation system'. That is a fact that should be kept in mind by researchers who are interested in how innovation actually happens in practice. Looking at policies or innovation systems alone would miss important innovation activities. Second, this result only says that direct policy influences were hardly seen; however, indirect influences might have existed but were not discovered by the methods. To find out about indirect influences would require more detailed and refined analyses of innovative milieus or learning regions. Assuming that only direct policy influences are important would reflect a traditional linear thinking about innovations and a too instrumental understanding of policies. In this instrumental understanding, typical interventions are direct support measures such as research or pilot financing. Systemic thinking, in contrast, includes wider dimensions such as education, cultural factors and the influence of multiple policy fields related to environment, employment or quality of life. Systemic interventions address complex relationships in innovation governance and are thus not directly visible by the actors in the field – and by research approaches that do not go deep enough. Profound analyses of innovation processes in rural areas require better financial resources and employ approaches from the fields of regional governance, regional innovation systems, creative milieus or learning regions. Not too many such studies seem to exist so far but they could significantly contribute to an important and promising research field. The empirical results of the book were not able to evaluate comprehensively the influence of relevant policies, but at least they did confirm the importance of the institutional dimension, of stakeholder relations, and of policies, a field that demands further research.

18.5.2 Systemic innovation support needs to be further developed

In sum, it can be stated that most relevant policies, such as regional and rural development policies, innovation policies and the sectoral policies from forestry and forest industry, do not all strongly affect innovations in the sector, and if they do, they mostly apply a traditional linear thinking instead of systemic approaches (Rametsteiner and Weiss, 2006; Weiss *et al.*, 2010). Systemic policies are the exception rather than the rule. Good examples are the creation and activities of forest owners' associations, cluster initiatives and the Leader approach. Forest owners' associations (FOA) have the required capacities in innovation support that small forest owners do not have (Mendes *et al.*, Chapter 5). Long-established FOA have developed important services for forest owners and have supported innovations such as in the fields of timber marketing or bio-energy. Newly established organizations such as in southern or eastern European countries need further political support in order to develop their potentials. It has, however, also to be remarked that sectoral agencies often have a pronounced focus on traditional forest-based products and that they would have to develop a greater openness towards new societal demands, which might offer interesting business opportunities such as in the well-known fields of climate change mitigation, bio-energy, recreation or biodiversity conservation.

Cluster initiatives have often been successful systemic interventions, but in practice limitations are also observed. Clusters as a policy means are typically part of regional development policies, aim at nurturing the strengths and potentials of certain selected localized industries and strive to create favourable institutional environments for their innovation activities. The cases that were investigated in this book (Rimmler *et al.*, Chapter 7; Ollonqvist *et al.*, Chapter 14) have different legal status and financing structures and apply the cluster concept in different ways, but all aim at the coordination of relevant policies and to enhance the cluster industries' knowledge base. The operational structure and capacities of the studied cluster initiatives differ in practice and their success in mobilizing firms to join the cluster activities varies. Companies are often sceptical of cooperation if they do not see clear benefits. In fact, instead of focusing only on the institutional setting of competing firms, the initiatives typically also aim at a closer cooperation of the enterprises. In that they follow rather innovation networks and innovation systems ideas than that of Porter's cluster model. The term cluster is often a catch word for quite different policy support approaches.

The Leader instrument is another prototypical example for a systemic innovation support even if deficiencies in its implementation are also reported (Feliciano *et al.*, Chapter 6). In the original set-up, it contains all the features for systemic support of an integrated multifunctional rural development. As such it has brought about highly interesting and innovative projects across Europe. In many practical cases, however, the innovative character of the instrument is compromised. One important success was the installation of regional development agencies in many regions. Those agencies develop specific know-how in regional governance and project support and offer services in innovation and project management and networking. With that they comprehensively offer innovation system functions and have become platforms for regional innovation governance. To what extent the risk and innovation-orientation, area-based and bottom-up approach, multi-sectoral principle, and networking and inter-regional cooperation have been realized strongly depends on the institutional framework in which the Leader structure was embedded and that strongly differs across countries or even provinces. Forestry and forest-based industries were often not included in the activity focus of Leader groups, and forestry sector agencies and firms did not successfully claim their themes to be considered, mostly because they simply did not know about the instrument. With the mainstreaming of the instrument, much of the innovative and distinct systemic features are getting lost as established groups increasingly capture the benefits. An innovation-oriented policy should try to reclaim the cross-sectorally, risk- and innovation-oriented characteristic of the programme and ensure its effective implementation.

18.5.3 Integrated rural development has many dimensions

What does integrated rural development actually mean? How does it work? The

Leader approach is a good example from policy practice of what integrated rural development and systemic innovation support could look like (Böcher, 2008). The concept of integrated rural development is closely linked to the regional innovation systems (Asheim and Gertler, 2005) and regional governance models (Adamaschek and Pröhl, 2003). There are many relevant dimensions that all together make what can be called integrated regional development:

- the intersectoral communication and cooperation between actors in the public and private spheres;
- the multi-level coordination of policies and actors;
- the bottom-up principle, which allows for the participation of all concerned interest groups, civil society initiatives and the population;
- the territory-based principle, which connects policy intervention to the local specificities;
- the multi-instrumental policy mix of measures that optimize mutual synergies but avoid contradictory incentives;
- the process-oriented view, which supports all stages of innovation, development and diffusion, knowledge generation and application;
- the systemic approach, which considers all relevant public and private actors and institutional frameworks;
- the innovation- and risk-friendly logic, which is interested in the new and unusual and which also allows the trying out of new things that have no guaranteed success in the short run;
- the connection of the local and the global that brings together regional and external values, experiences and knowledge sources;
- the interactive learning that brings together different types of knowledge and creates an innovative and creative milieu of trust and social capital;
- the undogmatic but pragmatic exploration and utilization of state, market and community mechanisms; and

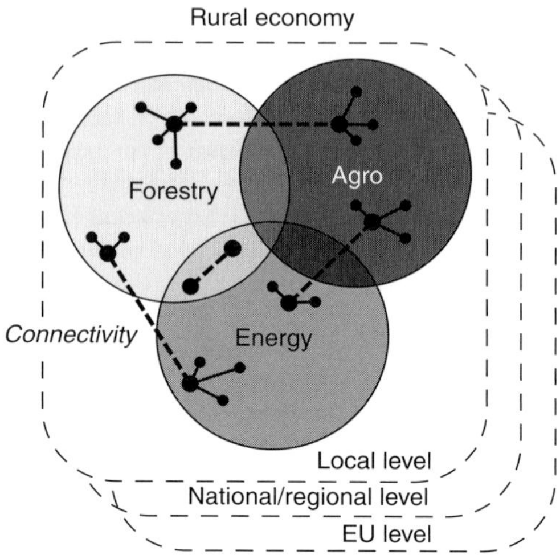

Fig. 18.3. A 'landscape approach' to integrated rural development policy.

- the search for a balance of economic, ecological and social interests with a view to enable a sustainable regional development.

Some of these aspects are pictured in Fig. 18.3, which illustrates some of the multiple sectors that play relevant roles in the use and development of the landscape. It pictures the importance of the connectivity between sectors and administrative levels.

18.6 Research Needs on Innovation Governance in the Forest Sector

18.6.1 Innovation, international competitiveness and sustainable regional development

Research needs result from the particularities of the production fields in question, whereby the territorial goods and services and the wood value chains have a number of characteristics in common but also differ in a number of basic features. Common challenges are related to their connection to rural backgrounds and traditional industries, which is challenged by societal changes and resulting new demands in regard to the sustainable European and local economies. The differences lie, on the one hand, in the specific connection of many ecosystem services to their territory where they are produced and, on the other hand, in the implications of globally traded commodities. Further research needs are related to the required institutional settings and policy actions, their implementation and impacts.

Some of the manifold research questions may be the following:

- compatibility of sustainability and innovation concepts in the forest sector and rural development;
- scenario or foresight studies on the coming challenges and structural changes in the wood-based sectors and rural areas;
- innovation support infrastructures suitable to connect small enterprises in the wood industries along the wood chain, within regional clusters, with research organizations and specific knowledge providers, and/or across countries and regions on international level;
- regional production and marketing concepts and innovation system models for the support of territorial goods and services of forests, such as territorial marketing, regional development agencies, land owner or producer associations or cooperatives, etc.; and
- preconditions for innovation processes and effective policies in eastern and south-eastern European post-communist countries.

18.6.2 Mechanisms for the delivery of territorial goods and services of forests

As many ecosystem services have partly public good characteristics and are often provided as externalities without markets, the question arises through which mechanisms they can optimally be secured. These mechanisms include public policies, market exchange and self-organized delivery mechanisms. They cover a broad range of options that are mostly not yet well understood.

- Forest policies are in general well studied, but for a number of promising new options of intervention there is hardly any knowledge. Policy options include regulations but market-based instruments as well. The range of options has been presented many times (Mavsar *et al.*, 2008), but they are often not understood in detail yet; for instance, tendering schemes for ecosystem services, eco-taxes, the creation of new markets, innovation support structures, etc.
- Market mechanisms could be used much more than is actually the case, including for goods and services with partly public good characteristics, for instance, through add-on services, eco-sponsoring, labelling, territorial marketing, etc.
- Self-regulated mechanisms for common-pool resources such as through common property regimes are hardly studied yet because of a different property rights tradition in Europe. These solutions, therefore, mostly require institutional innovations, including user or producer associations, etc.

18.6.3 A better understanding of innovation processes in a rural development context

We have reported on research results about innovations in traditional industries and small businesses. They show how important these sectors are but how the general approaches of innovation research and innovation policies do not fit the realities. Much more effort is needed to provide knowledge about the factors for innovation and the needs of the innovating small enterprises in traditional sectors such as forestry and the forest-based industries. Innovation research in these sectors has practically only started recently because the forest sector has mostly been ignored so far by innovation research (Hansen *et al.*, 2006). The COST Action has certainly triggered further initiatives, but there is a need to expand the current research capacities. Important questions refer, for instance, to the important role of institutional actors that provide complementary innovation enabling capabilities to small businesses such as in transforming local and global knowledge sources. From the experiences of the present work, the following themes would merit research attention:

- innovation processes in rural areas apart from established innovation systems;
- influence of sectoral and regional innovation systems on the generation and diffusion of innovations for integrated rural development;
- the role of market-oriented and social entrepreneurship for innovations for sustainable regional development;
- the role of intermediary actors as support infrastructures for innovations in rural small businesses, providing knowledge transfer, networking and financial incentives; and
- sources of knowledge for innovating firms in the forest sector, including the absorption of scientific knowledge and knowledge from other sectors.

18.6.4 Analysis of systemic innovation support policies

This book reports on the high awareness and interest of policy makers to apply more systemic innovation support but also on many deficiencies in the availability and application of systemic policy measures. There is a deficit of knowledge about the proper design of such measures in order to support their successful implementation. Our research gives an account of many gaps, but new knowledge is needed to understand the mechanisms in detail and to provide practical recommendations for improvement. Even well-established instruments – such as cluster initiatives, regional innovation systems or the Leader instrument – are not well understood in their practical application. Comparative studies across countries would enable the understanding of the influence of regional contexts on the implementation and also how these need to be considered in the policy design. Specific questions should address the barriers to innovation that stem from market factors (demand structures, lock-in situations, etc.), power relations (interests of powerful groups, relations between private and state sectors, etc.) or conventional mindsets and traditional ways of thinking. Comprehensive European-wide analyses of cluster programmes or the Leader scheme would be highly needed research programmes. Possible research topics could be as follows:

- comparative international analyses of the strengths and weaknesses of cluster policies in wood-based value chains;
- experiences with different forms of systemic policy measures aimed at the support of wood-based value chains;
- natural, economic, political or social barriers to wood sector innovations that would contribute to sustainable development;
- comparative international analyses of the institutional frameworks and implementation designs within national and regional Leader programmes;
- experiences with different forms of systemic policy measures aimed at integrated rural development; and
- natural, economic, political or social barriers to innovations for sustainable rural development through multifunctional land uses.

18.7 Territorial and Value Chain Approaches for Innovation Support: Summary Insights and Policy Recommendations

After studying innovation and innovation policies in various forestry and forest industry contexts and from multiple analytical perspectives, what have we learnt? Actually, there seems to be a fast growing political interest in the manifold contributions that forests and wood products make to the sustainable development in Europe: these aspects range from ecosystem services to the possible reduction of greenhouse gas emissions through the use of wood in construction and for other purposes. Political programmes express the high value that forest biodiversity, drinking water supply or recreational services have in Europe. Support programmes furthermore aim at the benefits of renewable materials and energy sources that stem from forests. Interestingly, policy measures related to forest ecosystem services are hardly targeted by innovation support while this is rather the case in the field of new timber products ('bio-based economy'). The discussions at the COST Action E51 conference, as reported by Aggestam and Weiss (Chapter 17), showed that it is much easier to talk about innovation in the wood-based industry sector than it is in the field of territorial goods and services or even in relation to rural development. Innovation, however, seems of equal importance here and there.

Innovation is supported but, in our view, it is still not sufficient. Forest-based environmental and social services are partly addressed in the EU Rural Development Programme and the Leader instrument in particular would be a suitable instrument to support innovative projects: if (i) there were more resources dedicated; it were implemented to be innovation-oriented; and the instrument would be used more often for projects with a forestry connection. In the field of wood value chains, it is often policies that drive innovations for sustainable development – the use of wood as a renewable energy source, or as a material for green building. Important new uses are seen for bio-based products in the chemical or fibre industries. The EU Lead Markets Initiative, for instance, has a remarkably strong relevance for the forest sector.

The results of this book strongly call for the further development of systemic policy measures. The case studies and the discussion with policy makers and stakeholders have consistently shown that a too narrow focus on research and development totally misses the needs of the sector enterprises in their innovation efforts: at least as important as research is the translation of the results to the companies. This knowledge transfer, however, is the bottleneck. Measures are most importantly needed to facilitate the (two-way) knowledge flow between researchers and firms and to support the whole innovation process. Innovation support infrastructures such as cluster organization, regional and rural development agencies and the like play important roles as intermediary agents and complementary knowledge and capacity providers. Furthermore, networks, education and training are measures that are of high relevance.

What needs to be repeatedly emphasized is to keep in mind the particularities of the different production fields: in territorial goods and services, the marketing possibilities are limited and narrowing down the policy efforts to market-oriented innovation would miss the target. Although we call for a better support of innovation for rural development, the limitations of the markets need to be remembered. All the more important are social, institutional and policy innovations. When it comes to wood value chains, the particular challenges are related to the specific problems in traditional industries and small businesses. Especially here, systemic approaches are necessary if the firms' innovation efforts should be best nurtured. Despite many policy initiatives – for instance, to support SMEs – small businesses are still disadvantaged by most industrial policies.

We have applied territorial and value chain approaches in theoretical–methodological terms, and we have also studied territorial and value chain approaches as they are applied in innovation policies. It is true for the research as well as the policy realm that the two approaches are both valuable, and even more so is the combination of both: they should be combined because goods and services have both spatial and sectoral dimensions, and they are dependent on regional as well as sectoral policies. Although it seems that regional innovation systems are especially suited for the support of territorial goods and services, it is also true that regional company networks and local milieus play crucial roles in fostering innovations within wood industries. The regional dimension of innovation with a view towards local networks and institutional support structures seems to require much more attention from policy makers in general. An integrated view of sustainable regional/rural development needs to support cross-sectoral interaction much more than is currently the case.

'Interaction' is perhaps the magic term for innovation policy in the sector: support of interaction across sectoral boundaries, regions and national borders, between the private and public, and between practice and research.

References

Adamaschek, B. and Pröhl, M. (eds) (2003) *Regionen erfolgreich steuern. Regional Governance – von der kommunalen zur regionalen Strategie.* Verlag Bertelsmann Stiftung, Gütersloh.

Asheim, B. and Gertler, M.S. (2005) The geography of innovation: regional innovation systems. In: Fagerberg, J., Mowery, D.C. and Nelson, R. (eds) *The Oxford Handbook of Innovation.* Oxford University Press, Oxford, pp. 291–317.

Aslesen, H. and Isaksen, A. (2007) New perspectives on knowledge-intensive services and innovation. *Geografiska Annaler* 89B(S1), 45–58.

Bjarnar, O. (2008) Global pipelines and diverging patterns of knowledge sharing in regional clusters. Working Paper 2008:1. Molde University College.

Böcher, M. (2008) Regional governance and rural development in Germany: the implementation of Leader+. *Sociologia Ruralis* 48, 372–388.

Böheim, M. (2008) Sectoral innovation systems in Europe: monitoring and analyzing trends and identifying challenges – textiles. Sector-specific analysis, Europe INNOVA.

Elands, B.H.M. and Wiersum, K.F. (2001) Forestry and rural development in Europe: an exploration of socio-political discourses. *Forest Policy and Economics* 3, 5–16.

European Commission (2004) Innovation management and the knowledge-driven economy. EC Directorate General for Enterprise and Industry, Brussels.

Gaboardi, E. and Monti, E. (2008) ICT and e-business impact in the furniture industry. Impact Study Report No. 03. European Commission, DG Enterprise & Industry.

Glück, P. (2000) Policy means for ensuring the full value of forests to society. *Land Use Policy* 17, 177–185.

Hansen, E., Korhonen, S., Rametsteiner, E. and Shook, S. (2006) Current state-of-knowledge: innovation research in the global forest sector. *Forest Products Journal* 3, 27.

Heanue, K. (2009) Aspects of the furniture industry in Ireland. PhD thesis, Dublin City University Business School, Dublin.

Hirsch-Kreinsen, H., Hahn, K. and Jacobson, D. (2008) The low-tech issue. In: Hirsch-Kreinsen, H. and Jacobson, D. (eds) *Innovation in Low-Tech Firms and Industries.* Edward Elgar, Cheltenham, UK, pp. 3–22.

Jenssen, J.I. and Jørgensen, G. (2004) How do corporate champions promote innovations? A literature review. *International Journal of Innovation Management* 8, 63–86.

Kennedy, J.J., Thomas, J.W. and Glück, P. (2001) Evolving forestry and rural development beliefs at midpoint and close of the 20th century. *Forest Policy and Economics* 3, 81–95.

Lorenzen, M. (1998) Information cost, learning and trust. Lessons from cooperation and higher-order capabilities amongst geographically proximate firms. DRUID Working Paper No. 98-21. Copenhagen.

Mantau, U., Merlo, M., Sekot, W. and Welcker, B. (2001) *Recreational and Environmental Markets for Forest Enterprises: A New Approach Towards Marketability of Public Goods.* CAB International, Wallingford, UK.

Mavsar, R., Ramcilovic, S., Palahi, M., Weiss, G., Rametsteiner, E., Tykkä, S., Apeldoorn, R. v., Vreke, J., Wijk, M. v. and Janse, G. (2008) Study on the development and marketing of non-market forest products and services (FORVALUE). Study Report for DG AGRI, Study Contract No. 30-CE-0162979/00-21.

Morris, M. (1998) *Entrepreneurial Intensity: Sustainable Advantages for Individuals, Organisations, and Societies.* Quorum Books, Westport, Connecticut.

Niskanen, A., Slee, B., Ollonqvist, P., Pettenella, D., Bouriaud, L. and Rametsteiner, E. (eds) (2007) *Entrepreneurship in the Forest Sector in Europe. Silva Carelica 52.* University of Joensuu, Faculty of Forestry, Joensuu.

Nybakk, E. (2009) *Innovation and Entrepreneurship in Small Firms: The Influence of Entrepreneurial Attitudes, External Relationships and Learning Orientation.* Norwegian University of Life Sciences, Ås.

Porter, M.E. (1990) *The Competitive Advantage of Nations.* McMillan, London.

Rametsteiner, E. and Weiss, G. (2006) Innovation and innovation policy in forestry: linking innovation process with systems models. *Forest Policy and Economics* 8, 691–703.

Rametsteiner, E., Weiss, G. and Kubeczko, K. (2005) *Innovation and Entrepreneurship in Forestry in Central Europe.* Brill, Leiden.

Tödtling, F., Lehner, P. and Kaufmann, A. (2009) Do different types of innovation rely on specific kinds of knowledge interactions? *Technovation* 29, 59–71.

Tunzelmann, N.v. and Acha, V. (2005) Innovation in low-tech industries. In: Fagerberg, J., Mowery, D.C. and Nelson, R. (eds) *The Oxford Handbook of Innovation.* Oxford University Press, Oxford, pp. 407–432.

Weiss, G., Ollonqvist, P., Rametsteiner, E. and Slee, B. (2010) Summary results and conclusions. In: Rametsteiner, E., Weiss, G., Ollonqvist, P. and Slee, B. (eds) *Policy Integration and Coordination: the Case of Innovation and the Forest Sector in Europe.* OPOCE, Brussels.

Index

Page numbers in **bold** type refer to figures, tables and boxed text.

acid rain 131
air quality 131–132
alliances, strategic *see* networks, business
Assigned Amount Units (emissions AAUs) 133
  interaction with voluntary market 146
  prices and trading 136
Austria
  cluster policy and structures 108–110
    funding sources 109
    Styrian Wood Cluster initiative 110, 112
  forest owners' associations, functions and influence 74
  land use, population and forest cover **90**
  Leader+ funding, forest-related projects 93, **93**, 97
  'social partnership' political structure 73
  timber construction enterprises, eco-friendly houses 212, 214
  tourism enterprises
    Almliesl forest cottages 174, **175**
    Sauwald canopy walkway (Baumkronenweg) 97, 174, **175**, 178
  wood biomass heating plants 27
    Güssing, regional cluster **198**

bio-energy *see* energy generation
biodiversity 298, 300, 304
birch sap products, new market opportunities **125**
buildings
  design and build (D&B) businesses
    cost-efficient modular multi-storey, Norway 218–219
    energy-efficient homes, Austria 212, 214
    local government building, Finland 215–217
    panelling and garden buildings, Romania 221–223
    residential multi-storey, Estonia 214–215
    service quality and reliability, Scotland 219–220
  regulations
    EU harmonization 205, 206
    variation between countries 205, 225–227
Bulgaria, traditional forest ownership and management 78–79
businesses
  alternative enterprise strategies 239–240
  formal and informal networks 37–39, **38**, **40**, 266–267
    importance of social capital 43
    participation objectives 41
  innovative capabilities 13–14, 48
    competence areas 195
    competitive advantage in traditional industries 192–194
    related to successful performance 271–272

businesses (*continued*)
market position 279
stakeholders, direct and indirect influence 171–172
see also micro-enterprises; small and medium sized enterprises

carbon sequestration
European forests
and emissions trading 141–144, 147
potential amount, CO_2 equivalents 132–133, 137
monetary value of emission allowances 134, **136**, 136–137
regulated carbon markets, Kyoto framework 133–134, **135**, 139–141
impact of rules on project costs 140–141
tradeable property rights establishment 131–132
voluntary
carbon market trading 144–145, 147–148
as corporate social responsibility 121, 144
pioneer forestry sequestration projects 145–147
TreeSmart program (Australia) **127**
chestnuts, marketing innovation **45**
Clean Development Mechanism (CDM)
accepted afforestation/reforestation activities 134, 140
Certified Emission Reduction (CER) units 133
project costs and approval procedures 140–141
climate change
deforestation impact estimates 143
energy sector path dependency 306
mitigation
forests, potential carbon sink estimates 132–133
impact on forest sector 129, 137–139
statutory measures 263
see also carbon sequestration
clusters
characteristics and benefits 15
formal organization and management 111, 112
inward investment attraction 104, 111–112
knowledge resource sharing 103–104, 195–196, **199**, 242
training and skills recruitment 112, 238
transaction cost savings 103
as framework for policy action 101–102, 194, 209, 243, 249–250
policy direction recommendations 113
regional definition and national coordination 111, 307–308
implementation, variation between countries 110–112, 113
innovation research model 17, 22–23, 102
policy initiatives 23, 245, **248**, 313
Austria 108–110
Croatia 104–105
Finland 107–108
Scotland 106–107
wood-based industries 191, **192**
Common Agricultural Policy (CAP)
incorporation of environmental services 300
reforms 3–4
thematic axes and pillars 88, 305
Community Innovation Survey (CIS) programme 28
Finland/Estonia wood sector businesses data **56**, 56–57
competition
competitive advantage (CA) 189
categories of CA sources 239
global spread of computer-aided systems 234, 235
sustainability 193, 194, 289
global (international) 2, 4, 194, 236–238
in industry clusters 15, 50, 234
information exchange between competitors 52–53
position of forest operations contractors 277
construction industry
characteristics and culture 204–205
adaptation of model to EU context 208–209, **209**
project structured learning 209–210, **210**
types and degrees of innovation 210–211, **211**
Winch's CoPS model 206–207, **207**
design, interactive 3D modelling 200, 307
market share, timber structures 191, 205
overcoming path dependency **201**, 228–229
policy frameworks
environment and infrastructure, individual countries **213**
EU Construction Product Directive (CPD) 205, 206, 228
knowledge and policy fragmentation 207–208
national systems, interaction with EU 208

recommendations for innovation support 229–230
variation between countries, effect on innovation 227–228
timber construction innovations, case study analysis **224**, **227**
actors, business focus and origins 225
micro-cluster establishment, Finland **199**
policy and regulation differences 225–227
recommendations for improvement 228–229
types of innovation 223, 225
see also timber-frame housing
consultants 53
COST Action E51
aims and objectives 6
conference (2010) discussion forum 294–301, 317
findings, bio-energy sector 256
scope and definition of forestry 91
crafts, wood products 91, 123
Croatia
cluster-based Competitiveness Initiative 104–105
North-West Timber Cluster 105, **244**
Forest Extension Service activities 77–78
furniture production enterprise clusters **244**, **246**, **248**
customers
business-to-business (BtoB) procurement markets 221, 225, 279
inventory management for 193–194
as knowledge source 52, 194
needs analysis, in furniture design 242, 245
public bodies as customers 229
public image of nature-based products/services 157, 170

DUI (doing, using and interacting) activities
innovation capability, SMEs 190, 228–229
knowledge base 194

economic development policies, EU
building regulations harmonization 205, 206, 228
'Europe 2020' strategy 1–2, 5
globalization, responses to 4, 5, 208–209, 234–235
information technology focus, catching-up economies 55
knowledge-based, human resources 124
Lisbon Competitiveness and Innovation Framework Programme (2006) 204, 206
Lisbon Strategy for Growth and Jobs (2000) 1, 54
research investment targets 2, 54, 55
transformations in eastern Europe 27
ecosystem services 119, 121, 123, 305–306
Emissions Trading Scheme (ETS)
European Emission Allowances (EUA), prices 136, **136**
auctions 136, 143
size and scope of market 141–143
energy generation
biomass pricing and demand 144, 254
transaction costs 255, 273
wood chips as novel harvesting market 281–282
wood fibre demand/supply 106, 191, 261
external networks and partnerships **264–265**
strong networking ties 263, 266–267
weak/underdeveloped linkages 267
fuel wood products 91
chips, pellets and briquettes 260, 261
waste (pallets, bark, sawdust) 261, 266, 273
innovation types and drivers
business system innovations 260–261
case study descriptions 257, **258–259**, 260
incremental process innovations 260, 261–262
personal and environmental drivers 262–263
product innovations 261
institutional and policy impacts 267–271, **268–269**
policy variations and research needs 271–273
research study, methods and cases 256–257, **257**
summary of findings **271**
wood biomass heating plants, Austria 27
energy self-sufficiency, Güssing **198**
entrepreneurs
creative urban migrants 124, **125**, 125–126
informal cooperative links 160, 267
management skills 193, 194
personal qualities, innovation champions 13, 262, 272, 303
potential and limitations 303–304
social milieu and social capital 43, 183
Estonia
data on business innovation 56–57
forest tourism

Estonia (*continued*)
    Liego tourism farm **175**, 178
    Woods Crone family park (tourism) **175**, 178–179
  KBE innovation strategy 55–56
  research and education 58–59
  RMK (State Forest Management Centre) 286, 291
  state energy policies 267, **268–269**, 270
  wood-related sector
    combined heat and power plant **258**
    financial support analysis 62, **62**
    forest operations, innovative machinery uses 286–287
    fuel wood production **258–259**, 262
    furniture production enterprise clusters **244**, **246**, **248**
    timber residential construction elements 214–215
European forest research initiatives
  COST Action E51 6, 90, 256, 294
  INNOFORCE Project Centre 5, 13, 20
  multifunctional forest use projects 305

FDI (foreign direct investment) 58
Finland
  Forest Management Associations 72
  innovation policies and institutions 54–55
    bio-energy R&D support **269**
    data on business innovation 56–57
    spending on innovations by SMEs 58
  land use, population and forest cover **90**
  Leader+ funding, forest-related projects 93, **93**, 97
  National Forest Program (FNFP), research 54
  regional development strategic focus
    autonomy and cluster centres 107–108
    industrial park micro-cluster, Satakunta **199**
    Lahti region industrial cluster 108, 112, **245**
  territorial goods and services
    horseback riding tours **176**, 179
    hunting activities enterprise (Finnhunt Oy) **176**, 179
    LUPPO nature-based Lapp rural development project 160–162, **163**
    Nature-based Entrepreneurship Association 162, 164–166, **165–166**
  timber-related industries
    furniture industry, design innovation support **197**, **244–245**, **246–247**, **248**
    modular housing construction, firm Z 215–217
    wood-based bio-energy cooperatives **259**, 260, **265**
forest operations
  case study analysis of innovation
    case firms, overview **280**, 280–281
    data collection methods 279–280
    harvesting and forwarding mechanization 282–283, 285–286
    logging tractor design improvement, Slovakia 283–284
    marketing of bio-fuel, Germany 281–282
    new uses for machinery, Estonia 286–287
    timber storage packing, Slovakia 284–285
  enterprises, position in forest-based value chain 276–278
    timber procurement market 279
    types of forestry contractor firms 287–288
  innovation encouragement 278–279, 288–292
  see also timber harvesting
forest owners
  growth in private ownership, Europe 68, 155–156
    local community Private Forest Districts (Romania) 79
    variation between countries 156
  size of holdings 69, 156
  tolerance of entrepreneurial activity 185
    market-oriented or social entrepreneurship 304
  see also owners' associations
forest sector innovation processes
  matching to Leader Initiative themes 89–90
  sector-specific patterns 24–25
    actors and relationships 25–26, **26**, 185, 297–298
    institutions 26–27
    'low-tech'/traditional designation 27–28
    small enterprise structure impacts 28–30
    technologies 25, 190
  spatial dimension 21–22
    regional innovation systems (RIS) 18, 23–24, 191
  timber-based and NWFP&S, changing focus 35, 304–305
    degree of innovation in core and fringe activities 112
    resource conflicts 298–299

Forestry Commission, UK
cluster strategy involvement 106, 107
mountain biking trail development **127**, 180
renewable energy networking 266
wildlife interpretation centres 180
France
green-headed wood, sawmill product innovation 217–218
pulp manufacturing and bio-energy production **258**, 261
financial and network support **264**, 266, **268**
FTP (Forest-based Sector Technology Platform)
forest sector research 305
Innovation Task Force, proactive approach 296
Strategic Research Agenda (SRA) 3
furniture industry
business characteristics and structures, Europe 233–235, **237**
business performance 236, **253**
enterprise strategy types 239–240
outsourcing 237–238, 240
product segments 236–237, 238–239
regional and local clusters 236, **236**, 238
innovation approaches
computer-aided technology adoption 240–241, 307
knowledge availability and competences 241–242
policy flexibility 249–250
potential success factors 248–250
market changes and innovation support
cluster cooperation **197**, 242
customer base changes 243
market distribution 242–243
policy support 243
market share, wood materials 191, 249
research objectives and case studies 235
classifications **248**
infrastructures and activities 244, **244–245**
innovation activities and drivers 244–245, **246–247**, 248

Germany, bio-fuel harvesting 281–282
governance
concept (political science) 20
regional 21, 128, 173
related to innovation systems 20–21, 314–316
greenhouse gas (GHG) emissions *see* carbon sequestration; Kyoto Protocol

hazards, natural
flood management 306
forest fires 42, 70
protection plans, Portugal 80, 81, 97
storm damage 281, 285, 289
hiking trails (long distance), organizational challenges **123**, 304

imitation and spillovers
new product protection, by branding 157
opportunities in networks 43, 158, 234
innovation
advantages and difficulties of small scale 29, 157, 271, 277–278
administrative burdens 297
barriers
in forest sector 126–128
political 267, 270
regulatory 291
research needs 316
structures, skills shortage and perceived risks 206–207, 297
in traditional timber-based industries 205, 229
definitions and scope **11**, 11–12, 89, 255, 278
contexts 123–124
types and degrees 12, 210–211, **211**, 223, 225
drivers
'champions' (individual drivers) 13, 262, 266, 272, 303
environmental scanning 262
social and economic circumstances 121, 144, 262–263, 288–289
sector characteristics 24–25, 189–190
characteristic mode of innovation 194, 277–278, 306–307
need for collaborative learning 297–298, 307
potential, in wood material uses 191
relevant research questions 314–316
support strategies
infrastructure provision 23, 304
institutions 3, 69, 308
internal capacity building (IEC) 52, 192–193, 195, 240
knowledge and education provision 54, 55, 59–61, 195–196, 291
local cluster promotion 103–104, 196–197, **199**, 228, 249–250
project or process basis 296, 301
regulatory environment 225–227, 263
small-scale project funding (Leader+) 96–99, 298

innovation (*continued*)
stakeholder cooperation
encouragement 164, 167, 170
tax and subsidy support 272–273
systemic approach, interactive learning 190–191, 206, 208–209, 279
research on policy influence 312
institutional innovation
creation of new markets 132
forest owners' associations 69, 313
industry cluster initiatives (CIs) 102–104, 313
public interest group involvement 304
technology coordination networks 3
IPCC (Intergovernmental Panel on Climate Change) 132, 141
Italy
Asiago Adventure Park 160, **161**
Borgotaro, *Boletus* mushroom-picking **121**, 158–160, **159**
Garfagnana (Tuscany), chestnut marketing **45**

Joint Implementation (JI), climate protection
approval rules and project costs 140–141
Emission Reduction Units (ERUs) 133
project specifications 134, 140

knowledge flows
access and transmission methods 25, 53
comparison study, Finland and Estonia 54–65
regional network model 17–18, 48–49
role in innovation 16–17, 43, 190–191
demand and supply 196–197, 200, 241–242
knowledge base components 124, 195
sources 49–54, **51**, 194–195
competitors 52–53, 196–197
consultants and KIBS 53, 190, 196, 200
customers 52
exhibitions and conferences 57–58
global (technology-related) 309
industrial clusters (CIs) 103–104, 107, 234, 308–309
internal 50, 52
suppliers 50
universities and research institutions 53–54, 196
knowledge intensive business services (KIBS)
growth, as response to globalization 309
regional and local supply 196, 200, 310
uses in innovation 53, 190, 242
Kyoto Protocol (KP)
forest sector implications 137–139, **138**
framework and commitments
CDM and JI mechanisms 140–141
first commitment period, agreement features 134, **135**
GHG emissions limitation 133–134
progress and prospects 139

Land Use, Land Use Change and Forestry (LULUCF) activities 133
carbon accounting procedures 139
compatibility with ETS targets 142–143, 147
registered forestry projects 141
landscape amenities, business development 178
Lead Market Initiative (LMI) 2, 5
Leader initiative (EU rural development programmes)
forest sector involvement 89–90, 96, 313, 317
risks and regulations 185–186
value and awareness 98–99
launch and objectives **88**, 88–89
Leader+ pilot projects, aims and timing 88, 89, 90
study
Leader+ projects, national analyses **92**, 92–93, 97
rural employment consequences 95, **95**, 97–98
sources of innovative ideas 93–94, **94**
types of innovation **94**, 94–95
user assessment of effectiveness 94, **95**, 95–96, **96**, 98
Lisbon Strategy
Competitiveness and Innovation Framework (2006) 204, 206
for Growth and Jobs (2000) 1, 54
literature (as innovation information source) 263
Lithuania
forest owners' associations and cooperatives 74–75
harvester machinery purchase 283
sawdust fuel briquette production **258**, 260
EU financial support **268**
LMT (low and medium technology) industries
creative capabilities of staff 50, 193
importance of vocational colleges 59
innovation potential 189–190, 234–235, 306–307
international knowledge/competency links 309

marketing innovations 12
conservation storage during over-supply 284–285
e-business adoption 241, 249
new market discovery 260, 281–282

packaging 160–162, **163**
promotion 43, **45**
retail distribution methods 242–243
MCPFE (Ministerial Conference on the Protection of Forests in Europe), Vienna Resolution 3, 89
micro-enterprises
assistance networks, Finland 161–162
definition 29
passive house-building, Austria 212, 214
Millennium Ecosystem Assessment (MEA) 119
montado ecosystem (Portugal) 95, 97, **126**
mountain biking
business opportunities **177**, 180
use of public infrastructure 126, **127**
multifunctional land use
political context 119, 300
research projects, forestry 5, 305
see also territorial goods and services
mushrooms, Borgotaro *Boletus* trail **121**, 158–160, **159**

nature conservation, business opportunities 178, 180
networks, business
characteristics and types 14–15
evolution 37, 39, **40**
factors affecting position of members 37
international cluster groups 112
spatial and social proximity 48–49, 103, 263, 266–267, 309
subcontracting 182, **199**, 225, 290
types of cooperative arrangement 37, **38**, **158**, 166–167
definitions **36**, 36–37
economic basis 35–36
association membership fees 158, 159
importance of startup public finding 162
Leader+ supported new development 95
outcomes and advantages
benefit distribution 41
economic costs and benefits 39–40, 162
resources access and exchange 40–41, 52–53, 164
role in innovation 41–45, **42**
forestry sector examples 42, **45**, 158–166
implications for policy makers 45–46, 272, 306
theoretical analysis, instruments 36–37
industrial district and cluster theories **39**, 48
learning regions 171
see also clusters; stakeholders
NIMBYism 126
non-wood forest products and services (NWFP&S)
importance of networking and cooperation 35, 45, **45**, 155–157, 166–167
analysis by case studies 158, **158**
Asiago Adventure Park 160, **161**
Borgotaro mushroom trail 158–160, **159**
Finnish nature-based entrepreneurship association 162, 164–166, **165–166**
nature-based Lapp products, gift packages 160–162, **163**
protected origin designations 120, 158–159
sector growth and business characteristics 154–155, 167
types of product 91, 154, 157
recreation and sports 160
regionally specific food and drink 120–121
see also tourism
Norway
forest owners' associations 72–73
timber construction businesses 218–219
wood fuel pellet production **258**, 260
innovation champion 262
local network links 263, **264**, 266
public funding support **268**

OECD (Organisation for Economic Co-operation and Development)
classifications
rural areas and forest cover 90
types of innovation 12, 278
low/high-tech sector definitions 27, 233
Oslo Manual 28, 190
reports on science research and innovation policies 4–5
offsetting (carbon) **127**, 144–147
organizational (in-firm) innovations 12
business linkage architecture 223
bio-energy cooperatives 260
external partnership relations 157
recruitment policy 53
service orientation 193–194
Oslo Manual (OECD), innovation data interpretation 28, 190
owners' associations
country case studies 72–81
critical mass for maintenance 71, 72
directions for future research 83
evolution and path dependence 71, 82–83

owners' associations (*continued*)
existing oligopsonic markets 81
land restitution after forest nationalization 78–79
weaknesses of top-down approach 80
'free riding' problems, coping measures 70–71, 82
forest road construction/repair 78
legal support and financial arrangements 72, 77
securing property rights 79
waterway-based geographical organization 73
importance of collective organization 69
counterbalance to buyers' market power 72
for fire protection measures 80–81
forest management coordination 76–78
influence and leverage 71, 82
international support projects 77
networking cooperation 74–75
support from semi-public institutions 73–74, 313
vertical integration 72, 73
theoretical reasons for emergence 69–71

passive (energy-efficient) building design 212, 214
patents, investment motivation 41–42
PGI (Protected Geographical Indication) certification 120, 158–159, **159**
Poland
biomass-fuelled energy generation **258**, 261–262
government subsidies **268**, 270
stakeholder network support 263, **264**
furniture production enterprise clusters **245**, **247**, **248**
timber forwarder purchase 282–283
policy frameworks
analysis, COST Action E51 programme 5, 6
climate 132
EU Emissions Trading Scheme 141–144
public funding and national legislation, bio-energy 267–271, **268–269**
sustainable energy use, buildings 206
UNFCCC/Kyoto Protocol 133–134, 140–141, 266
cross-sectoral integration/coordination 4–5, 54, 101–102, 172–173, 313–314
development policies 3–4, 281
influence on cooperative relationships 183–184, **184**, 186
regional 54, 61, 102
rural 54, 87–90, 128, **156**, 299–300
forestry sector 3, 26
forest-based industries 3, 277, 291
innovation-enabling conditions 299, 300, 312
integration, private and state-owned forests 76–77
renewable energy 2, 4
sensitivity to market/consumer demands 305–306
innovation policies (IP) 1, 4
cluster approach 102–104, 194, 234–235, 243
evolution and range 20–21
Finland and Estonia 54–56, **64**, 65
Lead Market Initiative (LMI) 2, 5, 317
national systems (NIS) 208
recommendations for forestry support 316–318
relationship to science and technology policies 207–208
small businesses, attitudes to 28–30, 49, 295–296
see also Common Agricultural Policy
Porter, M.E.
cluster concept, diamond model 15, 17, 102
five-force competition and innovation model 278, 279
Portugal
fire risks, forest 81
forest owners' associations, establishment 80–81, 97
land use, population and forest cover **90**
Leader+ funding, forest-related projects 93, **93**, 97
montado ecosystem and cork products 95, 97, **126**
process innovations
definition 12, 41
examples 95
expenditure (Finland) 58
industrialized (factory) production 205, 206, 238
production methods/machinery
bio-energy power plants 260–261
computer-aided design (furniture) 240–241, 307
novel uses of existing machinery 286–287
timber harvesting and transport mechanization 282–283, 285–286, 288
wood fuel pellet manufacture 260
product innovations 12
components and materials 261
design improvement 105, 283–284

novel products 156–157, 223, 225
territoriality attributes 120–121, 158–159
types, in forestry sector 41–42, 94–95
professional associations
Association of Forest Entrepreneurs (Poland) 282, 290–291
carpenters' guilds 196, 234
property rights
definition difficulty, for public goods 132
designated territorial products 120
economic value establishment 136–137
emissions allowances 133–134
restitution after state ownership 78–80, 156
public goods
air quality 131–132
'green tax' funding 78
marketability 131–132, 298, 304
role of forest owners' associations 70–71, 83
scope for innovation 127
pulp and paper industries 27, 49, 261

R&D institutes *see* research organizations, public sector
recreational services
adventure activities 157, 160, **161**, 179, 180
EU research projects 5
hunting 179
public access rights 304
see also tourism
regional development
cluster coordination 111, 307–308
governance 21
importance of network organization 170–171, 229–230
policy frameworks 54, 61, 102, 313
local tailoring 296
regional innovation systems (RIS) 18, 234
contrasted with cluster approach 23
interaction potential among actors 193
local grassroots and regionally networked 23–24
position of R&D activities 191
renewable energy sources (RES) *see* energy generation
research organizations, public sector
interactions with forestry industries 50, 195, 196
networking encouragement 226–227, 301
Romania
international cooperation, 'Sawdust 2000' project **265**, 267–268, **269**
private owners, establishment of associations 79–80
problems of forest mismanagement 79
territorial goods and services
Calimani National Park tourism **176**, 179–180
Dorna Adventure, recreation activities **176**, 179
timber-related industries
construction product opportunities 221–223
wood-based bio-energy generation **259**, 260–261
rural development
contributions from forest-based industries 2–3, 128–129, 299–300
potential, research needs 315–316
employment opportunities, forest sector 2, 95, **95**
policy frameworks
EU strategic guidelines 156, **156**, 172
horizontal approach, Finland 54, 162, 166
integration of private sector 74–77
landscape approach to system integration 298–299, **299**, 300–301, **301**, 313–314
Leader instrument 5, 24, **88**, 88–90, 313
rural area problems and revival actions 87–88
diversification 154–155, 169–170
innovation driven by incomers **125**, 125–126
knowledge and competence supply 196–197, **198**
post-productivist economic aim 119, 120

Schumpeter, J.A.
definition of innovation 11
Mark I and Mark II patterns of innovation 17, 24
science parks/technopoles 23, 196
Scotland
economic development strategy 106
land use, population and forest cover **90**
Leader+ funding, forest-related projects 93, **93**, 97
Scottish Enterprise forest industries cluster 106–107
Sitka spruce knowledge base partnership 107
territorial goods and services
Huntly peregrine wild watch project **177**, 180
mountain biking 126, **127**, **177**, 180
tourist marketing iconography 121, **122**
timber-related industries

Scotland (*continued*)
timber-frame construction engineering 219–220
wood pellets for bio-energy **259**, 260, 263, 266, **269**
Serbia, legislative support for forest owners 76–77
Slovakia
forest operations innovation
custom-made logging tractor 283–284
logging tractor purchase 285–286
timber storage, new packing method 284–285
post-socialist forest owners' associations 75–76
tourism development
Velky Klížforests **177**, 180–181
Vydrovská valley **177**, 181
small and medium sized enterprises (SMEs)
advantages of collaborative networks 17, 40–41, 45–46, 155, 166
innovation sharing 42–43, 242
production and marketing 43, 45, 157, 162
quality and sustainability support 164
innovation capacity and support 29–30, 52, 157
access to customers and natural resources 170, 179, 180
access to knowledge resources 53–54, 59, 60, 64, 195–196
'doing, using and interacting' (DUI) activities 190, 194
Innovation Enabling Capabilities (IEC) 192–193, 195, 240
use of tacit and local knowledge sources 194, 277–278, 308–309
status in EU economies 28–29
business culture, wood-based enterprises 191, 200
effectiveness of innovation policies 295–296
social capital
definition 43, **44**
strong rural bonding 125, 128
stakeholders
beneficiaries of Leader+ funding 93
COST Action E51 conference participants 295, 302
in forestry sector innovation systems 25–26, **26**
cooperative relationships 181–183
influence on business performance 171–172
Local Action Groups (LAGs) 87, **88**, 89, **92**
system approach to cooperation study (Freeman's) 15, 256
suppliers
bio-energy supply chain innovation stimuli 260–261, 262, 273
equipment manufacturers 50
knowledge base 58
timber procurement, bargaining power 279, 289
sustainability
contribution of forest sector 2, 126, 304
policy instrument support 306
impact assessment research 5
Sweden
furniture production enterprise clusters **245**, **247**, **248**
timber-frame housing
innovation support **201**
wood product development 220–221

technology
characteristics of traditional sectors 25, 50, 190, 193, 200
information and communication (ICT)
management and marketing uses 241
requirements of wood-based value chains 307
skills transfer links 195
lock-in (path-dependent) situations 17, 200
low- and high-tech industries, comparison 27–28
machinery
age variation, timber harvesting 277
computer-controlled (CNC), furniture industry 240
see also LMT industries
technopoles *see* science parks/technopoles
Tekes (Finnish Agency for Technology and Innovation) 54, 97, 216, **248**
territorial goods and services
innovation (forest-related)
conflicts of interest with commodity production 127–128
drawdown of corporate profits 121
obstacles from ownership complexity 126–127, **127**
opportunities 126, **126**
public goods, innovation scope 127, **127**, 305–306
provision contexts 119–120
related to policy and economic climate 118–119
balance of multifunctional outputs 304–305
integration with innovation policies 128–129

scope and interaction of products 22
definition 91, 120
delivery mechanisms 315
multiple environmental services 121, **123**
place-specific regional goods 120–121, **121**, **122**, 158–159
related to forests and woodland 123
see also carbon sequestration; ecosystem services; non-wood forest products; tourism
timber-frame housing
analysis of innovations 223–227, **224**, **227**
business opportunities, recent growth 205
national and local development drivers, Sweden **201**
design and marketing
garden houses and gluelam elements, Romania 221–223
homes for the elderly, Finland 215–217
quality timber-frame engineering, Scotland 219–220
modular passive (eco-friendly) houses, Austria 212, 214
multi-storey/apartment buildings
cost-efficient modular units, Norway 218–219
timber use in construction elements, Estonia 214–215
pre-fabrication of components 205, 206
timber harvesting
diesel fuel regulations 284, 286, 287, 291
European forests, use and age of machinery 277
innovative uses and designs 283–284, 286–287
investments in mechanization 282–283, 285–286, 288
manpower 289
chainsaw operators 282, 283
value, compared to carbon sequestration 138–139
tourism
economic benefits of forests, external spillover effect 118–119
local network cooperation 160
importance of personal relationships 183, 185
role in innovation 171, 184–185
structured and unstructured 182–183, 185
typical characteristics 181–182
nature-based
Austrian case studies 174, **175**, 178
characteristics and growth 169–170
Estonian case studies **175**, 178–179
Finnish case studies **176**, 179
key stakeholders 171–172
policy support 172–173, 183–184, **184**
research methods, case studies 173–174
Romanian case studies **176**, 179–180
Scottish case studies **177**, 180
Slovakian case studies **177**, 180–181
value of territorial and cultural attributes 121
TreeSmart carbon sequestration program **127**

United Nations Framework Convention on Climate Change (UNFCCC) 132–134, 139–142
universities
education and technology transfer approaches 53, 65
international knowledge distribution 61
knowledge access for SMEs 53–54, 59, 60, 64

value chains
flow of market information 307
positioning, modes of change 240
relationship to traditional clusters 194
sectoral innovation systems (research approach) 18
vertical, in wood-based industries 191, **192**
see also construction industry; energy generation; furniture industry; timber harvesting
VER (Voluntary/Verified Emission Reduction) units 144–146

wood industries
business performance
forest service operations 277
furniture **253**
timber construction 204–205
wood fuel bio-energy 254–255
communication and trust across value chains 297
knowledge creation and transfer 309–311, **310**, **311**
economic returns 156, 193
enterprise types 56, 91
industry clusters and interfaces 191, **192**
technological characteristics 50, 193
see also energy generation